NUMERICAL LINEAR ALGEBRA

NUMERICAL LINEAR ALGEBRA

LLOYD N. TREFETHEN
Cornell University
Ithaca, New York

DAVID BAU, III
Microsoft Corporation
Redmond, Washington

Society for Industrial and Applied Mathematics
Philadelphia

Library of Congress Cataloging-in-Publication Data

Trefethen, Lloyd N. (Lloyd Nicholas)
    Numerical linear algebra / Lloyd N. Trefethen, David Bau III.
        p. cm.
    Includes bibliographical references and index.
    ISBN 0-89871-361-7 (pbk.)
    1. Algebras, Linear. 2. Numerical calculations. I. Bau, David.
  II. Title.
  QA184.T74 1997
  512'.5--dc21   96-52458

Cover Illustration. The four curves reminiscent of water drops are polynomial lemniscates in the complex plane associated with steps 5,6,7,8 of an Arnoldi iteration. The small dots are the eigenvalues of the underlying matrix A, and the large dots are the Ritz values of the Arnoldi iteration. As the iteration proceeds, the lemniscate first reaches out to engulf one of the eigenvalues λ, then pinches off and shrinks steadily to a point. The Ritz value inside it thus converges geometrically to λ. See Figure 34.3 on p. 263.

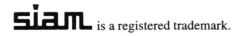 is a registered trademark.

To our parents
Florence and Lloyd MacG. Trefethen
and
Rachel and Paul Bau

Contents

Preface   ix

Acknowledgments   xi

I   Fundamentals   1
    Lecture 1   Matrix-Vector Multiplication   3
    Lecture 2   Orthogonal Vectors and Matrices   11
    Lecture 3   Norms   17
    Lecture 4   The Singular Value Decomposition   25
    Lecture 5   More on the SVD   32

II   QR Factorization and Least Squares   39
    Lecture 6   Projectors   41
    Lecture 7   QR Factorization   48
    Lecture 8   Gram–Schmidt Orthogonalization   56
    Lecture 9   MATLAB   63
    Lecture 10   Householder Triangularization   69
    Lecture 11   Least Squares Problems   77

III Conditioning and Stability   87
    Lecture 12   Conditioning and Condition Numbers   89
    Lecture 13   Floating Point Arithmetic   97
    Lecture 14   Stability   102
    Lecture 15   More on Stability   108
    Lecture 16   Stability of Householder Triangularization   114
    Lecture 17   Stability of Back Substitution   121
    Lecture 18   Conditioning of Least Squares Problems   129
    Lecture 19   Stability of Least Squares Algorithms   137

IV   Systems of Equations   145
✓Lecture 20   Gaussian Elimination 147
✓Lecture 21   Pivoting 155
  Lecture 22   Stability of Gaussian Elimination 163
  Lecture 23   Cholesky Factorization 172

V   Eigenvalues   179
  Lecture 24   Eigenvalue Problems 181
✓Lecture 25   Overview of Eigenvalue Algorithms 190
✓Lecture 26   Reduction to Hessenberg or Tridiagonal Form 196
  Lecture 27   Rayleigh Quotient, Inverse Iteration 202
  Lecture 28   QR Algorithm without Shifts 211
  Lecture 29   QR Algorithm with Shifts 219
  Lecture 30   Other Eigenvalue Algorithms 225
  Lecture 31   Computing the SVD 234

VI   Iterative Methods   241
  Lecture 32   Overview of Iterative Methods 243
  Lecture 33   The Arnoldi Iteration 250
  Lecture 34   How Arnoldi Locates Eigenvalues 257
  Lecture 35   GMRES 266
  Lecture 36   The Lanczos Iteration 276
  Lecture 37   From Lanczos to Gauss Quadrature 285
  Lecture 38   Conjugate Gradients 293
  Lecture 39   Biorthogonalization Methods 303
  Lecture 40   Preconditioning 313

Appendix   The Definition of Numerical Analysis   321
Notes   329
Bibliography   343
Index   353

Preface

Since the early 1980s, the first author has taught a graduate course in numerical linear algebra at MIT and Cornell. The alumni of this course, now numbering in the hundreds, have been graduate students in all fields of engineering and the physical sciences. This book is an attempt to put this course on paper.

In the field of numerical linear algebra, there is already an encyclopedic treatment on the market: *Matrix Computations*, by Golub and Van Loan, now in its third edition. This book is in no way an attempt to duplicate that one. It is small, scaled to the size of one university semester. Its aim is to present fundamental ideas in as elegant a fashion as possible. We hope that every reader of this book will have access also to Golub and Van Loan for the pursuit of further details and additional topics, and for its extensive references to the research literature. Two other important recent books are those of Higham and Demmel, described in the Notes at the end (p. 329).

The field of numerical linear algebra is more beautiful, and more fundamental, than its rather dull name may suggest. More beautiful, because it is full of powerful ideas that are quite unlike those normally emphasized in a linear algebra course in a mathematics department. (At the end of the semester, students invariably comment that there is more to this subject than they ever imagined.) More fundamental, because, thanks to a trick of history, "numerical" linear algebra is really *applied* linear algebra. It is here that one finds the essential ideas that every mathematical scientist needs to work effectively with vectors and matrices. In fact, our subject is more than just vectors and matrices, for virtually everything we do carries over to functions and operators. Numerical linear algebra is really functional analysis, but with the emphasis always on practical algorithmic ideas rather than mathematical technicalities.

The book is divided into forty lectures. We have tried to build each lecture around one or two central ideas, emphasizing the unity between topics and never getting lost in details. In many places our treatment is nonstandard. This is not the place to list all of these points (see the Notes), but we will mention one unusual aspect of this book. We have departed from the customary practice by not starting with Gaussian elimination. That algorithm is atypical of numerical linear algebra, exceptionally difficult to analyze, yet at the same time tediously familiar to every student entering a course like this. Instead, we begin with the QR factorization, which is more important, less complicated, and a fresher idea to most students. The QR factorization is the thread that connects most of the algorithms of numerical linear algebra, including methods for least squares, eigenvalue, and singular value problems, as well as iterative methods for all of these and also for systems of equations. Since the 1970s, iterative methods have moved to center stage in scientific computing, and to them we devote the last part of the book.

We hope the reader will come to share our view that if any other mathematical topic is as fundamental to the mathematical sciences as calculus and differential equations, it is numerical linear algebra.

Acknowledgments

We could not have written this book without help from many people. We must begin by thanking the hundreds of graduate students at MIT (Math 335) and Cornell (CS 621) whose enthusiasm and advice over a period of ten years guided the choice of topics and the style of presentation. About seventy of these students at Cornell worked from drafts of the book itself and contributed numerous suggestions. The number of typos caught by Keith Sollers alone was astonishing.

Most of Trefethen's own graduate students during the period of writing read the text from beginning to end—sometimes on short notice and under a gun. Thanks for numerous constructive suggestions go to Jeff Baggett, Toby Driscoll, Vicki Howle, Gudbjorn Jonsson, Kim Toh, and Divakar Viswanath. It is a privilege to have students, then colleagues, like these.

Working with the publications staff at SIAM has been a pleasure; there can be few organizations that match SIAM's combination of flexibility and professionalism. We are grateful to the half-dozen SIAM editorial, production, and design staff whose combined efforts have made this book attractive, and in particular, to Beth Gallagher, whose contributions begin with first-rate copy editing but go a long way beyond.

No institution on earth is more supportive of numerical linear algebra—or produces more books on the subject!—than the Computer Science Department at Cornell. The other three department faculty members with interests in this area are Tom Coleman, Charlie Van Loan, and Steve Vavasis, and we would like to thank them for making Cornell such an attractive center of scientific computing. Vavasis read a draft of the book in its entirety and made many valuable suggestions, and Van Loan was the one who brought Trefethen to Cornell in the first place. Among our non-numerical colleagues, we thank Dexter Kozen for providing the model on which this book was based: *The Design and Analysis of Algorithms*, also in the form of forty brief lectures. Among the department's support staff, we have depended especially on the professionalism, hard work, and good spirits of Rebekah Personius.

Outside Cornell, though a frequent and welcome visitor, another colleague who provided extensive suggestions on the text was Anne Greenbaum, one of the deepest thinkers about numerical linear algebra whom we know.

From September 1995 to December 1996, a number of our colleagues taught courses from drafts of this book and contributed their own and their students' suggestions. Among these were Gene Golub (Stanford), Bob Lynch (Purdue), Suely Oliveira (Texas A & M), Michael Overton (New York University), Haesun Park and Ahmed Sameh (University of Minnesota), Irwin Pressmann (Carleton University), Bob Russell and Manfred Trummer (Simon Fraser University), Peter Schmid (University of Washington), Daniel Szyld (Temple University), and Hong Zhang and Bill Moss (Clemson University). The record-breakers in the group were Lynch and Overton, each of whom provided long lists of detailed suggestions. Though eager to dot the last i, we found these contributions too sensible to ignore, and there are now hundreds of places in the book where the exposition is better because of Lynch or Overton.

Most important of all, when it comes to substantive help in making this a better book, we owe a debt that cannot be repaid (he refuses to consider it) to Nick Higham of the University of Manchester, whose creativity and scholarly attention to detail have inspired numerical analysts from half his age to twice it. At short notice and with characteristic good will, Higham read a draft of this book carefully and contributed many pages of technical suggestions, some of which changed the book significantly.

For decades, numerical linear algebra has been a model of a friendly and socially cohesive field. Trefethen would like in particular to acknowledge the three "father figures" whose classroom lectures first attracted him to the subject: Gene Golub, Cleve Moler, and Jim Wilkinson.

Still, it takes more than numerical linear algebra to make life worth living. For this, the first author thanks Anne, Emma (5), and Jacob (3) Trefethen, and the second thanks Heidi Yeh.

Part I

Fundamentals

Lecture 1. Matrix-Vector Multiplication

You already know the formula for matrix-vector multiplication. Nevertheless, the purpose of this first lecture is to describe a way of interpreting such products that may be less familiar. If $b = Ax$, then b *is a linear combination of the columns of A.*

Familiar Definitions

Let x be an n-dimensional column vector and let A be an $m \times n$ matrix (m rows, n columns). Then the matrix-vector product $b = Ax$ is the m-dimensional column vector defined as follows:

$$b_i = \sum_{j=1}^{n} a_{ij}x_j, \qquad i = 1, \ldots, m. \qquad (1.1)$$

Here b_i denotes the ith entry of b, a_{ij} denotes the i,j entry of A (ith row, jth column), and x_j denotes the jth entry of x. For simplicity, we assume in all but a few lectures of this book that quantities such as these belong to $\mathbb{C}$, the field of complex numbers. The space of m-vectors is $\mathbb{C}^m$, and the space of $m \times n$ matrices is $\mathbb{C}^{m \times n}$.

The map $x \mapsto Ax$ is *linear*, which means that, for any $x, y \in \mathbb{C}^n$ and any $\alpha \in \mathbb{C}$,

$$A(x + y) = Ax + Ay,$$
$$A(\alpha x) = \alpha Ax.$$

Conversely, every linear map from $\mathbb{C}^n$ to $\mathbb{C}^m$ can be expressed as multiplication by an $m \times n$ matrix.

A Matrix Times a Vector

Let a_j denote the jth column of A, an m-vector. Then (1.1) can be rewritten

$$b = Ax = \sum_{j=1}^{n} x_j a_j. \tag{1.2}$$

This equation can be displayed schematically as follows:

$$\begin{bmatrix} \\ b \\ \\ \end{bmatrix} = \begin{bmatrix} & & & \\ a_1 & a_2 & \cdots & a_n \\ & & & \end{bmatrix} \begin{bmatrix} x_1 \\ x_2 \\ \vdots \\ x_n \end{bmatrix} = x_1 \begin{bmatrix} \\ a_1 \\ \\ \end{bmatrix} + x_2 \begin{bmatrix} \\ a_2 \\ \\ \end{bmatrix} + \cdots + x_n \begin{bmatrix} \\ a_n \\ \\ \end{bmatrix}.$$

In (1.2), b is expressed as a linear combination of the columns a_j. Nothing but a slight change of notation has occurred in going from (1.1) to (1.2). Yet thinking of Ax in terms of the form (1.2) is essential for a proper understanding of the algorithms of numerical linear algebra.

    We can summarize these different descriptions of matrix-vector products in the following way. As mathematicians, we are used to viewing the formula $Ax = b$ as a statement that A acts on x to produce b. The formula (1.2), by contrast, suggests the interpretation that x acts on A to produce b.

Example 1.1. Vandermonde Matrix. Fix a sequence of numbers $\{x_1, x_2, \ldots, x_m\}$. If p and q are polynomials of degree $< n$ and α is a scalar, then $p+q$ and αp are also polynomials of degree $< n$. Moreover, the values of these polynomials at the points x_i satisfy the following linearity properties:

$$\begin{aligned} (p + q)(x_i) &= p(x_i) + q(x_i), \\ (\alpha p)(x_i) &= \alpha(p(x_i)). \end{aligned}$$

Thus the map from vectors of coefficients of polynomials p of degree $< n$ to vectors $(p(x_1), p(x_2), \ldots, p(x_m))$ of sampled polynomial values is linear. Any linear map can be expressed as multiplication by a matrix; this is an example. In fact, it is expressed by an $m \times n$ *Vandermonde matrix*

$$A = \begin{bmatrix} 1 & x_1 & x_1^2 & \cdots & x_1^{n-1} \\ 1 & x_2 & x_2^2 & \cdots & x_2^{n-1} \\ \vdots & \vdots & \vdots & & \vdots \\ 1 & x_m & x_m^2 & \cdots & x_m^{n-1} \end{bmatrix}.$$

If c is the column vector of coefficients of p,

$$c = \begin{bmatrix} c_0 \\ c_1 \\ c_2 \\ \vdots \\ c_{n-1} \end{bmatrix}, \qquad p(x) = c_0 + c_1 x + c_2 x^2 + \cdots + c_{n-1} x^{n-1},$$

then the product Ac gives the sampled polynomial values. That is, for each i from 1 to m, we have

$$(Ac)_i = c_0 + c_1 x_i + c_2 x_i^2 + \cdots + c_{n-1} x_i^{n-1} = p(x_i). \qquad (1.3)$$

In this example, it is clear that the matrix-vector product Ac need not be thought of as m distinct scalar summations, each giving a different linear combination of the entries of c, as (1.1) might suggest. Instead, A can be viewed as a matrix of columns, each giving sampled values of a monomial,

$$A = \begin{bmatrix} 1 & x & x^2 & \cdots & x^{n-1} \end{bmatrix}, \qquad (1.4)$$

and the product Ac should be understood as a single vector summation in the form of (1.2) that at once gives a linear combination of these monomials,

$$Ac = c_0 + c_1 x + c_2 x^2 + \cdots + c_{n-1} x^{n-1} = p(x). \qquad \square$$

The remainder of this lecture will review some fundamental concepts in linear algebra from the point of view of (1.2).

A Matrix Times a Matrix

For the matrix-matrix product $B = AC$, *each column of B is a linear combination of the columns of A.* To derive this fact, we begin with the usual formula for matrix products. If A is $\ell \times m$ and C is $m \times n$, then B is $\ell \times n$, with entries defined by

$$b_{ij} = \sum_{k=1}^{m} a_{ik} c_{kj}. \qquad (1.5)$$

Here b_{ij}, a_{ik}, and c_{kj} are entries of B, A, and C, respectively. Written in terms of columns, the product is

$$\begin{bmatrix} b_1 & b_2 & \cdots & b_n \end{bmatrix} = \begin{bmatrix} a_1 & a_2 & \cdots & a_m \end{bmatrix} \begin{bmatrix} c_1 & c_2 & \cdots & c_n \end{bmatrix},$$

and (1.5) becomes

$$b_j = Ac_j = \sum_{k=1}^{m} c_{kj} a_k. \qquad (1.6)$$

Thus b_j is a linear combination of the columns a_k with coefficients c_{kj}.

Example 1.2. Outer Product. A simple example of a matrix-matrix product is the *outer product*. This is the product of an m-dimensional column vector u with an n-dimensional row vector v; the result is an $m \times n$ matrix of rank 1. The outer product can be written

$$\begin{bmatrix} \\ u \\ \\ \end{bmatrix} \begin{bmatrix} v_1 & v_2 & \cdots & v_n \end{bmatrix} = \begin{bmatrix} & & & \\ v_1 u & v_2 u & \cdots & v_n u \\ & & & \end{bmatrix} = \begin{bmatrix} v_1 u_1 & \cdots & v_n u_1 \\ \vdots & & \vdots \\ v_1 u_m & \cdots & v_n u_m \end{bmatrix}.$$

The columns are all multiples of the same vector u, and similarly, the rows are all multiples of the same vector v.   □

Example 1.3. As a second illustration, consider $B = AR$, where R is the upper-triangular $n \times n$ matrix with entries $r_{ij} = 1$ for $i \leq j$ and $r_{ij} = 0$ for $i > j$. This product can be written

$$\begin{bmatrix} & & \\ b_1 & \cdots & b_n \\ & & \end{bmatrix} = \begin{bmatrix} & & \\ a_1 & \cdots & a_n \\ & & \end{bmatrix} \begin{bmatrix} 1 & \cdots & 1 \\ & \ddots & \vdots \\ & & 1 \end{bmatrix}.$$

The column formula (1.6) now gives

$$b_j = Ar_j = \sum_{k=1}^{j} a_k. \qquad (1.7)$$

That is, the jth column of B is the sum of the first j columns of A. The matrix R is a discrete analogue of an indefinite integral operator.   □

Range and Nullspace

The *range* of a matrix A, written range(A), is the set of vectors that can be expressed as Ax for some x. The formula (1.2) leads naturally to the following characterization of range(A).

Theorem 1.1. range(A) *is the space spanned by the columns of A.*

Proof. By (1.2), any Ax is a linear combination of the columns of A. Conversely, any vector y in the space spanned by the columns of A can be written as a linear combination of the columns, $y = \sum_{j=1}^{n} x_j a_j$. Forming a vector x out of the coefficients x_j, we have $y = Ax$, and thus y is in the range of A. □

In view of Theorem 1.1, the range of a matrix A is also called the *column space* of A.

The *nullspace* of $A \in \mathbb{C}^{m \times n}$, written null$(A)$, is the set of vectors x that satisfy $Ax = 0$, where 0 is the 0-vector in $\mathbb{C}^m$. The entries of each vector $x \in$ null(A) give the coefficients of an expansion of zero as a linear combination of columns of A: $0 = x_1 a_1 + x_2 a_2 + \cdots + x_n a_n$.

Rank

The *column rank* of a matrix is the dimension of its column space. Similarly, the *row rank* of a matrix is the dimension of the space spanned by its rows. Row rank always equals column rank (among other proofs, this is a corollary of the singular value decomposition, discussed in Lectures 4 and 5), so we refer to this number simply as the *rank* of a matrix.

An $m \times n$ matrix of *full rank* is one that has the maximal possible rank (the lesser of m and n). This means that a matrix of full rank with $m \geq n$ must have n linearly independent columns. Such a matrix can also be characterized by the property that the map it defines is one-to-one.

Theorem 1.2. *A matrix $A \in \mathbb{C}^{m \times n}$ with $m \geq n$ has full rank if and only if it maps no two distinct vectors to the same vector.*

Proof. ($\Longrightarrow$) If A is of full rank, its columns are linearly independent, so they form a basis for range(A). This means that every $b \in$ range(A) has a unique linear expansion in terms of the columns of A, and therefore, by (1.2), every $b \in$ range(A) has a unique x such that $b = Ax$. ($\Longleftarrow$) Conversely, if A is not of full rank, its columns a_j are dependent, and there is a nontrivial linear combination such that $\sum_{j=1}^{n} c_j a_j = 0$. The nonzero vector c formed from the coefficients c_j satisfies $Ac = 0$. But then A maps distinct vectors to the same vector since, for any x, $Ax = A(x + c)$. □

Inverse

A *nonsingular* or *invertible* matrix is a square matrix of full rank. Note that the m columns of a nonsingular $m \times m$ matrix A form a basis for the whole space $\mathbb{C}^m$. Therefore, we can uniquely express any vector as a linear combination of them. In particular, the canonical unit vector with 1 in the jth entry and zeros elsewhere, written e_j, can be expanded:

$$e_j = \sum_{i=1}^{m} z_{ij} a_i. \tag{1.8}$$

Let Z be the matrix with entries z_{ij}, and let z_j denote the jth column of Z. Then (1.8) can be written $e_j = A z_j$. This equation has the form (1.6); it can be written again, most concisely, as

$$\left[\begin{array}{c|c|c} & & \\ e_1 & \cdots & e_m \\ & & \end{array} \right] = I = AZ,$$

where I is the $m \times m$ matrix known as the *identity*. The matrix Z is the *inverse* of A. Any square nonsingular matrix A has a unique inverse, written A^{-1}, that satisfies $AA^{-1} = A^{-1}A = I$.

The following theorem records a number of equivalent conditions that hold when a square matrix is nonsingular. These conditions appear in linear algebra texts, and we shall not give a proof here. Concerning (f), see Lecture 5.

Theorem 1.3. *For $A \in \mathbb{C}^{m \times m}$, the following conditions are equivalent:*
(a) *A has an inverse A^{-1},*
(b) *$\mathrm{rank}(A) = m$,*
(c) *$\mathrm{range}(A) = \mathbb{C}^m$,*
(d) *$\mathrm{null}(A) = \{0\}$,*
(e) *0 is not an eigenvalue of A,*
(f) *0 is not a singular value of A,*
(g) *$\det(A) \neq 0$.*

Concerning (g), we mention that the determinant, though a convenient notion theoretically, rarely finds a useful role in numerical algorithms.

A Matrix Inverse Times a Vector

When writing the product $x = A^{-1}b$, it is important not to let the inverse-matrix notation obscure what is really going on! Rather than thinking of x as the result of applying A^{-1} to b, we should understand it as the unique vector that satisfies the equation $Ax = b$. By (1.2), this means that x is the vector of coefficients of the unique linear expansion of b in the basis of columns of A.

This point cannot be emphasized too much, so we repeat:

$A^{-1}b$ is the vector of coefficients of the expansion of b
in the basis of columns of A.

Multiplication by A^{-1} is a *change of basis* operation:

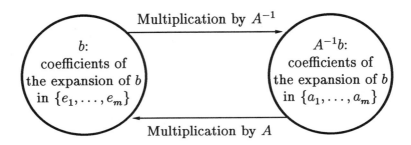

In this description we are being casual with terminology, using "b" in one instance to denote an m-tuple of numbers, and in another, as a point in an abstract vector space. The reader should think about these matters until he or she is comfortable with the distinction.

A Note on m and n

Throughout numerical linear algebra, it is customary to take a rectangular matrix to have dimensions $m \times n$. We follow this convention in this book.

What if the matrix is square? The usual convention is to give it dimensions $n \times n$, but in this book we shall generally take the other choice, $m \times m$. Many of our algorithms require us to look at rectangular submatrices formed by taking a subset of the columns of a square matrix. If the submatrix is to be $m \times n$, the original matrix had better be $m \times m$.

Exercises

1.1. Let B be a 4×4 matrix to which we apply the following operations:
    1. double column 1,
    2. halve row 3,
    3. add row 3 to row 1,
    4. interchange columns 1 and 4,
    5. subtract row 2 from each of the other rows,
    6. replace column 4 by column 3,
    7. delete column 1 (so that the column dimension is reduced by 1).

(a) Write the result as a product of eight matrices.

(b) Write it again as a product ABC (same B) of three matrices.

1.2. Suppose masses m_1, m_2, m_3, m_4 are located at positions x_1, x_2, x_3, x_4 in a line and connected by springs with spring constants k_{12}, k_{23}, k_{34} whose natural lengths of extension are $\ell_{12}, \ell_{23}, \ell_{34}$. Let f_1, f_2, f_3, f_4 denote the rightward forces on the masses, e.g., $f_1 = k_{12}(x_2 - x_1 - \ell_{12})$.

(a) Write the 4×4 matrix equation relating the column vectors f and x. Let K denote the matrix in this equation.

(b) What are the dimensions of the entries of K in the physics sense (e.g., mass times time, distance divided by mass, etc.)?

(c) What are the dimensions of $\det(K)$, again in the physics sense?

(d) Suppose K is given numerical values based on the units meters, kilograms, and seconds. Now the system is rewritten with a matrix K' based on centimeters, grams, and seconds. What is the relationship of K' to K? What is the relationship of $\det(K')$ to $\det(K)$?

1.3. Generalizing Example 1.3, we say that a square or rectangular matrix R with entries r_{ij} is *upper-triangular* if $r_{ij} = 0$ for $i > j$. By considering what space is spanned by the first n columns of R and using (1.8), show that if R is a nonsingular $m \times m$ upper-triangular matrix, then R^{-1} is also upper-triangular. (The analogous result also holds for lower-triangular matrices.)

1.4. Let $f_1, \ldots, f_8$ be a set of functions defined on the interval $[1, 8]$ with the property that for any numbers $d_1, \ldots, d_8$, there exists a set of coefficients $c_1, \ldots, c_8$ such that

$$\sum_{j=1}^{8} c_j f_j(i) = d_i, \qquad i = 1, \ldots, 8.$$

(a) Show by appealing to the theorems of this lecture that $d_1, \ldots, d_8$ determine $c_1, \ldots, c_8$ uniquely.

(b) Let A be the 8×8 matrix representing the linear mapping from data $d_1, \ldots, d_8$ to coefficients $c_1, \ldots, c_8$. What is the i, j entry of A^{-1}?

Lecture 2. Orthogonal Vectors and Matrices

Since the 1960s, many of the best algorithms of numerical linear algebra have been based in one way or another on orthogonality. In this lecture we present the ingredients: orthogonal vectors and orthogonal (unitary) matrices.

Adjoint

The *complex conjugate* of a scalar z, written $\overline{z}$ or z^*, is obtained by negating its imaginary part. For real z, $\overline{z} = z$.

The *hermitian conjugate* or *adjoint* of an $m \times n$ matrix A, written A^*, is the $n \times m$ matrix whose i,j entry is the complex conjugate of the j,i entry of A. For example,

$$A = \begin{bmatrix} a_{11} & a_{12} \\ a_{21} & a_{22} \\ a_{31} & a_{32} \end{bmatrix} \qquad \Longrightarrow \qquad A^* = \begin{bmatrix} \overline{a}_{11} & \overline{a}_{21} & \overline{a}_{31} \\ \overline{a}_{12} & \overline{a}_{22} & \overline{a}_{32} \end{bmatrix}.$$

If $A = A^*$, A is *hermitian*. By definition, a hermitian matrix must be square. For real A, the adjoint simply interchanges the rows and columns of A. In this case, the adjoint is also known as the *transpose*, and is written A^T. If a real matrix is hermitian, that is, $A = A^T$, then it is also said to be *symmetric*.

Most textbooks of numerical linear algebra assume that the matrices under discussion are real and thus principally use ^{T} instead of *. Since most of the ideas to be dealt with are not intrinsically restricted to the reals, however, we have followed the other course. Thus, for example, in this book a row vector

11

will usually be denoted by, say, a^* rather than a^T. The reader who prefers to imagine that all quantities are real and that * is a synonym for ^{T} will rarely get into trouble.

Inner Product

The *inner product* of two column vectors $x, y \in \mathbb{C}^m$ is the product of the adjoint of x by y:

$$x^* y = \sum_{i=1}^{m} \overline{x}_i y_i. \tag{2.1}$$

The Euclidean length of x may be written $\|x\|$ (vector norms such as this are discussed systematically in the next lecture), and can be defined as the square root of the inner product of x with itself:

$$\|x\| = \sqrt{x^* x} = \left(\sum_{i=1}^{m} |x_i|^2 \right)^{1/2}. \tag{2.2}$$

The cosine of the angle α between x and y can also be expressed in terms of the inner product:

$$\cos \alpha = \frac{x^* y}{\|x\| \, \|y\|}. \tag{2.3}$$

At various points of this book, as here, we mention geometric interpretations of algebraic formulas. For these geometric interpretations, the reader should think of the vectors as real rather than complex, although usually the interpretations can be carried over in one way or another to the complex case too.

The inner product is *bilinear*, which means that it is linear in each vector separately:

$$
\begin{aligned}
(x_1 + x_2)^* y &= x_1^* y + x_2^* y, \\
x^* (y_1 + y_2) &= x^* y_1 + x^* y_2, \\
(\alpha x)^* (\beta y) &= \overline{\alpha} \beta x^* y.
\end{aligned}
$$

We shall also frequently use the easily proved property that for any matrices or vectors A and B of compatible dimensions,

$$(AB)^* = B^* A^*. \tag{2.4}$$

This is analogous to the equally important formula for products of invertible square matrices,

$$(AB)^{-1} = B^{-1} A^{-1}. \tag{2.5}$$

The notation A^{-*} is a shorthand for $(A^*)^{-1}$ or $(A^{-1})^*$; these two are equal, as can be verified by applying (2.4) with $B = A^{-1}$.

Orthogonal Vectors

A pair of vectors x and y are *orthogonal* if $x^*y = 0$. If x and y are real, this means they lie at right angles to each other in $\mathbb{R}^m$. Two sets of vectors X and Y are orthogonal (also stated "X is orthogonal to Y") if every $x \in X$ is orthogonal to every $y \in Y$.

A set of nonzero vectors S is *orthogonal* if its elements are pairwise orthogonal, i.e., if for $x, y \in S$, $x \neq y \Rightarrow x^*y = 0$. A set of vectors is *orthonormal* if it is orthogonal and, in addition, every $x \in S$ has $\|x\| = 1$.

Theorem 2.1. *The vectors in an orthogonal set S are linearly independent.*

Proof. If the vectors in S are not independent, then some $v_k \in S$ can be expressed as a linear combination of other members $v_1, \ldots, v_n \in S$,

$$v_k = \sum_{\substack{i=1 \\ i \neq k}}^{n} c_i v_i.$$

Since $v_k \neq 0$, $v_k^* v_k = \|v_k\|^2 > 0$. Using the bilinearity of inner products and the orthogonality of S, we calculate

$$v_k^* v_k = \sum_{\substack{i=1 \\ i \neq k}}^{n} c_i v_k^* v_i = 0,$$

which contradicts the assumption that the vectors in S are nonzero. □

As a corollary of Theorem 2.1 it follows that if an orthogonal set $S \subseteq \mathbb{C}^m$ contains m vectors, then it is a basis for $\mathbb{C}^m$.

Components of a Vector

The most important idea to draw from the concepts of inner products and orthogonality is this: inner products can be used to decompose arbitrary vectors into orthogonal components.

For example, suppose that $\{q_1, q_2, \ldots, q_n\}$ is an orthonormal set, and let v be an arbitrary vector. The quantity $q_j^* v$ is a scalar. Utilizing these scalars as coordinates in an expansion, we find that the vector

$$r = v - (q_1^* v)q_1 - (q_2^* v)q_2 - \cdots - (q_n^* v)q_n \tag{2.6}$$

is orthogonal to $\{q_1, q_2, \ldots, q_n\}$. This can be verified by computing $q_i^* r$:

$$q_i^* r = q_i^* v - (q_1^* v)(q_i^* q_1) - \cdots - (q_n^* v)(q_i^* q_n).$$

This sum collapses, since $q_i^* q_j = 0$ for $i \neq j$:

$$q_i^* r = q_i^* v - (q_i^* v)(q_i^* q_i) = 0.$$

Thus we see that v can be decomposed into $n + 1$ orthogonal components:

$$v = r + \sum_{i=1}^{n}(q_i^* v)q_i = r + \sum_{i=1}^{n}(q_i q_i^*)v. \tag{2.7}$$

In this decomposition, r is the part of v orthogonal to the set of vectors $\{q_1, q_2, \ldots, q_n\}$, or, equivalently, to the subspace spanned by this set of vectors, and $(q_i^* v)q_i$ is the part of v in the direction of q_i.

If $\{q_i\}$ is a basis for $\mathbb{C}^m$, then n must be equal to m and r must be the zero vector, so v is completely decomposed into m orthogonal components in the directions of the q_i:

$$v = \sum_{i=1}^{m}(q_i^* v)q_i = \sum_{i=1}^{m}(q_i q_i^*)v. \tag{2.8}$$

In both (2.7) and (2.8) we have written the formula in two different ways, once with $(q_i^* v)q_i$ and again with $(q_i q_i^*)v$. These expressions are equal, but they have different interpretations. In the first case, we view v as a sum of coefficients $q_i^* v$ times vectors q_i. In the second, we view v as a sum of orthogonal projections of v onto the various directions q_i. The ith projection operation is achieved by the very special rank-one matrix $q_i q_i^*$. We shall discuss this and other projection processes in Lecture 6.

Unitary Matrices

A square matrix $Q \in \mathbb{C}^{m \times m}$ is *unitary* (in the real case, we also say *orthogonal*) if $Q^* = Q^{-1}$, i.e, if $Q^* Q = I$. In terms of the columns of Q, this product can be written

$$\begin{bmatrix} \underline{\quad q_1^* \quad} \\ \underline{\quad q_2^* \quad} \\ \vdots \\ \underline{\quad q_m^* \quad} \end{bmatrix} \begin{bmatrix} \big| & \big| & & \big| \\ q_1 & q_2 & \cdots & q_m \\ \big| & \big| & & \big| \end{bmatrix} = \begin{bmatrix} 1 & & & \\ & 1 & & \\ & & \ddots & \\ & & & 1 \end{bmatrix}.$$

In other words, $q_i^* q_j = \delta_{ij}$, and the columns of a unitary matrix Q form an orthonormal basis of $\mathbb{C}^m$. The symbol δ_{ij} is the *Kronecker delta*, equal to 1 if $i = j$ and 0 if $i \neq j$.

Multiplication by a Unitary Matrix

In the last lecture we discussed the interpretation of matrix-vector products Ax and $A^{-1}b$. If A is a unitary matrix Q, these products become Qx and Q^*b, and the same interpretations are of course still valid. As before, Qx is the linear combination of the columns of Q with coefficients x. Conversely,

Q^*b is the vector of coefficients of the expansion of b
in the basis of columns of Q.

Schematically, the situation looks like this:

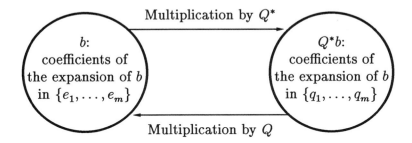

These processes of multiplication by a unitary matrix or its adjoint preserve geometric structure in the Euclidean sense, because inner products are preserved. That is, for unitary Q,

$$(Qx)^*(Qy) = x^*y, \qquad (2.9)$$

as is readily verified by (2.4). The invariance of inner products means that angles between vectors are preserved, and so are their lengths:

$$\|Qx\| = \|x\|. \qquad (2.10)$$

In the real case, multiplication by an orthogonal matrix Q corresponds to a rigid rotation (if $\det Q = 1$) or reflection (if $\det Q = -1$) of the vector space.

Exercises

2.1. Show that if a matrix A is both triangular and unitary, then it is diagonal.

2.2. The Pythagorean theorem asserts that for a set of n orthogonal vectors $\{x_i\}$,

$$\left\| \sum_{i=1}^{n} x_i \right\|^2 = \sum_{i=1}^{n} \|x_i\|^2.$$

(a) Prove this in the case $n = 2$ by an explicit computation of $\|x_1 + x_2\|^2$.
(b) Show that this computation also establishes the general case, by induction.

2.3. Let $A \in \mathbb{C}^{m \times m}$ be hermitian. An eigenvector of A is a nonzero vector $x \in \mathbb{C}^m$ such that $Ax = \lambda x$ for some $\lambda \in \mathbb{C}$, the corresponding eigenvalue.
(a) Prove that all eigenvalues of A are real.

(b) Prove that if x and y are eigenvectors corresponding to distinct eigenvalues, then x and y are orthogonal.

2.4. What can be said about the eigenvalues of a unitary matrix?

2.5. Let $S \in \mathbb{C}^{m \times m}$ be *skew-hermitian*, i.e., $S^* = -S$.
(a) Show by using Exercise 2.1 that the eigenvalues of S are pure imaginary.
(b) Show that $I - S$ is nonsingular.
(c) Show that the matrix $Q = (I-S)^{-1}(I+S)$, known as the *Cayley transform* of S, is unitary. (This is a matrix analogue of a linear fractional transformation $(1 + s)/(1 - s)$, which maps the left half of the complex s-plane conformally onto the unit disk.)

2.6. If u and v are m-vectors, the matrix $A = I + uv^*$ is known as a *rank-one perturbation of the identity*. Show that if A is nonsingular, then its inverse has the form $A^{-1} = I + \alpha uv^*$ for some scalar α, and give an expression for α. For what u and v is A singular? If it is singular, what is null(A)?

2.7. A *Hadamard matrix* is a matrix whose entries are all ± 1 and whose transpose is equal to its inverse times a constant factor. It is known that if A is a Hadamard matrix of dimension $m > 2$, then m is a multiple of 4. It is not known, however, whether there is a Hadamard matrix for every such m, though examples have been found for all cases $m \leq 424$.

Show that the following recursive description provides a Hadamard matrix of each dimension $m = 2^k$, $k = 0, 1, 2, \ldots$:

$$H_0 = \begin{bmatrix} 1 \end{bmatrix}, \qquad H_{k+1} = \begin{bmatrix} H_k & H_k \\ H_k & -H_k \end{bmatrix}.$$

Lecture 3. Norms

The essential notions of size and distance in a vector space are captured by norms. These are the yardsticks with which we measure approximations and convergence throughout numerical linear algebra.

Vector Norms

A *norm* is a function $\| \cdot \| : \mathbb{C}^m \to \mathbb{R}$ that assigns a real-valued length to each vector. In order to conform to a reasonable notion of length, a norm must satisfy the following three conditions. For all vectors x and y and for all scalars $\alpha \in \mathbb{C}$,

$$
\begin{aligned}
&(1) \ \ \|x\| \geq 0, \text{ and } \|x\| = 0 \text{ only if } x = 0, \\
&(2) \ \ \|x + y\| \leq \|x\| + \|y\|, \\
&(3) \ \ \|\alpha x\| = |\alpha| \, \|x\|.
\end{aligned}
\tag{3.1}
$$

In words, these conditions require that (1) the norm of a nonzero vector is positive, (2) the norm of a vector sum does not exceed the sum of the norms of its parts—the *triangle inequality*, and (3) scaling a vector scales its norm by the same amount.

In the last lecture, we used $\| \cdot \|$ to denote the Euclidean length function (the square root of the sum of the squares of the entries of a vector). However, the three conditions (3.1) allow for different notions of length, and at times it is useful to have this flexibility.

The most important class of vector norms, the p-norms, are defined below. The closed unit ball $\{x \in \mathbb{C}^m : \|x\| \leq 1\}$ corresponding to each norm is illustrated to the right for the case $m = 2$.

$$\|x\|_1 = \sum_{i=1}^{m} |x_i|,$$

$$\|x\|_2 = \left(\sum_{i=1}^{m} |x_i|^2\right)^{1/2} = \sqrt{x^*x},$$

$$\|x\|_\infty = \max_{1 \leq i \leq m} |x_i|, \qquad\qquad\qquad\qquad (3.2)$$

$$\|x\|_p = \left(\sum_{i=1}^{m} |x_i|^p\right)^{1/p} \quad (1 \leq p < \infty).$$

The 2-norm is the Euclidean length function; its unit ball is spherical. The 1-norm is used by airlines to define the maximal allowable size of a suitcase. The Sergel plaza in Stockholm, Sweden has the shape of the unit ball in the 4-norm; the Danish poet Piet Hein popularized this "superellipse" as a pleasing shape for objects such as conference tables.

Aside from the p-norms, the most useful norms are the *weighted p-norms*, where each of the coordinates of a vector space is given its own weight. In general, given any norm $\|\cdot\|$, a weighted norm can be written as

$$\|x\|_W = \|Wx\|. \qquad\qquad (3.3)$$

Here W is the diagonal matrix in which the ith diagonal entry is the weight $w_i \neq 0$. For example, a weighted 2-norm $\|\cdot\|_W$ on $\mathbb{C}^m$ is specified as follows:

$$\|x\|_W = \left(\sum_{i=1}^{m} |w_i x_i|^2\right)^{1/2}. \qquad\qquad (3.4)$$

One can also generalize the idea of weighted norms by allowing W to be an arbitrary nonsingular matrix, not necessarily diagonal (Exercise 3.1).

The most important norms in this book are the unweighted 2-norm and its induced matrix norm.

Matrix Norms Induced by Vector Norms

An $m \times n$ matrix can be viewed as a vector in an mn-dimensional space: each of the mn entries of the matrix is an independent coordinate. Any mn-dimensional norm can therefore be used for measuring the "size" of such a matrix.

However, in dealing with a space of matrices, certain special norms are more useful than the vector norms (3.2)–(3.3) already discussed. These are the *induced matrix norms*, defined in terms of the behavior of a matrix as an operator between its normed domain and range spaces.

Given vector norms $\| \cdot \|_{(n)}$ and $\| \cdot \|_{(m)}$ on the domain and the range of $A \in \mathbb{C}^{m \times n}$, respectively, the induced matrix norm $\|A\|_{(m,n)}$ is the smallest number C for which the following inequality holds for all $x \in \mathbb{C}^n$:

$$\|Ax\|_{(m)} \leq C\|x\|_{(n)}. \tag{3.5}$$

In other words, $\|A\|_{(m,n)}$ is the supremum of the ratios $\|Ax\|_{(m)}/\|x\|_{(n)}$ over all vectors $x \in \mathbb{C}^n$—the maximum factor by which A can "stretch" a vector x. We say that $\| \cdot \|_{(m,n)}$ is the matrix norm induced by $\| \cdot \|_{(m)}$ and $\| \cdot \|_{(n)}$.

Because of condition (3) of (3.1), the action of A is determined by its action on unit vectors. Therefore, the matrix norm can be defined equivalently in terms of the images of the unit vectors under A:

$$\|A\|_{(m,n)} = \sup_{\substack{x \in \mathbb{C}^n \\ x \neq 0}} \frac{\|Ax\|_{(m)}}{\|x\|_{(n)}} = \sup_{\substack{x \in \mathbb{C}^n \\ \|x\|_{(n)}=1}} \|Ax\|_{(m)}. \tag{3.6}$$

This form of the definition can be convenient for visualizing induced matrix norms, as in the sketches in (3.2) above.

Examples

Example 3.1. The matrix

$$A = \begin{bmatrix} 1 & 2 \\ 0 & 2 \end{bmatrix} \tag{3.7}$$

maps $\mathbb{C}^2$ to $\mathbb{C}^2$. It also maps $\mathbb{R}^2$ to $\mathbb{R}^2$, which is more convenient if we want to draw pictures and also (it can be shown) sufficient for determining matrix p-norms, since the coefficients of A are real.

Figure 3.1 depicts the action of A on the unit balls of $\mathbb{R}^2$ defined by the 1-, 2-, and ∞-norms. From this figure, one can see a graphical interpretation of these three norms of A. Regardless of the norm, A maps $e_1 = (1,0)^*$ to the first column of A, namely e_1 itself, and $e_2 = (0,1)^*$ to the second column of A, namely $(2,2)^*$. In the 1-norm, the unit vector x that is amplified most by A is $(0,1)^*$ (or its negative), and the amplification factor is 4. In the ∞-norm, the unit vector x that is amplified most by A is $(1,1)^*$ (or its negative), and the amplification factor is 3. In the 2-norm, the unit vector that is amplified most by A is the vector indicated by the dashed line in the figure (or its negative), and the amplification factor is approximately 2.9208. (Note that it must be at least $\sqrt{8} \approx 2.8284$, since $(0,1)^*$ maps to $(2,2)^*$.) We shall consider how to calculate such 2-norm results in Lecture 5.   □

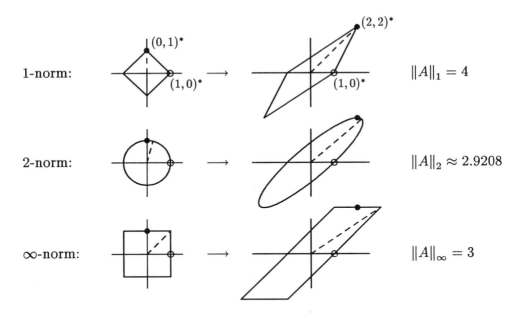

Figure 3.1. *On the left, the unit balls of $\mathbb{R}^2$ with respect to $\|\cdot\|_1$, $\|\cdot\|_2$, and $\|\cdot\|_\infty$. On the right, their images under the matrix A of (3.7). Dashed lines mark the vectors that are amplified most by A in each norm.*

Example 3.2. The p-Norm of a Diagonal Matrix. Let D be the diagonal matrix

$$D = \begin{bmatrix} d_1 & & & \\ & d_2 & & \\ & & \ddots & \\ & & & d_m \end{bmatrix}.$$

Then, as in the second row of Figure 3.1, the image of the 2-norm unit sphere under D is an m-dimensional ellipse whose semiaxis lengths are given by the numbers $|d_i|$. The unit vectors amplified most by D are those that are mapped to the longest semiaxis of the ellipse, of length $\max_i\{|d_i|\}$. Therefore, we have $\|D\|_2 = \max_{1\le i\le m}\{|d_i|\}$. In the next lecture we shall see that *every* matrix maps the 2-norm unit sphere to an ellipse—properly called a *hyperellipse* if $m > 2$—though the axes may be oriented arbitrarily.

This result for the 2-norm generalizes to any p: if D is diagonal, then $\|D\|_p = \max_{1\le i\le m}|d_i|$. $\square$

Example 3.3. The 1-Norm of a Matrix. If A is any $m \times n$ matrix, then $\|A\|_1$ is equal to the "maximum column sum" of A. We explain and derive this result as follows. Write A in terms of its columns

$$A = \left[\; a_1 \;\middle|\; \cdots \;\middle|\; a_n \;\right], \tag{3.8}$$

where each a_j is an m-vector. Consider the diamond-shaped 1-norm unit ball in $\mathbb{C}^n$, illustrated in (3.2). This is the set $\{x \in \mathbb{C}^n : \sum_{j=1}^n |x_j| \leq 1\}$. Any vector Ax in the image of this set satisfies

$$\|Ax\|_1 = \left\|\sum_{j=1}^n x_j a_j\right\|_1 \leq \sum_{j=1}^n |x_j|\,\|a_j\|_1 \leq \max_{1 \leq j \leq n} \|a_j\|_1.$$

Therefore, the induced matrix 1-norm satisfies $\|A\|_1 \leq \max_{1 \leq j \leq n} \|a_j\|_1$. By choosing $x = e_j$, where j maximizes $\|a_j\|_1$, we attain this bound, and thus the matrix norm is

$$\|A\|_1 = \max_{1 \leq j \leq n} \|a_j\|_1. \tag{3.9}$$

$\square$

Example 3.4. The ∞-Norm of a Matrix. By much the same argument, it can be shown that the ∞-norm of an $m \times n$ matrix is equal to the "maximum row sum,"

$$\|A\|_\infty = \max_{1 \leq i \leq m} \|a_i^*\|_1, \tag{3.10}$$

where a_i^* denotes the ith row of A. $\square$

Cauchy–Schwarz and Hölder Inequalities

Computing matrix p-norms with $p \neq 1, \infty$ is more difficult, and to approach this problem, we note that inner products can be bounded using p-norms. Let p and q satisfy $1/p + 1/q = 1$, with $1 \leq p, q \leq \infty$. Then the *Hölder inequality* states that, for any vectors x and y,

$$|x^*y| \leq \|x\|_p \|y\|_q. \tag{3.11}$$

The *Cauchy–Schwarz inequality* is the special case $p = q = 2$:

$$|x^*y| \leq \|x\|_2 \|y\|_2. \tag{3.12}$$

Derivations of these results can be found in linear algebra texts. Both bounds are tight in the sense that for certain choices of x and y, the inequalities become equalities.

Example 3.5. The 2-Norm of a Row Vector. Consider a matrix A containing a single row. This matrix can be written as $A = a^*$, where a is a column vector. The Cauchy–Schwarz inequality allows us to obtain the induced matrix 2-norm. For any x, we have $\|Ax\|_2 = |a^*x| \leq \|a\|_2 \|x\|_2$. This bound is tight: observe that $\|Aa\|_2 = \|a\|_2^2$. Therefore, we have

$$\|A\|_2 = \sup_{x \neq 0}\{\|Ax\|_2 / \|x\|_2\} = \|a\|_2. \qquad \square$$

Example 3.6. The 2-Norm of an Outer Product. More generally, consider the rank-one outer product $A = uv^*$, where u is an m-vector and v is an n-vector. For any n-vector x, we can bound $\|Ax\|_2$ as follows:

$$\|Ax\|_2 = \|uv^*x\|_2 = \|u\|_2|v^*x| \leq \|u\|_2\|v\|_2\|x\|_2. \qquad (3.13)$$

Therefore $\|A\|_2 \leq \|u\|_2\|v\|_2$. Again, this inequality is an equality: consider the case $x = v$. $\qquad\qquad\square$

Bounding $\|AB\|$ in an Induced Matrix Norm

The induced matrix norm of a matrix product can also be bounded. Let $\|\cdot\|_{(\ell)}$, $\|\cdot\|_{(m)}$, and $\|\cdot\|_{(n)}$ be norms on $\mathbb{C}^l$, $\mathbb{C}^m$, and $\mathbb{C}^n$, respectively, and let A be an $l \times m$ matrix and B an $m \times n$ matrix. For any $x \in \mathbb{C}^n$ we have

$$\|ABx\|_{(\ell)} \leq \|A\|_{(\ell,m)}\|Bx\|_{(m)} \leq \|A\|_{(\ell,m)}\|B\|_{(m,n)}\|x\|_{(n)}.$$

Therefore the induced norm of AB must satisfy

$$\|AB\|_{(\ell,n)} \leq \|A\|_{(\ell,m)}\|B\|_{(m,n)}. \qquad (3.14)$$

In general, this inequality is not an equality. For example, the inequality $\|A^n\| \leq \|A\|^n$ holds for any square matrix in any matrix norm induced by a vector norm, but $\|A^n\| = \|A\|^n$ does not hold in general for $n \geq 2$.

General Matrix Norms

As noted above, matrix norms do not have to be induced by vector norms. In general, a matrix norm must merely satisfy the three vector norm conditions (3.1) applied in the mn-dimensional vector space of matrices:

$$(1) \;\; \|A\| \geq 0, \text{ and } \|A\| = 0 \text{ only if } A = 0,$$
$$(2) \;\; \|A + B\| \leq \|A\| + \|B\|, \qquad\qquad\qquad (3.15)$$
$$(3) \;\; \|\alpha A\| = |\alpha|\,\|A\|.$$

The most important matrix norm which is not induced by a vector norm is the *Hilbert–Schmidt* or *Frobenius norm*, defined by

$$\|A\|_F = \left(\sum_{i=1}^{m}\sum_{j=1}^{n}|a_{ij}|^2\right)^{1/2}. \qquad (3.16)$$

Observe that this is the same as the 2-norm of the matrix when viewed as an mn-dimensional vector. The formula for the Frobenius norm can also be written in terms of individual rows or columns. For example, if a_j is the jth column of A, we have

$$\|A\|_F = \left(\sum_{j=1}^{n} \|a_j\|_2^{\,2} \right)^{1/2} \tag{3.17}$$

This identity, as well as the analogous result based on rows instead of columns, can be expressed compactly by the equation

$$\|A\|_F = \sqrt{\operatorname{tr}(A^*A)} = \sqrt{\operatorname{tr}(AA^*)}, \tag{3.18}$$

where $\operatorname{tr}(B)$ denotes the *trace* of B, the sum of its diagonal entries.

Like an induced matrix norm, the Frobenius norm can be used to bound products of matrices. Let $C = AB$ with entries c_{ik}, and let a_i^* denote the ith row of A and b_j the jth column of B. Then $c_{ij} = a_i^* b_j$, so by the Cauchy–Schwarz inequality we have $|c_{ij}| \leq \|a_i\|_2 \|b_j\|_2$. Squaring both sides and summing over all i, j, we obtain

$$
\begin{aligned}
\|AB\|_F^2 &= \sum_{i=1}^{n} \sum_{j=1}^{m} |c_{ij}|^2 \\
&\leq \sum_{i=1}^{n} \sum_{j=1}^{m} \left(\|a_i\|_2 \|b_j\|_2 \right)^2 \\
&= \sum_{i=1}^{n} \left(\|a_i\|_2 \right)^2 \sum_{j=1}^{m} \left(\|b_j\|_2 \right)^2 = \|A\|_F^2 \|B\|_F^2.
\end{aligned}
$$

Invariance under Unitary Multiplication

One of the many special properties of the matrix 2-norm is that, like the vector 2-norm, it is invariant under multiplication by unitary matrices. The same property holds for the Frobenius norm.

Theorem 3.1. *For any $A \in \mathbb{C}^{m \times n}$ and unitary $Q \in \mathbb{C}^{m \times m}$, we have*

$$\|QA\|_2 = \|A\|_2, \qquad \|QA\|_F = \|A\|_F.$$

Proof. Since $\|Qx\|_2 = \|x\|_2$ for every x, by (2.10), the invariance in the 2-norm follows from (3.6). For the Frobenius norm we may use (3.18). $\square$

Theorem 3.1 remains valid if Q is generalized to a rectangular matrix with orthonormal columns, that is, $Q \in \mathbb{C}^{p \times m}$ with $p > m$. Analogous identities also hold for multiplication by unitary matrices on the right, or more generally, by rectangular matrices with orthonormal rows.

Exercises

3.1. Prove that if W is an arbitrary nonsingular matrix, the function $\|\cdot\|_W$ defined by (3.3) is a vector norm.

3.2. Let $\|\cdot\|$ denote any norm on $\mathbb{C}^m$ and also the induced matrix norm on $\mathbb{C}^{m \times m}$. Show that $\rho(A) \leq \|A\|$, where $\rho(A)$ is the *spectral radius* of A, i.e., the largest absolute value $|\lambda|$ of an eigenvalue λ of A.

3.3. Vector and matrix p-norms are related by various inequalities, often involving the dimensions m or n. For each of the following, verify the inequality and give an example of a nonzero vector or matrix (for general m, n) for which equality is achieved. In this problem x is an m-vector and A is an $m \times n$ matrix.

(a) $\|x\|_\infty \leq \|x\|_2$,

(b) $\|x\|_2 \leq \sqrt{m}\,\|x\|_\infty$,

(c) $\|A\|_\infty \leq \sqrt{n}\,\|A\|_2$,

(d) $\|A\|_2 \leq \sqrt{m}\,\|A\|_\infty$.

3.4. Let A be an $m \times n$ matrix and let B be a submatrix of A, that is, a $\mu \times \nu$ matrix ($\mu \leq m$, $\nu \leq n$) obtained by selecting certain rows and columns of A.

(a) Explain how B can be obtained by multiplying A by certain row and column "deletion matrices" as in step 7 of Exercise 1.1.

(b) Using this product, show that $\|B\|_p \leq \|A\|_p$ for any p with $1 \leq p \leq \infty$.

3.5. Example 3.6 shows that if E is an outer product $E = uv^*$, then $\|E\|_2 = \|u\|_2\|v\|_2$. Is the same true for the Frobenius norm, i.e., $\|E\|_F = \|u\|_F\|v\|_F$? Prove it or give a counterexample.

3.6. Let $\|\cdot\|$ denote any norm on $\mathbb{C}^m$. The corresponding *dual norm* $\|\cdot\|'$ is defined by the formula $\|x\|' = \sup_{\|y\|=1} |y^*x|$.

(a) Prove that $\|\cdot\|'$ is a norm.

(b) Let $x, y \in \mathbb{C}^m$ with $\|x\| = \|y\| = 1$ be given. Show that there exists a rank-one matrix $B = yz^*$ such that $Bx = y$ and $\|B\| = 1$, where $\|B\|$ is the matrix norm of B induced by the vector norm $\|\cdot\|$. You may use the following lemma, without proof: given $x \in \mathbb{C}^m$, there exists a nonzero $z \in \mathbb{C}^m$ such that $|z^*x| = \|z\|'\|x\|$.

Lecture 4. The Singular Value Decomposition

奇異值分解

The singular value decomposition (SVD) is a matrix factorization whose computation is a step in many algorithms. Equally important is the use of the SVD for conceptual purposes. Many problems of linear algebra can be better understood if we first ask the question: what if we take the SVD?

A Geometric Observation

The SVD is motivated by the following geometric fact:

The image of the unit sphere under any $m \times n$ matrix is a hyperellipse.

The SVD is applicable to both real and complex matrices. However, in describing the geometric interpretation, we assume as usual that the matrix is real.

The term "hyperellipse" may be unfamiliar, but this is just the m-dimensional generalization of an ellipse. We may define a hyperellipse in $\mathbb{R}^m$ as the surface obtained by stretching the unit sphere in $\mathbb{R}^m$ by some factors $\sigma_1, \ldots, \sigma_m$ (possibly zero) in some orthogonal directions $u_1, \ldots, u_m \in \mathbb{R}^m$. For convenience, let us take the u_i to be unit vectors, i.e., $\|u_i\|_2 = 1$. The vectors $\{\sigma_i u_i\}$ are the *principal semiaxes* of the hyperellipse, with lengths $\sigma_1, \ldots, \sigma_m$. If A has rank r, exactly r of the lengths σ_i will turn out to be nonzero, and in particular, if $m \geq n$, at most n of them will be nonzero.

25

Our opening statement about the image of the unit sphere has the following meaning. By the unit sphere, we mean the usual Euclidean sphere in n-space, i.e., the unit sphere in the 2-norm; let us denote it by S. Then AS, the image of S under the mapping A, is a hyperellipse as just defined.

This geometric fact is not obvious. We shall restate it in the language of linear algebra and prove it later. For the moment, assume it is true.

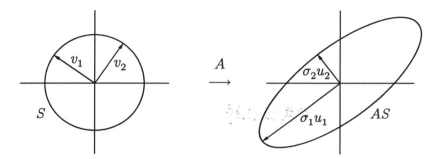

Figure 4.1. *SVD of a* 2×2 *matrix.*

Let S be the unit sphere in $\mathbb{R}^n$, and take any $A \in \mathbb{R}^{m \times n}$ with $m \geq n$. For simplicity, suppose for the moment that A has full rank n. The image AS is a hyperellipse in $\mathbb{R}^m$. We now define some properties of A in terms of the shape of AS. The key ideas are indicated in Figure 4.1.

First, we define the n *singular values* of A. These are the lengths of the n principal semiaxes of AS, written $\sigma_1, \sigma_2, \ldots, \sigma_n$. It is conventional to assume that the singular values are numbered in descending order, $\sigma_1 \geq \sigma_2 \geq \cdots \geq \sigma_n > 0$.

Next, we define the n *left singular vectors* of A. These are the unit vectors $\{u_1, u_2, \ldots, u_n\}$ oriented in the directions of the principal semiaxes of AS, numbered to correspond with the singular values. Thus the vector $\sigma_i u_i$ is the ith largest principal semiaxis of AS.

Finally, we define the n *right singular vectors* of A. These are the unit vectors $\{v_1, v_2, \ldots, v_n\} \in S$ that are the preimages of the principal semiaxes of AS, numbered so that $Av_j = \sigma_j u_j$.

The terms "left" and "right" in the definitions above are decidedly awkward. They come from the positions of the factors U and V in (4.2) and (4.3), below. What is awkward is that in a sketch like Figure 4.1, the left singular vectors correspond to the space on the right, and the right singular vectors correspond to the space on the left! One could resolve this problem by interchanging the two halves of the figure, with the map A pointing from right to left, but that would go against deeply ingrained habits.

Reduced SVD

We have just mentioned that the equations relating right singular vectors $\{v_j\}$ and left singular vectors $\{u_j\}$ can be written

$$Av_j = \sigma_j u_j, \qquad 1 \le j \le n. \tag{4.1}$$

This collection of vector equations can be expressed as a matrix equation,

$$\begin{bmatrix} & \\ & A & \\ & \end{bmatrix} \begin{bmatrix} v_1 & v_2 & \cdots & v_n \end{bmatrix} = \begin{bmatrix} u_1 & u_2 & \cdots & u_n \end{bmatrix} \begin{bmatrix} \sigma_1 & & & \\ & \sigma_2 & & \\ & & \ddots & \\ & & & \sigma_n \end{bmatrix},$$

or, more compactly, $AV = \hat{U}\hat{\Sigma}$. In this matrix equation, $\hat{\Sigma}$ is an $n \times n$ diagonal matrix with positive real entries (since A was assumed to have full rank n), $\hat{U}$ is an $m \times n$ matrix with orthonormal columns, and V is an $n \times n$ matrix with orthonormal columns. Thus V is unitary, and we can multiply on the right by its inverse V^* to obtain

$$A = \hat{U}\hat{\Sigma}V^*. \tag{4.2}$$

This factorization of A is called a *reduced singular value decomposition*, or *reduced SVD*, of A. Schematically, it looks like this:

Reduced SVD ($m \ge n$)

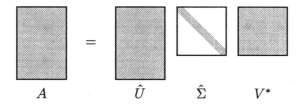

$$A \qquad\qquad \hat{U} \qquad \hat{\Sigma} \qquad V^*$$

Full SVD

In most applications, the SVD is used in exactly the form just described. However, this is not the way in which the idea of an SVD is usually formulated in textbooks. We have introduced the term "reduced" and the hats on U and Σ in order to distinguish the factorization (4.2) from the more standard "full" SVD. This "reduced" vs. "full" terminology and hatted notation will be maintained throughout the book, and we shall make a similar distinction between reduced and full QR factorizations. Reminders of these conventions are printed on the inside front cover.

The idea is as follows. The columns of $\hat{U}$ are n orthonormal vectors in the m-dimensional space $\mathbb{C}^m$. Unless $m = n$, they do not form a basis of $\mathbb{C}^m$, nor is $\hat{U}$ a unitary matrix. However, by adjoining an additional $m - n$ orthonormal columns, $\hat{U}$ can be extended to a unitary matrix. Let us do this in an arbitrary fashion, and call the result U.

If $\hat{U}$ is replaced by U in (4.2), then $\hat{\Sigma}$ will have to change too. For the product to remain unaltered, the last $m - n$ columns of U should be multiplied by zero. Accordingly, let Σ be the $m \times n$ matrix consisting of $\hat{\Sigma}$ in the upper $n \times n$ block together with $m - n$ rows of zeros below. We now have a new factorization, the *full SVD* of A:

$$A = U\Sigma V^*. \tag{4.3}$$

Here U is $m \times m$ and unitary, V is $n \times n$ and unitary, and Σ is $m \times n$ and diagonal with positive real entries. Schematically:

Full SVD $(m \geq n)$

$$A \qquad\qquad U \qquad\qquad \Sigma \qquad V^*$$

The dashed lines indicate the "silent" columns of U and rows of Σ that are discarded in passing from (4.3) to (4.2).

Having described the full SVD, we can now discard the simplifying assumption that A has full rank. If A is rank-deficient, the factorization (4.3) is still appropriate. All that changes is that now not n but only r of the left singular vectors of A are determined by the geometry of the hyperellipse. To construct the unitary matrix U, we introduce $m - r$ instead of just $m - n$ additional arbitrary orthonormal columns. The matrix V will also need $n - r$ arbitrary orthonormal columns to extend the r columns determined by the geometry. The matrix Σ will now have r positive diagonal entries, with the remaining $n - r$ equal to zero.

By the same token, the reduced SVD (4.2) also makes sense for matrices A of less than full rank. One can take $\hat{U}$ to be $m \times n$, with $\hat{\Sigma}$ of dimensions $n \times n$ with some zeros on the diagonal, or further compress the representation so that $\hat{U}$ is $m \times r$ and $\hat{\Sigma}$ is $r \times r$ and strictly positive on the diagonal.

Formal Definition

Let m and n be arbitrary; we do not require $m \geq n$. Given $A \in \mathbb{C}^{m \times n}$, not necessarily of full rank, a *singular value decomposition* (SVD) of A is a factorization

$$A = U\Sigma V^* \tag{4.4}$$

where

$$U \in \mathbb{C}^{m \times m} \quad \text{is unitary,}$$
$$V \in \mathbb{C}^{n \times n} \quad \text{is unitary,}$$
$$\Sigma \in \mathbb{R}^{m \times n} \quad \text{is diagonal.}$$

In addition, it is assumed that the diagonal entries σ_j of Σ are nonnegative and in nonincreasing order; that is, $\sigma_1 \geq \sigma_2 \geq \cdots \geq \sigma_p \geq 0$, where $p = \min(m, n)$.

Note that the diagonal matrix Σ has the same shape as A even when A is not square, but U and V are always square unitary matrices.

It is clear that the image of the unit sphere in $\mathbb{R}^n$ under a map $A = U\Sigma V^*$ must be a hyperellipse in $\mathbb{R}^m$. The unitary map V^* preserves the sphere, the diagonal matrix Σ stretches the sphere into a hyperellipse aligned with the canonical basis, and the final unitary map U rotates or reflects the hyperellipse without changing its shape. Thus, if we can prove that every matrix has an SVD, we shall have proved that the image of the unit sphere under any linear map is a hyperellipse, as claimed at the outset of this lecture.

Existence and Uniqueness

Theorem 4.1. *Every matrix $A \in \mathbb{C}^{m \times n}$ has a singular value decomposition (4.4). Furthermore, the singular values $\{\sigma_j\}$ are uniquely determined, and, if A is square and the σ_j are distinct, the left and right singular vectors $\{u_j\}$ and $\{v_j\}$ are uniquely determined up to complex signs (i.e., complex scalar factors of absolute value 1).*

Proof. To prove existence of the SVD, we isolate the direction of the largest action of A, and then proceed by induction on the dimension of A.

Set $\sigma_1 = \|A\|_2$. By a compactness argument, there must be a vector $v_1 \in \mathbb{C}^n$ with $\|v_1\|_2 = 1$ and $\|u_1\|_2 = \sigma_1$, where $u_1 = Av_1$. Consider any extensions of v_1 to an orthonormal basis $\{v_j\}$ of $\mathbb{C}^n$ and of u_1 to an orthonormal basis $\{u_j\}$ of $\mathbb{C}^m$, and let U_1 and V_1 denote the unitary matrices with columns u_j and v_j, respectively. Then we have

$$U_1^* A V_1 = S = \begin{bmatrix} \sigma_1 & w^* \\ 0 & B \end{bmatrix}, \tag{4.5}$$

where 0 is a column vector of dimension $m-1$, w^* is a row vector of dimension $n - 1$, and B has dimensions $(m - 1) \times (n - 1)$. Furthermore,

$$\left\| \begin{bmatrix} \sigma_1 & w^* \\ 0 & B \end{bmatrix} \begin{bmatrix} \sigma_1 \\ w \end{bmatrix} \right\|_2 \geq \sigma_1^2 + w^* w = (\sigma_1^2 + w^* w)^{1/2} \left\| \begin{bmatrix} \sigma_1 \\ w \end{bmatrix} \right\|_2,$$

implying $\|S\|_2 \geq (\sigma_1^2 + w^*w)^{1/2}$. Since U_1 and V_1 are unitary, we know that $\|S\|_2 = \|A\|_2 = \sigma_1$, so this implies $w = 0$.

If $n = 1$ or $m = 1$, we are done. Otherwise, the submatrix B describes the action of A on the subspace orthogonal to v_1. By the induction hypothesis, B has an SVD $B = U_2 \Sigma_2 V_2^*$. Now it is easily verified that

$$A = U_1 \begin{bmatrix} 1 & 0 \\ 0 & U_2 \end{bmatrix} \begin{bmatrix} \sigma_1 & 0 \\ 0 & \Sigma_2 \end{bmatrix} \begin{bmatrix} 1 & 0 \\ 0 & V_2 \end{bmatrix}^* V_1^*$$

is an SVD of A, completing the proof of existence.

For the uniqueness claim, the geometric justification is straightforward: if the semiaxis lengths of a hyperellipse are distinct, then the semiaxes themselves are determined by the geometry, up to signs. Algebraically, we can argue as follows. First we note that σ_1 is uniquely determined by the condition that it is equal to $\|A\|_2$, as follows from (4.4). Now suppose that in addition to v_1, there is another linearly independent vector w with $\|w\|_2 = 1$ and $\|Aw\|_2 = \sigma_1$. Define a unit vector v_2, orthogonal to v_1, as a linear combination of v_1 and w,

$$v_2 = \frac{w - (v_1^*w)v_1}{\|w - (v_1^*w)v_1\|_2}.$$

Since $\|A\|_2 = \sigma_1$, $\|Av_2\|_2 \leq \sigma_1$; but this must be an equality, for otherwise, since $w = v_1 c + v_2 s$ for some constants c and s with $|c|^2 + |s|^2 = 1$, we would have $\|Aw\|_2 < \sigma_1$. This vector v_2 is a second right singular vector of A corresponding to the singular value σ_1; it will lead to the appearance of a vector y (equal to the last $n - 1$ components of $V_1^*v_2$) with $\|y\|_2 = 1$ and $\|By\|_2 = \sigma_1$. We conclude that, if the singular vector v_1 is not unique, then the corresponding singular value σ_1 is not simple. To complete the uniqueness proof we note that, as indicated above, once σ_1, v_1, and u_1 are determined, the remainder of the SVD is determined by the action of A on the space orthogonal to v_1. Since v_1 is unique up to sign, this orthogonal space is uniquely defined, and the uniqueness of the remaining singular values and vectors now follows by induction. $\square$

Exercises

4.1. Determine SVDs of the following matrices (by hand calculation):

(a) $\begin{bmatrix} 3 & 0 \\ 0 & -2 \end{bmatrix}$, (b) $\begin{bmatrix} 2 & 0 \\ 0 & 3 \end{bmatrix}$, (c) $\begin{bmatrix} 0 & 2 \\ 0 & 0 \\ 0 & 0 \end{bmatrix}$, (d) $\begin{bmatrix} 1 & 1 \\ 0 & 0 \end{bmatrix}$, (e) $\begin{bmatrix} 1 & 1 \\ 1 & 1 \end{bmatrix}$.

4.2. Suppose A is an $m \times n$ matrix and B is the $n \times m$ matrix obtained by rotating A ninety degrees clockwise on paper (not exactly a standard mathematical transformation!). Do A and B have the same singular values? Prove that the answer is yes or give a counterexample.

4.3. Write a MATLAB program (see Lecture 9) which, given a real 2×2 matrix A, plots the right singular vectors v_1 and v_2 in the unit circle and also the left singular vectors u_1 and u_2 in the appropriate ellipse, as in Figure 4.1. Apply your program to the matrix (3.7) and also to the 2×2 matrices of Exercise 4.1.

4.4. Two matrices $A, B \in \mathbb{C}^{m \times m}$ are *unitarily equivalent* if $A = QBQ^*$ for some unitary $Q \in \mathbb{C}^{m \times m}$. Is it true or false that A and B are unitarily equivalent if and only if they have the same singular values?

4.5. Theorem 4.1 asserts that every $A \in \mathbb{C}^{m \times n}$ has an SVD $A = U\Sigma V^*$. Show that if A is real, then it has a real SVD ($U \in \mathbb{R}^{m \times m}$, $V \in \mathbb{R}^{n \times n}$).

Lecture 5. More on the SVD

We continue our discussion of the singular value decomposition, emphasizing its connection with low-rank approximation of matrices in the 2-norm and the Frobenius norm.

A Change of Bases

The SVD makes it possible for us to say that every matrix is diagonal—if only one uses the proper bases for the domain and range spaces.

Here is how the change of bases works. Any $b \in \mathbb{C}^m$ can be expanded in the basis of left singular vectors of A (columns of U), and any $x \in \mathbb{C}^n$ can be expanded in the basis of right singular vectors of A (columns of V). The coordinate vectors for these expansions are

$$b' = U^*b, \qquad x' = V^*x.$$

By (4.3), the relation $b = Ax$ can be expressed in terms of b' and x':

$$b = Ax \quad \Longleftrightarrow \quad U^*b = U^*Ax = U^*U\Sigma V^*x \quad \Longleftrightarrow \quad b' = \Sigma x'.$$

Whenever $b = Ax$, we have $b' = \Sigma x'$. Thus A reduces to the diagonal matrix Σ when the range is expressed in the basis of columns of U and the domain is expressed in the basis of columns of V.

SVD vs. Eigenvalue Decomposition

The theme of diagonalizing a matrix by expressing it in terms of a new basis also underlies the study of eigenvalues. A nondefective square matrix A can be expressed as a diagonal matrix of eigenvalues Λ, if the range and domain are represented in a basis of eigenvectors.

If the columns of a matrix $X \in \mathbb{C}^{m \times m}$ contain linearly independent eigenvectors of $A \in \mathbb{C}^{m \times m}$, the *eigenvalue decomposition* of A is

$$A = X \Lambda X^{-1}, \tag{5.1}$$

where Λ is an $m \times m$ diagonal matrix whose entries are the eigenvalues of A. This implies that if we define, for $b, x \in \mathbb{C}^m$ satisfying $b = Ax$,

$$b' = X^{-1}b, \qquad x' = X^{-1}x,$$

then the newly expanded vectors b' and x' satisfy $b' = \Lambda x'$. Eigenvalues are treated systematically in Lecture 24.

There are fundamental differences between the SVD and the eigenvalue decomposition. One is that the SVD uses two different bases (the sets of left and right singular vectors), whereas the eigenvalue decomposition uses just one (the eigenvectors). Another is that the SVD uses orthonormal bases, whereas the eigenvalue decomposition uses a basis that generally is not orthogonal. A third is that not all matrices (even square ones) have an eigenvalue decomposition, but all matrices (even rectangular ones) have a singular value decomposition, as we established in Theorem 4.1. In applications, eigenvalues tend to be relevant to problems involving the behavior of iterated forms of A, such as matrix powers A^k or exponentials e^{tA}, whereas singular vectors tend to be relevant to problems involving the behavior of A itself, or its inverse.

Matrix Properties via the SVD

The power of the SVD becomes apparent as we begin to catalogue its connections with other fundamental topics of linear algebra. For the following theorems, assume that A has dimensions $m \times n$. Let p be the minimum of m and n, let $r \leq p$ denote the number of nonzero singular values of A, and let $\langle x, y, \ldots, z \rangle$ denote the space spanned by the vectors $x, y, \ldots, z$.

Theorem 5.1. *The rank of A is r, the number of nonzero singular values.*

Proof. The rank of a diagonal matrix is equal to the number of its nonzero entries, and in the decomposition $A = U \Sigma V^*$, U and V are of full rank. Therefore $\mathrm{rank}(A) = \mathrm{rank}(\Sigma) = r$. $\square$

Theorem 5.2. $\mathrm{range}(A) = \langle u_1, \ldots, u_r \rangle$ *and* $\mathrm{null}(A) = \langle v_{r+1}, \ldots, v_n \rangle$.

Proof. This is a consequence of the fact that $\mathrm{range}(\Sigma) = \langle e_1, \ldots, e_r \rangle \subseteq \mathbb{C}^m$ and $\mathrm{null}(\Sigma) = \langle e_{r+1}, \ldots, e_n \rangle \subseteq \mathbb{C}^n$. $\square$

Theorem 5.3. $\|A\|_2 = \sigma_1$ *and* $\|A\|_F = \sqrt{\sigma_1^2 + \sigma_2^2 + \cdots + \sigma_r^2}$.

Proof. The first result was already established in the proof of Theorem 4.1: since $A = U\Sigma V^*$ with unitary U and V, $\|A\|_2 = \|\Sigma\|_2 = \max\{|\sigma_j|\} = \sigma_1$, by Theorem 3.1. For the second, note that by Theorem 3.1 and the remark following, the Frobenius norm is invariant under unitary multiplication, so $\|A\|_F = \|\Sigma\|_F$, and by (3.16), this is given by the stated formula.   □

Theorem 5.4. *The nonzero singular values of A are the square roots of the nonzero eigenvalues of A^*A or AA^*. (These matrices have the same nonzero eigenvalues.)*

Proof. From the calculation

$$A^*A = (U\Sigma V^*)^*(U\Sigma V^*) = V\Sigma^*U^*U\Sigma V^* = V(\Sigma^*\Sigma)V^*,$$

we see that A^*A is similar to $\Sigma^*\Sigma$ and hence has the same n eigenvalues (see Lecture 24). The eigenvalues of the diagonal matrix $\Sigma^*\Sigma$ are $\sigma_1^2, \sigma_2^2, \ldots, \sigma_p^2$, with $n - p$ additional zero eigenvalues if $n > p$. A similar calculation applies to the m eigenvalues of AA^*.   □

Theorem 5.5. *If $A = A^*$, then the singular values of A are the absolute values of the eigenvalues of A.*

Proof. As is well known (see Exercise 2.3), a hermitian matrix has a complete set of orthogonal eigenvectors, and all of the eigenvalues are real. An equivalent statement is that (5.1) holds with X equal to some unitary matrix Q and Λ a real diagonal matrix. But then we can write

$$A = Q\Lambda Q^* = Q|\Lambda|\text{sign}(\Lambda)Q^*, \qquad (5.2)$$

where $|\Lambda|$ and $\text{sign}(\Lambda)$ denote the diagonal matrices whose entries are the numbers $|\lambda_j|$ and $\text{sign}(\lambda_j)$, respectively. (We could equally well have put the factor $\text{sign}(\Lambda)$ on the left of $|\Lambda|$ instead of the right.) Since $\text{sign}(\Lambda)Q^*$ is unitary whenever Q is unitary, (5.2) is an SVD of A, with the singular values equal to the diagonal entries of $|\Lambda|$, $|\lambda_j|$. If desired, these numbers can be put into nonincreasing order by inserting suitable permutation matrices as factors in the left-hand unitary matrix of (5.2), Q, and the right-hand unitary matrix, $\text{sign}(\Lambda)Q^*$.   □

Theorem 5.6. *For $A \in \mathbb{C}^{m \times m}$, $|\det(A)| = \prod_{i=1}^{m} \sigma_i$.*

Proof. The determinant of a product of square matrices is the product of the determinants of the factors. Furthermore, the determinant of a unitary matrix is always 1 in absolute value; this follows from the formula $U^*U = I$ and the property $\det(U^*) = (\det(U))^*$. Therefore,

$$|\det(A)| = |\det(U\Sigma V^*)| = |\det(U)|\,|\det(\Sigma)|\,|\det(V^*)| = |\det(\Sigma)| = \prod_{i=1}^{m}\sigma_i.$$

$\square$

Low-Rank Approximations

But what *is* the SVD? Another approach to an explanation is to consider how a matrix A might be represented as a sum of rank-one matrices.

Theorem 5.7. *A is the sum of r rank-one matrices:*

$$A = \sum_{j=1}^{r}\sigma_j u_j v_j^*. \tag{5.3}$$

Proof. If we write Σ as a sum of r matrices Σ_j, where $\Sigma_j = \mathrm{diag}(0,\dots,0,\sigma_j,0, \dots,0)$, then (5.3) follows from (4.3). $\square$

There are many ways to express an $m \times n$ matrix A as a sum of rank-one matrices. For example, A could be written as the sum of its m rows, or its n columns, or its mn entries. For another example, Gaussian elimination reduces A to the sum of a full rank-one matrix, a rank-one matrix whose first row and column are zero, a rank-one matrix whose first two rows and columns are zero, and so on.

Formula (5.3), however, represents a decomposition into rank-one matrices with a deeper property: *the νth partial sum captures as much of the energy of A as possible.* This statement holds with "energy" defined by either the 2-norm or the Frobenius norm. We can make it precise by formulating a problem of best approximation of a matrix A by matrices of lower rank.

Theorem 5.8. *For any ν with $0 \le \nu \le r$, define*

$$A_\nu = \sum_{j=1}^{\nu}\sigma_j u_j v_j^*; \tag{5.4}$$

if $\nu = p = \min\{m,n\}$, define $\sigma_{\nu+1} = 0$. Then

$$\|A - A_\nu\|_2 = \inf_{\substack{B\in \mathbb{C}^{m\times n}\\ \mathrm{rank}(B)\le\nu}} \|A - B\|_2 = \sigma_{\nu+1}.$$

Proof. Suppose there is some B with rank$(B) \leq \nu$ such that $\|A - B\|_2 < \|A - A_\nu\|_2 = \sigma_{\nu+1}$. Then there is an $(n-\nu)$-dimensional subspace $W \subseteq \mathbb{C}^n$ such that $w \in W \Rightarrow Bw = 0$. Accordingly, for any $w \in W$, we have $Aw = (A-B)w$ and

$$\|Aw\|_2 = \|(A - B)w\|_2 \leq \|A - B\|_2 \|w\|_2 < \sigma_{\nu+1}\|w\|_2.$$

Thus W is an $(n - \nu)$-dimensional subspace where $\|Aw\| < \sigma_{\nu+1}\|w\|$. But there is a $(\nu + 1)$-dimensional subspace where $\|Aw\| \geq \sigma_{\nu+1}\|w\|$, namely the space spanned by the first $\nu + 1$ right singular vectors of A. Since the sum of the dimensions of these spaces exceeds n, there must be a nonzero vector lying in both, and this is a contradiction. $\qquad\square$

Theorem 5.8 has a geometric interpretation. What is the best approximation of a hyperellipsoid by a line segment? Take the line segment to be the longest axis. What is the best approximation by a two-dimensional ellipsoid? Take the ellipsoid spanned by the longest and the second-longest axis. Continuing in this fashion, at each step we improve the approximation by adding into our approximation the largest axis of the hyperellipsoid not yet included. After r steps, we have captured all of A. This idea has ramifications in areas as disparate as image compression (see Exercise 9.3) and functional analysis.

We state the analogous result for the Frobenius norm without proof.

Theorem 5.9. *For any ν with $0 \leq \nu \leq r$, the matrix A_ν of* (5.4) *also satisfies*

$$\|A - A_\nu\|_F = \inf_{\substack{B \in \mathbb{C}^{m \times n} \\ \text{rank}(B) \leq \nu}} \|A - B\|_F = \sqrt{\sigma_{\nu+1}^2 + \cdots + \sigma_r^2}.$$

Computation of the SVD

In this and the previous lecture, we have examined the properties of the SVD but not considered how it can be computed. As it happens, the computation of the SVD is a fascinating subject. The best methods are variants of algorithms used for computing eigenvalues, and we shall discuss them in Lecture 31.

Once one can compute it, the SVD can be used as a tool for all kinds of problems. In fact, most of the theorems of this lecture have computational consequences. The best method for determining the rank of a matrix is to count the number of singular values greater than a judiciously chosen tolerance (Theorem 5.1). The most accurate method for finding an orthonormal basis of a range or a nullspace is via Theorem 5.2. (For both of these examples, QR factorization provides alternative algorithms that are faster but not always as accurate.) Theorem 5.3 represents the standard method for computing $\|A\|_2$, and Theorems 5.8 and 5.9, the standards for computing low-rank approximations with respect to $\|\cdot\|_2$ and $\|\cdot\|_F$. Besides these examples, the SVD is also an ingredient in robust algorithms for least squares fitting, intersection of subspaces, regularization, and numerous other problems.

Exercises

5.1. In Example 3.1 we considered the matrix (3.7) and asserted, among other things, that its 2-norm is approximately 2.9208. Using the SVD, work out (on paper) the exact values of $\sigma_{\min}(A)$ and $\sigma_{\max}(A)$ for this matrix.

5.2. Using the SVD, prove that any matrix in $\mathbb{C}^{m \times n}$ is the limit of a sequence of matrices of full rank. In other words, prove that the set of full-rank matrices is a dense subset of $\mathbb{C}^{m \times n}$. Use the 2-norm for your proof. (The norm doesn't matter, since all norms on a finite-dimensional space are equivalent.)

5.3. Consider the matrix

$$A = \begin{bmatrix} -2 & 11 \\ -10 & 5 \end{bmatrix}.$$

(a) Determine, on paper, a real SVD of A in the form $A = U\Sigma V^T$. The SVD is not unique, so find the one that has the minimal number of minus signs in U and V.

(b) List the singular values, left singular vectors, and right singular vectors of A. Draw a careful, labeled picture of the unit ball in $\mathbb{R}^2$ and its image under A, together with the singular vectors, with the coordinates of their vertices marked.

(c) What are the 1-, 2-, ∞-, and Frobenius norms of A?

(d) Find A^{-1} not directly, but via the SVD.

(e) Find the eigenvalues λ_1, λ_2 of A.

(f) Verify that $\det A = \lambda_1 \lambda_2$ and $|\det A| = \sigma_1 \sigma_2$.

(g) What is the area of the ellipsoid onto which A maps the unit ball of $\mathbb{R}^2$?

5.4. Suppose $A \in \mathbb{C}^{m \times m}$ has an SVD $A = U\Sigma V^*$. Find an eigenvalue decomposition (5.1) of the $2m \times 2m$ hermitian matrix

$$\begin{bmatrix} 0 & A^* \\ A & 0 \end{bmatrix}.$$

Part II

QR Factorization and Least Squares

Lecture 6. Projectors

We now enter the second part of the book, whose theme is orthogonality. We begin with the fundamental tool of projection matrices, or projectors, both orthogonal and nonorthogonal.

Projectors

A *projector* is a square matrix P that satisfies

$$P^2 = P. \tag{6.1}$$

(Such a matrix is also said to be *idempotent*.) This definition includes both orthogonal projectors, to be discussed in a moment, and nonorthogonal ones. To avoid confusion one may use the term *oblique projector* in the nonorthogonal case.

The term projector might be thought of as arising from the notion that if one were to shine a light onto the subspace range(P) from just the right direction, then Pv would be the shadow projected by the vector v. We shall carry this physical picture forward for a moment.

Observe that if $v \in$ range(P), then it lies exactly on its own shadow, and applying the projector results in v itself. Mathematically, we have $v = Px$ for some x and

$$Pv = P^2x = Px = v.$$

From what direction does the light shine when $v \neq Pv$? In general the answer depends on v, but for any particular v, it is easily deduced by drawing the

41

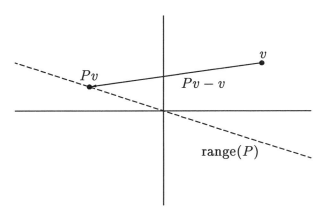

Figure 6.1. *An oblique projection.*

line from v to Pv, $Pv - v$ (Figure 6.1). Applying the projector to this vector gives a zero result:

$$P(Pv - v) = P^2 v - Pv = 0.$$

This means that $Pv - v \in \text{null}(P)$. That is, the direction of the light may be different for different v, but it is always described by a vector in $\text{null}(P)$.

Complementary Projectors

If P is a projector, $I - P$ is also a projector, for it is also idempotent:

$$(I - P)^2 = I - 2P + P^2 = I - P.$$

The matrix $I - P$ is called the *complementary projector* to P.

Onto what space does $I - P$ project? Exactly the nullspace of P! We know that $\text{range}(I - P) \supseteq \text{null}(P)$, because if $Pv = 0$, we have $(I - P)v = v$. Conversely, we know that $\text{range}(I - P) \subseteq \text{null}(P)$, because for any v, we have $(I - P)v = v - Pv \in \text{null}(P)$. Therefore, for any projector P,

$$\text{range}(I - P) = \text{null}(P). \tag{6.2}$$

By writing $P = I - (I - P)$ we derive the complementary fact

$$\text{null}(I - P) = \text{range}(P). \tag{6.3}$$

We can also see that $\text{null}(I - P) \cap \text{null}(P) = \{0\}$: any vector v in both sets satisfies $v = v - Pv = (I - P)v = 0$. Another way of stating this fact is

$$\text{range}(P) \cap \text{null}(P) = \{0\}. \tag{6.4}$$

These computations show that *a projector separates* $\mathbb{C}^m$ *into two spaces.* Conversely, let S_1 and S_2 be two subspaces of $\mathbb{C}^m$ such that $S_1 \cap S_2 = \{0\}$

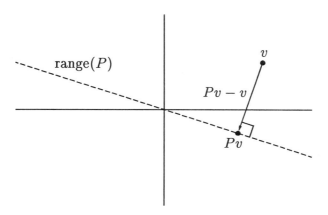

Figure 6.2. *An orthogonal projection.*

and $S_1 + S_2 = \mathbb{C}^m$, where $S_1 + S_2$ denotes the span of S_1 and S_2, that is, the set of vectors $s_1 + s_2$ with $s_1 \in S_1$ and $s_2 \in S_2$. (Such a pair are said to be *complementary subspaces.*) Then there is a projector P such that range$(P) = S_1$ and null$(P) = S_2$. We say that P is the projector *onto S_1 along S_2*. This projector and its complement can be seen as the unique solution to the following problem:

Given v, find vectors $v_1 \in S_1$ and $v_2 \in S_2$ such that $v_1 + v_2 = v$.

The projection Pv gives v_1, and the complementary projection $(I - P)v$ gives v_2. These vectors are unique because all solutions must be of the form

$$(Pv + v_3) + ((I - P)v - v_3) \ = \ v,$$

where it is clear that v_3 must be in both S_1 and S_2, i.e., $v_3 = 0$.

One context in which projectors and their complements arise is particularly familiar. Suppose an $m \times m$ matrix A has a complete set of eigenvectors $\{v_j\}$, as in (5.1), meaning that $\{v_j\}$ is a basis of $\mathbb{C}^m$. We are frequently concerned with problems associated with expansions of vectors in this basis. Given $x \in \mathbb{C}^m$, for example, what is the component of x in the direction of a particular eigenvector v? The answer is Px, where P is a certain rank-one projector. Rather than give details here, however, we turn now to the special class of projectors that will be of primary interest to us in this book.

Orthogonal Projectors

An *orthogonal projector* (Figure 6.2) is one that projects onto a subspace S_1 along a space S_2, where S_1 and S_2 are orthogonal. (Warning: orthogonal projectors are not orthogonal matrices!)

There is also an algebraic definition: an orthogonal projector is any projector that is hermitian, satisfying $P^* = P$ as well as (6.1). Of course, we must establish that this definition is equivalent to the first.

Theorem 6.1. *A projector P is orthogonal if and only if $P = P^*$.*

Proof. If $P = P^*$, then the inner product between a vector $Px \in S_1$ and a vector $(I - P)y \in S_2$ is zero:

$$x^* P^* (I - P)y = x^* (P - P^2)y = 0.$$

Thus the projector is orthogonal, providing the proof in the "if" direction.

For "only if," we can use the SVD. Suppose P projects onto S_1 along S_2, where $S_1 \perp S_2$ and S_1 has dimension n. Then an SVD of P can be constructed as follows. Let $\{q_1, q_2, \ldots, q_m\}$ be an orthonormal basis for $\mathbb{C}^m$, where $\{q_1, \ldots, q_n\}$ is a basis for S_1 and $\{q_{n+1}, \ldots, q_m\}$ is a basis for S_2. For $j \leq n$, we have $Pq_j = q_j$, and for $j > n$, we have $Pq_j = 0$. Now let Q be the unitary matrix whose jth column is q_j. We then have

$$PQ = \begin{bmatrix} q_1 & \cdots & q_n & 0 & \cdots \end{bmatrix},$$

so that

$$Q^* P Q = \begin{bmatrix} 1 & & & & \\ & \ddots & & & \\ & & 1 & & \\ & & & 0 & \\ & & & & \ddots \end{bmatrix} = \Sigma,$$

a diagonal matrix with ones in the first n entries and zeros everywhere else. Thus we have constructed a singular value decomposition of P:

$$P = Q \Sigma Q^*. \tag{6.5}$$

(Note that this is also an eigenvalue decomposition (5.1).) From here we see that P is hermitian, since $P^* = (Q\Sigma Q^*)^* = Q\Sigma^* Q^* = Q\Sigma Q^* = P$. $\square$

Projection with an Orthonormal Basis

Since an orthogonal projector has some singular values equal to zero (except in the trivial case $P = I$), it is natural to drop the silent columns of Q in (6.5) and use the reduced rather than the full SVD. We obtain the marvelously simple expression

$$P = \hat{Q}\hat{Q}^*, \tag{6.6}$$

where the columns of $\hat{Q}$ are orthonormal.

In (6.6), the matrix $\hat{Q}$ need not come from an SVD. Let $\{q_1, \ldots, q_n\}$ be any set of n orthonormal vectors in $\mathbb{C}^m$, and let $\hat{Q}$ be the corresponding $m \times n$ matrix. From (2.7) we know that

$$v = r + \sum_{i=1}^{n}(q_i q_i^*)v$$

represents a decomposition of a vector $v \in \mathbb{C}^m$ into a component in the column space of $\hat{Q}$ plus a component in the orthogonal space. Thus the map

$$v \mapsto \sum_{i=1}^{n}(q_i q_i^*)v \tag{6.7}$$

is an orthogonal projector onto range($\hat{Q}$), and in matrix form, it may be written $y = \hat{Q}\hat{Q}^*v$:

Thus any product $\hat{Q}\hat{Q}^*$ is always a projector onto the column space of $\hat{Q}$, regardless of how $\hat{Q}$ was obtained, as long as its columns are orthonormal. Perhaps $\hat{Q}$ was obtained by dropping some columns and rows from a full factorization $v = QQ^*v$ of the identity,

and perhaps it was not.

The complement of an orthogonal projector is also an orthogonal projector (proof: $I - \hat{Q}\hat{Q}^*$ is hermitian). The complement projects onto the space orthogonal to range($\hat{Q}$).

An important special case of orthogonal projectors is the rank-one orthogonal projector that isolates the component in a single direction q, which can be written

$$P_q = qq^*. \tag{6.8}$$

These are the pieces from which higher-rank projectors can be made, as in (6.7). Their complements are the rank $m - 1$ orthogonal projectors that eliminate the component in the direction of q:

$$P_{\perp q} = I - qq^*. \tag{6.9}$$

Equations (6.8) and (6.9) assume that q is a unit vector. For arbitrary nonzero vectors a, the analogous formulas are

$$P_a = \frac{aa^*}{a^*a}, \tag{6.10}$$

$$P_{\perp a} = I - \frac{aa^*}{a^*a}. \tag{6.11}$$

Projection with an Arbitrary Basis

An orthogonal projector onto a subspace of $\mathbb{C}^m$ can also be constructed beginning with an arbitrary basis, not necessarily orthogonal. Suppose that the subspace is spanned by the linearly independent vectors $\{a_1, \ldots, a_n\}$, and let A be the $m \times n$ matrix whose jth column is a_j.

In passing from v to its orthogonal projection $y \in \text{range}(A)$, the difference $y - v$ must be orthogonal to $\text{range}(A)$. This is equivalent to the statement that y must satisfy $a_j^*(y - v) = 0$ for every j. Since $y \in \text{range}(A)$, we can set $y = Ax$ and write this condition as $a_j^*(Ax - v) = 0$ for each j, or equivalently, $A^*(Ax - v) = 0$ or $A^*Ax = A^*v$. It is easily shown that since A has full rank, A^*A is nonsingular (Exercise 6.3). Therefore

$$x = (A^*A)^{-1}A^*v. \tag{6.12}$$

Finally, the projection of v, $y = Ax$, is $y = A(A^*A)^{-1}A^*v$. Thus the orthogonal projector onto $\text{range}(A)$ can be expressed by the formula

$$P = A(A^*A)^{-1}A^*. \tag{6.13}$$

Note that this is a multidimensional generalization of (6.10). In the orthonormal case $A = \hat{Q}$, the term in parentheses collapses to the identity and we recover (6.6).

Exercises

6.1. If P is an orthogonal projector, then $I - 2P$ is unitary. Prove this algebraically, and give a geometric interpretation.

6.2. Let E be the $m \times m$ matrix that extracts the "even part" of an m-vector: $Ex = (x + Fx)/2$, where F is the $m \times m$ matrix that flips $(x_1, \ldots, x_m)^*$ to $(x_m, \ldots, x_1)^*$. Is E an orthogonal projector, an oblique projector, or not a projector at all? What are its entries?

6.3. Given $A \in \mathbb{C}^{m \times n}$ with $m \geq n$, show that A^*A is nonsingular if and only if A has full rank.

6.4. Consider the matrices

$$A = \begin{bmatrix} 1 & 0 \\ 0 & 1 \\ 1 & 0 \end{bmatrix}, \qquad B = \begin{bmatrix} 1 & 2 \\ 0 & 1 \\ 1 & 0 \end{bmatrix}.$$

Answer the following questions by hand calculation.

(a) What is the orthogonal projector P onto range(A), and what is the image under P of the vector $(1, 2, 3)^*$?

(b) Same questions for B.

6.5. Let $P \in \mathbb{C}^{m \times m}$ be a nonzero projector. Show that $\|P\|_2 \geq 1$, with equality if and only if P is an orthogonal projector.

Lecture 7. QR Factorization

One algorithmic idea in numerical linear algebra is more important than all the others: QR factorization.

Reduced QR Factorization

For many applications, we find ourselves interested in the column spaces of a matrix A. Note the plural: these are the *successive* spaces spanned by the columns $a_1, a_2, \ldots$ of A:

$$\langle a_1 \rangle \subseteq \langle a_1, a_2 \rangle \subseteq \langle a_1, a_2, a_3 \rangle \subseteq \cdots .$$

Here, as in Lecture 5 and throughout the book, the notation $\langle \cdots \rangle$ indicates the subspace spanned by whatever vectors are included in the brackets. Thus $\langle a_1 \rangle$ is the one-dimensional space spanned by a_1, $\langle a_1, a_2 \rangle$ is the two-dimensional space spanned by a_1 and a_2, and so on. The idea of QR factorization is the construction of a sequence of orthonormal vectors $q_1, q_2, \ldots$ that span these successive spaces.

To be precise, assume for the moment that $A \in \mathbb{C}^{m \times n}$ $(m \geq n)$ has full rank n. We want the sequence $q_1, q_2, \ldots$ to have the property

$$\langle q_1, q_2, \ldots, q_j \rangle = \langle a_1, a_2, \ldots, a_j \rangle, \qquad j = 1, \ldots, n. \tag{7.1}$$

From the observations of Lecture 1, it is not hard to see that this amounts to the condition

$$
\begin{bmatrix} | & | & & | \\ a_1 & a_2 & \cdots & a_n \\ | & | & & | \end{bmatrix} = \begin{bmatrix} | & | & & | \\ q_1 & q_2 & \cdots & q_n \\ | & | & & | \end{bmatrix} \begin{bmatrix} r_{11} & r_{12} & \cdots & r_{1n} \\ & r_{22} & & \vdots \\ & & \ddots & \\ & & & r_{nn} \end{bmatrix}, \quad (7.2)
$$

where the diagonal entries r_{kk} are nonzero—for if (7.2) holds, then $a_1, \ldots, a_k$ can be expressed as linear combinations of $q_1, \ldots, q_k$, and the invertibility of the upper-left $k \times k$ block of the triangular matrix implies that, conversely, $q_1, \ldots, q_k$ can be expressed as linear combinations of $a_1, \ldots, a_k$. Written out, these equations take the form

$$
\begin{aligned}
a_1 &= r_{11}q_1, \\
a_2 &= r_{12}q_1 + r_{22}q_2, \\
a_3 &= r_{13}q_1 + r_{23}q_2 + r_{33}q_3, \\
&\;\;\vdots \\
a_n &= r_{1n}q_1 + r_{2n}q_2 + \cdots + r_{nn}q_n.
\end{aligned} \quad (7.3)
$$

As a matrix formula, we have

$$
A = \hat{Q}\hat{R}, \quad (7.4)
$$

where $\hat{Q}$ is $m \times n$ with orthonormal columns and $\hat{R}$ is $n \times n$ and upper-triangular. Such a factorization is called a *reduced QR factorization of A*.

Full QR Factorization

A *full QR factorization* of $A \in \mathbb{C}^{m \times n}$ ($m \geq n$) goes further, appending an additional $m - n$ orthonormal columns to $\hat{Q}$ so that it becomes an $m \times m$ unitary matrix Q. This is analogous to the passage from the reduced to the full SVD described in Lecture 4. In the process, rows of zeros are appended to $\hat{R}$ so that it becomes an $m \times n$ matrix R, still upper-triangular. The relationship between the full and reduced QR factorizations is as follows.

Full QR Factorization ($m \geq n$)

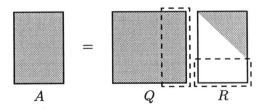

A $\qquad$ Q $\qquad$ R

In the full QR factorization, Q is $m \times m$, R is $m \times n$, and the last $m-n$ columns of Q are multiplied by zeros in R (enclosed by dashes). In the reduced QR factorization, the silent columns and rows are removed. Now $\hat{Q}$ is $m \times n$, $\hat{R}$ is $n \times n$, and none of the rows of $\hat{R}$ are necessarily zero.

Reduced QR Factorization $(m \geq n)$

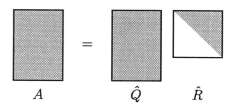

A $\hat{Q}$ $\hat{R}$

Notice that in the full QR factorization, the columns q_j for $j > n$ are orthogonal to range(A). Assuming A is of full rank n, they constitute an orthonormal basis for range$(A)^\perp$ (the space orthogonal to range(A)), or equivalently, for null(A^*).

Gram–Schmidt Orthogonalization

Equations (7.3) suggest a method for computing reduced QR factorizations. Given $a_1, a_2, \ldots$, we can construct the vectors $q_1, q_2, \ldots$ and entries r_{ij} by a process of successive orthogonalization. This is an old idea, known as *Gram–Schmidt orthogonalization*.

The process works like this. At the jth step, we wish to find a unit vector $q_j \in \langle a_1, \ldots, a_j \rangle$ that is orthogonal to $q_1, \ldots, q_{j-1}$. As it happens, we have already considered the necessary orthogonalization technique in (2.6). From that equation, we see that

$$v_j = a_j - (q_1^* a_j)q_1 - (q_2^* a_j)q_2 - \cdots - (q_{j-1}^* a_j)q_{j-1} \qquad (7.5)$$

is a vector of the kind required, except that it is not yet normalized. If we divide by $\|v_j\|_2$, the result is a suitable vector q_j.

With this in mind, let us rewrite (7.3) in the form

$$q_1 = \frac{a_1}{r_{11}},$$

$$q_2 = \frac{a_2 - r_{12}q_1}{r_{22}},$$

$$q_3 = \frac{a_3 - r_{13}q_1 - r_{23}q_2}{r_{33}}, \qquad (7.6)$$

$$\vdots$$

$$q_n = \frac{a_n - \sum_{i=1}^{n-1} r_{in}q_i}{r_{nn}}.$$

From (7.5) it is evident that an appropriate definition for the coefficients r_{ij} in the numerators of (7.6) is

$$r_{ij} = q_i^* a_j \qquad (i \neq j). \tag{7.7}$$

The coefficients r_{jj} in the denominators are chosen for normalization:

$$|r_{jj}| = \left\| a_j - \sum_{i=1}^{j-1} r_{ij} q_i \right\|_2. \tag{7.8}$$

Note that the sign of r_{jj} is not determined. Arbitrarily, we may choose $r_{jj} > 0$, in which case we shall finish with a factorization $A = \hat{Q}\hat{R}$ in which $\hat{R}$ has positive entries along the diagonal.

The algorithm embodied in (7.6)–(7.8) is the Gram–Schmidt iteration. Mathematically, it offers a simple route to understanding and proving various properties of QR factorizations. Numerically, it turns out to be unstable because of rounding errors on a computer. To emphasize the instability, numerical analysts refer to this as the *classical Gram–Schmidt iteration*, as opposed to the *modified Gram–Schmidt iteration*, discussed in the next lecture.

Algorithm 7.1. Classical Gram–Schmidt (unstable)

for $j = 1$ **to** n
$\quad v_j = a_j$
$\quad$**for** $i = 1$ **to** $j - 1$
$\quad\quad r_{ij} = q_i^* a_j$
$\quad\quad v_j = v_j - r_{ij} q_i$
$\quad r_{jj} = \|v_j\|_2$
$\quad q_j = v_j / r_{jj}$

Existence and Uniqueness

All matrices have QR factorizations, and under suitable restrictions, they are unique. We state first the existence result.

Theorem 7.1. *Every $A \in \mathbb{C}^{m \times n}$ ($m \geq n$) has a full QR factorization, hence also a reduced QR factorization.*

Proof. Suppose first that A has full rank and that we want just a reduced QR factorization. In this case, a proof of existence is provided by the Gram–Schmidt algorithm itself. By construction, this process generates orthonormal columns of $\hat{Q}$ and entries of $\hat{R}$ such that (7.4) holds. Failure can occur only if at some step, v_j is zero and thus cannot be normalized to produce q_j.

However, this would imply $a_j \in \langle q_1, \ldots, q_{j-1} \rangle = \langle a_1, \ldots, a_{j-1} \rangle$, contradicting the assumption that A has full rank.

Now suppose that A does not have full rank. Then at one or more steps j, we shall find that (7.5) gives $v_j = 0$, as just mentioned. At this moment, we simply pick q_j arbitrarily to be any normalized vector orthogonal to $\langle q_1, \ldots, q_{j-1} \rangle$, and then continue the Gram–Schmidt process.

Finally, the full, rather than reduced, QR factorization of an $m \times n$ matrix with $m > n$ can be constructed by introducing arbitrary orthonormal vectors in the same fashion. We follow the Gram–Schmidt process through step n, then continue on an additional $m - n$ steps, introducing vectors q_j at each step.

The issues discussed in the last two paragraphs came up already in Lecture 4, in our discussion of the SVD.   □

We turn now to uniqueness. Suppose $A = \hat{Q}\hat{R}$ is a reduced QR factorization. If the ith column of $\hat{Q}$ is multiplied by z and the ith row of $\hat{R}$ is multiplied by z^{-1} for some scalar z with $|z| = 1$, we obtain another QR factorization of A. The next theorem asserts that if A has full rank, this is the only way to obtain distinct reduced QR factorizations.

Theorem 7.2. *Each $A \in \mathbb{C}^{m \times n}$ ($m \geq n$) of full rank has a unique reduced QR factorization $A = \hat{Q}\hat{R}$ with $r_{jj} > 0$.*

Proof. Again, the proof is provided by the Gram–Schmidt iteration. From (7.4), the orthonormality of the columns of $\hat{Q}$, and the upper-triangularity of $\hat{R}$, it follows that any reduced QR factorization of A must satisfy (7.6)–(7.8). By the assumption of full rank, the denominators (7.8) of (7.6) are nonzero, and thus at each successive step j, these formulas determine r_{ij} and q_j fully, except in one place: the sign of r_{jj}, not specified in (7.8). Once this is fixed by the condition $r_{jj} > 0$, as in Algorithm 7.1, the factorization is completely determined.   □

When Vectors Become Continuous Functions

The QR factorization has an analogue for orthonormal expansions of functions rather than vectors.

Suppose we replace $\mathbb{C}^m$ by $L^2[-1, 1]$, a vector space of complex-valued functions on $[-1, 1]$. We shall not introduce the properties of this space formally; suffice it to say that the inner product of f and g now takes the form

$$(f, g) = \int_{-1}^{1} \overline{f(x)}\, g(x)\, dx. \tag{7.9}$$

Consider, for example, the following "matrix" whose "columns" are the monomials x^j:

$$A = \left[\begin{array}{c|c|c|c|c} & & & & \\ 1 & x & x^2 & \cdots & x^{n-1} \\ & & & & \end{array} \right]. \tag{7.10}$$

Each column is a function in $L^2[-1,1]$, and thus, whereas A is discrete as usual in the horizontal direction, it is continuous in the vertical direction. It is a continuous analogue of the Vandermonde matrix (1.4) of Example 1.1.

The "continuous QR factorization" of A takes the form

$$A = QR = \left[\begin{array}{c|c|c|c} & & & \\ q_0(x) & q_1(x) & \cdots & q_{n-1}(x) \\ & & & \end{array} \right] \left[\begin{array}{cccc} r_{11} & r_{12} & \cdots & r_{1n} \\ & r_{22} & & \vdots \\ & & \ddots & \\ & & & r_{nn} \end{array} \right],$$

where the columns of Q are functions of x, orthonormal with respect to the inner product (7.9):

$$\int_{-1}^{1} \overline{q_i(x)}\, q_j(x)\, dx \;=\; \delta_{ij} \;=\; \left\{ \begin{array}{ll} 1 & \text{if } i = j, \\ 0 & \text{if } i \neq j. \end{array} \right.$$

From the Gram–Schmidt construction we can see that q_j is a polynomial of degree j. These polynomials are scalar multiples of what are known as the *Legendre polynomials*, P_j, which are conventionally normalized so that $P_j(1) = 1$. The first few P_j are

$$P_0(x) = 1, \quad P_1(x) = x, \quad P_2(x) = \tfrac{3}{2}x^2 - \tfrac{1}{2}, \quad P_3(x) = \tfrac{5}{2}x^3 - \tfrac{3}{2}x; \tag{7.11}$$

see Figure 7.1. Like the monomials $1, x, x^2, \ldots$, this sequence of polynomials spans the spaces of polynomials of successively higher degree. However, $P_0(x), P_1(x), P_2(x), \ldots$ have the advantage that they are orthogonal, making them far better suited for certain computations. In fact, computations with such polynomials form the basis of *spectral methods*, one of the most powerful techniques for the numerical solution of partial differential equations.

What is the "projection matrix" $\hat{Q}\hat{Q}^*$ (6.6) associated with $\hat{Q}$? It is a "$[-1,1] \times [-1,1]$ matrix," that is, an integral operator

$$f(\cdot) \;\mapsto\; \sum_{j=0}^{n-1} q_j(\cdot) \int_{-1}^{1} \overline{q_j(x)}\, f(x)\, dx \tag{7.12}$$

mapping functions in $L^2[-1,1]$ to functions in $L^2[-1,1]$.

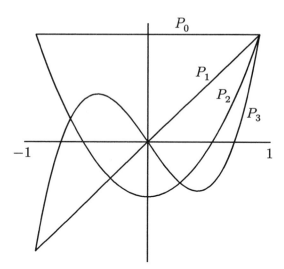

Figure 7.1. *The first four Legendre polynomials* (7.11). *Apart from scale factors, these can be interpreted as the columns of* $\hat{Q}$ *in a reduced QR factorization of the "*$[-1, 1] \times 4$ *matrix"* $[1, x, x^2, x^3]$.

Solution of $Ax = b$ by QR Factorization

In closing this lecture we return for a moment to discrete, finite matrices. Suppose we wish to solve $Ax = b$ for x, where $A \in \mathbb{C}^{m \times m}$ is nonsingular. If $A = QR$ is a QR factorization, then we can write $QRx = b$, or

$$Rx = Q^*b. \tag{7.13}$$

The right-hand side of this equation is easy to compute, if Q is known, and the system of linear equations implicit in the left-hand side is also easy to solve because it is triangular. This suggests the following method for computing the solution to $Ax = b$:

1. Compute a QR factorization $A = QR$.

2. Compute $y = Q^*b$.

3. Solve $Rx = y$ for x.

In later lectures we shall present algorithms for each of these steps.

The combination 1–3 is an excellent method for solving linear systems of equations; in Lecture 16, we shall prove this. However, it is not the standard method for such problems. Gaussian elimination is the algorithm generally used in practice, since it requires only half as many numerical operations.

Exercises

7.1. Consider again the matrices A and B of Exercise 6.4.

(a) Using any method you like, determine (on paper) a reduced QR factorization $A = \hat{Q}\hat{R}$ and a full QR factorization $A = QR$.

(b) Again using any method you like, determine reduced and full QR factorizations $B = \hat{Q}\hat{R}$ and $B = QR$.

7.2. Let A be a matrix with the property that columns $1, 3, 5, 7, \ldots$ are orthogonal to columns $2, 4, 6, 8, \ldots$. In a reduced QR factorization $A = \hat{Q}\hat{R}$, what special structure does $\hat{R}$ possess?

7.3. Let A be an $m \times m$ matrix, and let a_j be its jth column. Give an algebraic proof of *Hadamard's inequality*:

$$|\det A| \leq \prod_{j=1}^{m} \|a_j\|_2.$$

Also give a geometric interpretation of this result, making use of the fact that the determinant equals the volume of a parallelepiped.

7.4. Let $x^{(1)}$, $y^{(1)}$, $x^{(2)}$, and $y^{(2)}$ be nonzero vectors in $\mathbb{R}^3$ with the property that $x^{(1)}$ and $y^{(1)}$ are linearly independent and so are $x^{(2)}$ and $y^{(2)}$. Consider the two planes in $\mathbb{R}^3$,

$$P^{(1)} = \langle x^{(1)}, y^{(1)} \rangle, \qquad P^{(2)} = \langle x^{(2)}, y^{(2)} \rangle.$$

Suppose we wish to find a nonzero vector $v \in \mathbb{R}^3$ that lies in the intersection $P = P^{(1)} \cap P^{(2)}$. Devise a method for solving this problem by reducing it to the computation of QR factorizations of three 3×2 matrices.

7.5. Let A be an $m \times n$ matrix $(m \geq n)$, and let $A = \hat{Q}\hat{R}$ be a reduced QR factorization.

(a) Show that A has rank n if and only if all the diagonal entries of $\hat{R}$ are nonzero.

(b) Suppose $\hat{R}$ has k nonzero diagonal entries for some k with $0 \leq k < n$. What does this imply about the rank of A? Exactly k? At least k? At most k? Give a precise answer, and prove it.

Lecture 8. Gram–Schmidt Orthogonalization

The Gram–Schmidt iteration is the basis of one of the two principal numerical algorithms for computing QR factorizations. It is a process of "triangular orthogonalization," making the columns of a matrix orthonormal via a sequence of matrix operations that can be interpreted as multiplication on the right by upper-triangular matrices.

Gram–Schmidt Projections

In the last lecture we presented the Gram–Schmidt iteration in its classical form. To begin this lecture, we describe the same algorithm again in another way, using orthogonal projectors.

Let $A \in \mathbb{C}^{m \times n}$, $m \geq n$, be a matrix of full rank with columns $\{a_j\}$. Before, we expressed the Gram–Schmidt iteration by the formulas (7.6)–(7.8). Consider now the sequence of formulas

$$ q_1 = \frac{P_1 a_1}{\|P_1 a_1\|}, \quad q_2 = \frac{P_2 a_2}{\|P_2 a_2\|}, \quad \ldots, \quad q_n = \frac{P_n a_n}{\|P_n a_n\|}. \tag{8.1} $$

In these formulas, each P_j denotes an orthogonal projector. Specifically, P_j is the $m \times m$ matrix of rank $m - (j - 1)$ that projects $\mathbb{C}^m$ orthogonally onto the space orthogonal to $\langle q_1, \ldots, q_{j-1} \rangle$. (In the case $j = 1$, this prescription reduces to the identity: $P_1 = I$.) Now, observe that q_j as defined by (8.1) is orthogonal to $q_1, \ldots, q_{j-1}$, lies in the space $\langle a_1, \ldots, a_j \rangle$, and has norm 1. Thus we see that (8.1) is equivalent to (7.6)–(7.8) and hence to Algorithm 7.1.

The projector P_j can be represented explicitly. Let $\hat{Q}_{j-1}$ denote the $m \times (j-1)$ matrix containing the first $j-1$ columns of $\hat{Q}$,

$$\hat{Q}_{j-1} = \left[\begin{array}{c|c|c|c} q_1 & q_2 & \cdots & q_{j-1} \end{array} \right]. \tag{8.2}$$

Then P_j is given by

$$P_j = I - \hat{Q}_{j-1}\hat{Q}_{j-1}^*. \tag{8.3}$$

By now, the reader may be familiar enough with our notation and with orthogonality ideas to see at a glance that (8.3) represents the operator applied to a_j in (7.5).

Modified Gram–Schmidt Algorithm

In practice, the Gram–Schmidt formulas are not applied as we have indicated in Algorithm 7.1 and in (8.1), for this sequence of calculations turns out to be numerically unstable. Fortunately, there is a simple modification that improves matters. We have not discussed numerical stability yet; this will come in the next lecture and then systematically beginning in Lecture 14. For the moment, it is enough to know that a stable algorithm is one that is not too sensitive to the effects of rounding errors on a computer.

For each value of j, Algorithm 7.1 computes a single orthogonal projection of rank $m - (j-1)$,

$$v_j = P_j a_j. \tag{8.4}$$

In contrast, the modified Gram–Schmidt algorithm computes the same result by a sequence of $j-1$ projections of rank $m-1$. Recall from (6.9) that $P_{\perp q}$ denotes the rank $m-1$ orthogonal projector onto the space orthogonal to a nonzero vector $q \in \mathbb{C}^m$. By the definition of P_j, it is not difficult to see that

$$P_j = P_{\perp q_{j-1}} \cdots P_{\perp q_2} P_{\perp q_1}, \tag{8.5}$$

again with $P_1 = I$. Thus an equivalent statement to (8.4) is

$$v_j = P_{\perp q_{j-1}} \cdots P_{\perp q_2} P_{\perp q_1} a_j. \tag{8.6}$$

The modified Gram–Schmidt algorithm is based on the use of (8.6) instead of (8.4).

Mathematically, (8.6) and (8.4) are equivalent. However, the sequences of arithmetic operations implied by these formulas are different. The modified algorithm calculates v_j by evaluating the following formulas in order:

$$
\begin{aligned}
v_j^{(1)} &= a_j, \\
v_j^{(2)} &= P_{\perp q_1} v_j^{(1)} && = v_j^{(1)} - q_1 q_1^* v_j^{(1)}, \\
v_j^{(3)} &= P_{\perp q_2} v_j^{(2)} && = v_j^{(2)} - q_2 q_2^* v_j^{(2)}, \\
&\ \ \vdots && \ \ \vdots \\
v_j = v_j^{(j)} &= P_{\perp q_{j-1}} v_j^{(j-1)} && = v_j^{(j-1)} - q_{j-1} q_{j-1}^* v_j^{(j-1)}.
\end{aligned}
\tag{8.7}
$$

In finite precision computer arithmetic, we shall see that (8.7) introduces smaller errors than (8.4).

When the algorithm is implemented, the projector $P_{\perp q_i}$ can be conveniently applied to $v_j^{(i)}$ for each $j > i$ immediately after q_i is known. This is done in the description below.

Algorithm 8.1. Modified Gram–Schmidt

for $i = 1$ **to** n

$\qquad v_i = a_i$

for $i = 1$ **to** n

$\qquad r_{ii} = \|v_i\|$

$\qquad q_i = v_i / r_{ii}$

$\qquad$ **for** $j = i + 1$ **to** n

$\qquad\qquad r_{ij} = q_i^* v_j$

$\qquad\qquad v_j = v_j - r_{ij} q_i$

In practice, it is common to let v_i overwrite a_i and q_i overwrite v_i in order to save storage.

The reader should compare Algorithms 7.1 and 8.1 until he or she is confident of their equivalence.

Operation Count

The Gram–Schmidt algorithm is the first algorithm we have presented in this book, and with any algorithm, it is important to assess its cost. To do so, throughout the book we follow the classical route and count the number of floating point operations—*"flops"*—that the algorithm requires. Each addition, subtraction, multiplication, division, or square root counts as one flop.

We make no distinction between real and complex arithmetic, although in practice on most computers there is a sizable difference.

In fact, there is much more to the cost of an algorithm than operation counts. On a single-processor computer, the execution time is affected by the movement of data between elements of the memory hierarchy and by competing jobs running on the same processor. On multiprocessor machines the situation becomes more complex, with communication between processors sometimes taking on an importance much greater than that of actual "computation." With some regret, we shall ignore these important considerations, because this book is deliberately classical in style, focusing on algorithmic foundations.

For both variants of the Gram–Schmidt iteration, here is the classical result.

Theorem 8.1. *Algorithms 7.1 and 8.1 require $\sim 2mn^2$ flops to compute a QR factorization of an $m \times n$ matrix.*

Note that the theorem expresses only the leading term of the flop count. The symbol "$\sim$" has its usual asymptotic meaning:

$$\lim_{m,n\to\infty} \frac{\text{number of flops}}{2mn^2} = 1.$$

In discussing operation counts for algorithms, it is standard to discard lower-order terms as we have done here, since they are usually of little significance unless m and n are small.

Theorem 8.1 can be established as follows. To be definite, consider the modified Gram–Schmidt algorithm, Algorithm 8.1. When m and n are large, the work is dominated by the operations in the innermost loop:

$$r_{ij} = q_i^* v_j,$$
$$v_j = v_j - r_{ij} q_i.$$

The first line computes an inner product $q_i^* v_j$, requiring m multiplications and $m-1$ additions, and the second computes $v_j - r_{ij}q_i$, requiring m multiplications and m subtractions. The total work involved in a single inner iteration is consequently $\sim 4m$ flops, or 4 flops per column vector element. All together, the number of flops required by the algorithm is asymptotic to

$$\sum_{i=1}^{n}\sum_{j=i+1}^{n} 4m \sim \sum_{i=1}^{n}(i)4m \sim 2mn^2. \tag{8.8}$$

Counting Operations Geometrically

Operation counts can always be determined algebraically as in (8.8), and this is the standard procedure in the numerical analysis literature. However, it is also enlightening to take a different, geometrical route to the same conclusion. The argument goes like this. At the first step of the outer loop, Algorithm 8.1 operates on the whole matrix, subtracting a multiple of column 1 from the other columns. At the second step, it operates on a submatrix, subtracting a multiple of column 2 from columns $3, \ldots, n$. Continuing on in this way, at each step the column dimension shrinks by 1 until at the final step, only column n is modified. This process can be represented by the following diagram:

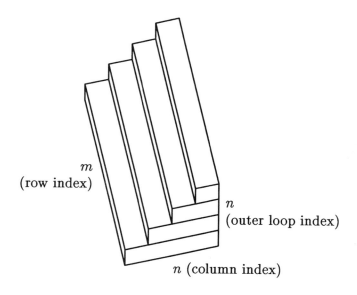

The $m \times n$ rectangle at the bottom corresponds to the first pass through the outer loop, the $m \times (n-1)$ rectangle above it to the second pass, and so on.

To leading order as $m, n \to \infty$, then, the operation count for Gram–Schmidt orthogonalization is proportional to the volume of the figure above. The constant of proportionality is four flops, because as noted above, the two steps of the inner loop correspond to four operations at each matrix location. Now as $m, n \to \infty$, the figure converges to a right triangular prism, with volume $mn^2/2$. Multiplying by four flops per unit volume gives, again,

Work for Gram–Schmidt orthogonalization: $\sim 2mn^2$ flops.   (8.9)

In this book we generally record operation counts in the format (8.9), without stating them as theorems. We often derive these results via figures like the one above, although algebraic derivations are also possible. One reason we do this is that a figure of this kind, besides being a route to an operation count, also serves as a reminder of the structure of an algorithm. For pictures of algorithms with different structures, see pp. 75 and 176.

Gram–Schmidt as Triangular Orthogonalization

Each outer step of the modified Gram–Schmidt algorithm can be interpreted as a right-multiplication by a square upper-triangular matrix. For example, beginning with A, the first iteration multiplies the first column a_1 by $1/r_{11}$ and then subtracts r_{1j} times the result from each of the remaining columns a_j. This is equivalent to right-multiplication by a matrix R_1:

$$
\begin{bmatrix} \\ v_1 & v_2 & \cdots & v_n \\ \\ \end{bmatrix}
\begin{bmatrix} \dfrac{1}{r_{11}} & \dfrac{-r_{12}}{r_{11}} & \dfrac{-r_{13}}{r_{11}} & \cdots \\ & 1 & \\ & & 1 \\ & & & \ddots \end{bmatrix}
=
\begin{bmatrix} \\ q_1 & v_2^{(2)} & \cdots & v_n^{(2)} \\ \\ \end{bmatrix}.
$$

In general, step i of Algorithm 8.1 subtracts r_{ij}/r_{ii} times column i of the current A from columns $j > i$ and replaces column i by $1/r_{ii}$ times itself. This corresponds to multiplication by an upper-triangular matrix R_i:

$$
R_2 = \begin{bmatrix} 1 & & & \\ & \dfrac{1}{r_{22}} & \dfrac{-r_{23}}{r_{22}} & \cdots \\ & & 1 & \\ & & & \ddots \end{bmatrix}, \qquad
R_3 = \begin{bmatrix} 1 & & & \\ & 1 & & \\ & & \dfrac{1}{r_{33}} & \cdots \\ & & & \ddots \end{bmatrix}, \quad \ldots.
$$

At the end of the iteration we have

$$
A \underbrace{R_1 R_2 \cdots R_n}_{\hat{R}^{-1}} = \hat{Q}. \tag{8.10}
$$

This formulation demonstrates that the Gram–Schmidt algorithm is a method of *triangular orthogonalization*. It applies triangular operations on the right of a matrix to reduce it to a matrix with orthonormal columns. Of course, in practice, we do not form the matrices R_i and multiply them together explicitly. The purpose of mentioning them is to give insight into the structure of the Gram–Schmidt algorithm. In Lecture 20 we shall see that it bears a close resemblance to the structure of Gaussian elimination.

Exercises

8.1. Let A be an $m \times n$ matrix. Determine the exact numbers of floating point additions, subtractions, multiplications, and divisions involved in computing the factorization $A = \hat{Q}\hat{R}$ by Algorithm 8.1.

8.2. Write a MATLAB function [Q,R] = mgs(A) (see next lecture) that computes a reduced QR factorization $A = \hat{Q}\hat{R}$ of an $m \times n$ matrix A with $m \geq n$ using modified Gram–Schmidt orthogonalization. The output variables are a matrix $Q \in \mathbb{C}^{m \times n}$ with orthonormal columns and a triangular matrix $R \in \mathbb{C}^{n \times n}$.

8.3. Each upper-triangular matrix R_j of p. 61 can be interpreted as the product of a diagonal matrix and a unit upper-triangular matrix (i.e., an upper-triangular matrix with 1 on the diagonal). Explain exactly what these factors are, and which line of Algorithm 8.1 corresponds to each.

Lecture 9. MATLAB

To learn numerical linear algebra, one must make a habit of experimenting on the computer. There is no better way to do this than by using the problem-solving environment known as MATLAB®.* In this lecture we illustrate MATLAB experimentation by three examples. Along the way, we make some observations about the stability of Gram–Schmidt orthogonalization.

MATLAB

MATLAB is a language for mathematical computations whose fundamental data types are vectors and matrices. It is distinguished from languages like Fortran and C by operating at a higher mathematical level, including hundreds of operations such as matrix inversion, the singular value decomposition, and the fast Fourier transform as built-in commands. It is also a problem-solving environment, processing top-level comments by an interpreter rather than a compiler and providing in-line access to 2D and 3D graphics.

Since the 1980s, MATLAB has become a widespread tool among numerical analysts and engineers around the world. For many problems of large-scale scientific computing, and for virtually all small- and medium-scale experimentation in numerical linear algebra, it is the language of choice.

---

*MATLAB is a registered trademark of The MathWorks, Inc., 24 Prime Park Way, Natick, MA 01760, USA, tel. 508-647-7000, fax 508-647-7001, info@mathworks.com, http://www.mathworks.com.

In this book, we use MATLAB now and then to present certain numerical experiments, and in some exercises. We do not describe the language systematically, since the number of experiments we present is limited, and only a reading knowledge of MATLAB is needed to follow them.

Experiment 1: Discrete Legendre Polynomials

In Lecture 7 we considered the Vandermonde "matrix" with "columns" consisting of the monomials 1, x, x^2, and x^3 on the interval $[-1, 1]$. Suppose we now make this a true Vandermonde matrix by discretizing $[-1, 1]$ by 257 equally spaced points. The following lines of MATLAB construct this matrix and compute its reduced QR factorization.

`x = (-128:128)'/128;`	Set x to a discretization of $[-1, 1]$.
`A = [x.^0 x.^1 x.^2 x.^3];`	Construct Vandermonde matrix.
`[Q,R] = qr(A,0);`	Find its reduced QR factorization.

Here are a few remarks on these commands. In the first line, the prime ' converts (-128:128) from a row to a column vector. In the second line, the sequences .^ indicate *entrywise* powers. In the third line, qr is a built-in MATLAB function for computing QR factorizations; the argument 0 indicates that a reduced rather than full factorization is needed. The method used here is not Gram–Schmidt orthogonalization but Householder triangularization, discussed in the next lecture, but this is of no consequence for the present purpose. In all three lines, the semicolons at the end suppress the printed output that would otherwise be produced (x, A, Q, and R).

The columns of the matrix Q are essentially the first four Legendre polynomials of Figure 7.1. They differ slightly, by amounts close to plotting accuracy, because the continuous inner product on $[-1, 1]$ that defines the Legendre polynomials has been replaced by a discrete analogue. They also differ in normalization, since a Legendre polynomial should satisfy $P_k(1) = 1$. We can fix this by dividing each column of Q by its final entry. The following lines of MATLAB do this by a right-multiplication by a 4×4 diagonal matrix.

`scale = Q(257,:);`	Select last row of Q.
`Q = Q*diag(1 ./scale);`	Rescale columns by these numbers.
`plot(Q)`	Plot columns of rescaled Q.

The result of our computation is a plot that looks just like Figure 7.1 (not shown). In Fortran or C, this would have taken dozens of lines of code containing numerous loops and nested loops. In our six lines of MATLAB, not a single loop has appeared explicitly, though at least one loop is implicit in every line.

Experiment 2: Classical vs. Modified Gram–Schmidt

Our second example has more algorithmic substance. Its purpose is to explore the difference in numerical stability between the classical and modified Gram–Schmidt algorithms.

First, we construct a square matrix A with random singular vectors and widely varying singular values spaced by factors of 2 between 2^{-1} and 2^{-80}.

`[U,X] = qr(randn(80));`	Set U to a random orthogonal matrix.
`[V,X] = qr(randn(80));`	Set V to a random orthogonal matrix.
`S=diag(2.^(-1:-1:-80));`	Set S to a diagonal matrix with exponentially graded entries.
`A = U*S*V;`	Set A to a matrix with these entries as singular values.

Now, we use Algorithms 7.1 and 8.1 to compute QR factorizations of A. In the following code, the programs `clgs` and `mgs` are MATLAB implementations, not listed here, of Algorithms 7.1 and 8.1.

`[QC,RC] = clgs(A);`	Compute a factorization $Q^{(c)}R^{(c)}$ by classical Gram–Schmidt.
`[QM,RM] = mgs(A);`	Compute a factorization $Q^{(m)}R^{(m)}$ by modified Gram–Schmidt.

Finally, we plot the diagonal elements r_{jj} produced by both computations (MATLAB code not shown). Since $r_{jj} = \|P_j a_j\|$, this gives us a picture of the size of the projection at each step. The results are shown on a logarithmic scale in Figure 9.1.

The first thing one notices in the figure is a steady decrease of r_{jj} with j, closely matching the line 2^{-j}. Evidently r_{jj} is not exactly equal to the jth singular value of A, but it is a reasonably good approximation. This phenomenon can be roughly explained as follows. The SVD of A can be written in the form (5.3) as

$$A = 2^{-1}u_1 v_1^* + 2^{-2}u_2 v_2^* + 2^{-3}u_3 v_3^* + \cdots + 2^{-80}u_{80}v_{80}^*,$$

where $\{u_j\}$ and $\{v_j\}$ are the left and right singular vectors of A, respectively. In particular, the jth column of A has the form

$$a_j = 2^{-1}\overline{v}_{j1}u_1 + 2^{-2}\overline{v}_{j2}u_2 + 2^{-3}\overline{v}_{j3}u_3 + \cdots + 2^{-80}\overline{v}_{j,80}u_{80}.$$

Since the singular vectors are random, we can expect that the numbers $\overline{v}_{ji}$ are all of a similar magnitude, on the order of $80^{-1/2} \approx 0.1$. Now, when we take the QR factorization, it is evident that the first vector q_1 is likely to be

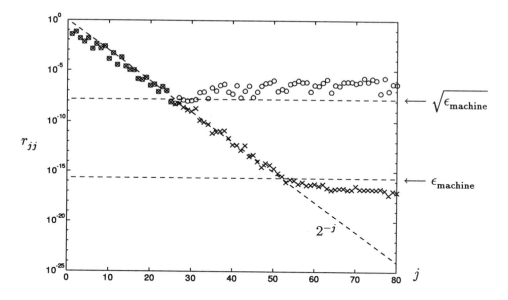

Figure 9.1. *Computed r_{jj} versus j for the QR factorization of a matrix with exponentially graded singular values. On this computer with about 16 digits of relative accuracy, the classical Gram–Schmidt algorithm produces the numbers represented by circles and the modified Gram–Schmidt algorithm produces the numbers represented by crosses.*

approximately equal to u_1, with r_{11} on the order of $2^{-1} \times 80^{-1/2}$. Orthogonalization at the next step will yield a second vector q_2 approximately equal to u_2, with r_{22} on the order of $2^{-2} \times 80^{-1/2}$—and so on.

The next thing one notices in Figure 9.1 is that the geometric decrease of r_{jj} does not continue all the way to $j = 80$. This is a consequence of rounding errors on the computer. With the classical Gram–Schmidt algorithm, the numbers never become smaller than about 10^{-8}. With the modified Gram–Schmidt algorithm, they shrink eight orders of magnitude further, down to the order of 10^{-16}, which is the level of *machine epsilon* for the computer used in this calculation. Machine epsilon is defined in Lecture 13.

Clearly, some algorithms are more stable than others. It is well established that the classical Gram–Schmidt process is one of the unstable ones. Consequently it is rarely used, except sometimes on parallel computers in situations where advantages related to communication may outweigh the disadvantage of instability.

Experiment 3: Numerical Loss of Orthogonality

At the risk of confusing the reader by presenting two instability phenomena in succession, we close this lecture by exhibiting another, different kind of instability that affects both the modified and classical Gram–Schmidt algorithms. In floating point arithmetic, these algorithms may produce vectors q_j that are far from orthogonal. The loss of orthogonality occurs when A is close to rank-deficient, and, like most instabilities, it can appear even in low dimensions.

Starting on paper rather than in MATLAB, consider the case of a matrix

$$A = \begin{bmatrix} 0.70000 & 0.70711 \\ 0.70001 & 0.70711 \end{bmatrix} \tag{9.1}$$

on a computer that rounds all computed results to five digits of relative accuracy (Lecture 13). The classical and modified algorithms are identical in the 2×2 case. At step $j = 1$, the first column is normalized, yielding

$$r_{11} = 0.98996, \qquad q_1 = a_1/r_{11} = \begin{bmatrix} 0.70000/0.98996 \\ 0.70001/0.98996 \end{bmatrix} = \begin{bmatrix} 0.70710 \\ 0.70711 \end{bmatrix}$$

in five-digit arithmetic. At step $j = 2$, the component of a_2 in the direction of q_1 is computed and subtracted out:

$$r_{12} = q_1^* a_2 = 0.70710 \times 0.70711 + 0.70711 \times 0.70711 = 1.0000,$$

$$v_2 = a_2 - r_{12}q_1 = \begin{bmatrix} 0.70711 \\ 0.70711 \end{bmatrix} - \begin{bmatrix} 0.70710 \\ 0.70711 \end{bmatrix} = \begin{bmatrix} 0.00001 \\ 0.00000 \end{bmatrix},$$

again with rounding to five digits. This computed v_2 is dominated by errors. The final computed Q is

$$Q = \begin{bmatrix} 0.70710 & 1.0000 \\ 0.70711 & 0.0000 \end{bmatrix},$$

which is not close to any orthogonal matrix.

On a computer with sixteen-digit precision, we still lose about five digits of orthogonality if we apply modified Gram–Schmidt to the matrix (9.1). Here is the MATLAB evidence. The "eye" function generates the identity of the indicated dimension.

`A = [.70000 .70711`	Define A.
`     .70001 .70711];`	
`[Q,R] = qr(A);`	Compute factor Q by Householder.
`   norm(Q'*Q-eye(2))`	Test orthogonality of Q.
`[Q,R] = mgs(A);`	Compute factor Q by modified G–S.
`   norm(Q'*Q-eye(2))`	Test orthogonality of Q.

The lines without semicolons produce the following printed output:

$$\text{ans} = 2.3515\text{e}{-16}, \qquad \text{ans} = 2.3014\text{e}{-11}.$$

Exercises

9.1. (a) Run the six-line MATLAB program of Experiment 1 to produce a plot of approximate Legendre polynomials.

(b) For $k = 0, 1, 2, 3$, plot the difference on the 257-point grid between these approximations and the exact polynomials (7.11). How big are the errors, and how are they distributed?

(c) Compare these results with what you get with grid spacings $\Delta x = 2^{-\nu}$ for other values of ν. What power of Δx appears to control the convergence?

9.2. In Experiment 2, the singular values of A match the diagonal elements of a QR factor R approximately. Consider now a very different example. Suppose $Q = I$ and $A = R$, the $m \times m$ matrix (a *Toeplitz matrix*) with 1 on the main diagonal, 2 on the first superdiagonal, and 0 everywhere else.

(a) What are the eigenvalues, determinant, and rank of A?

(b) What is A^{-1}?

(c) Give a nontrivial upper bound on σ_m, the mth singular value of A. You are welcome to use MATLAB for inspiration, but the bound you give should be justified analytically. (Hint: Use part (b).)

This problem illustrates that you cannot always infer much about the singular values of a matrix from its eigenvalues or from the diagonal entries of a QR factor R.

9.3. (a) Write a MATLAB program that sets up a 15×40 matrix with entries 0 everywhere except for the values 1 in the positions indicated in the picture below. The upper-leftmost 1 is in position $(2, 2)$, and the lower-rightmost 1 is in position $(13, 39)$. This picture was produced with the command spy(A).

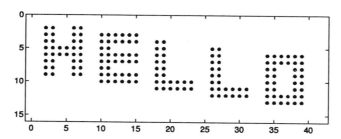

(b) Call svd to compute the singular values of A, and print the results. Plot these numbers using both plot and semilogy. What is the mathematically exact rank of A? How does this show up in the computed singular values?

(c) For each i from 1 to rank(A), construct the rank-i matrix B that is the best approximation to A in the 2-norm. Use the command pcolor(B) with colormap(gray) to create images of these various approximations.

Lecture 10. Householder Triangularization

The other principal method for computing QR factorizations is Householder triangularization, which is numerically more stable than Gram–Schmidt orthogonalization, though it lacks the latter's applicability as a basis for iterative methods. The Householder algorithm is a process of "orthogonal triangularization," making a matrix triangular by a sequence of unitary matrix operations.

Householder and Gram–Schmidt

As we saw in Lecture 8, the Gram–Schmidt iteration applies a succession of elementary triangular matrices R_k on the right of A, so that the resulting matrix

$$A \underbrace{R_1 R_2 \cdots R_n}_{\hat{R}^{-1}} = \hat{Q}$$

has orthonormal columns. The product $\hat{R} = R_n^{-1} \cdots R_2^{-1} R_1^{-1}$ is upper-triangular too, and thus $A = \hat{Q}\hat{R}$ is a reduced QR factorization of A.

In contrast, the Householder method applies a succession of elementary unitary matrices Q_k on the left of A, so that the resulting matrix

$$\underbrace{Q_n \cdots Q_2 Q_1}_{Q^*} A = R$$

is upper-triangular. The product $Q = Q_1^* Q_2^* \cdots Q_n^*$ is unitary too, and therefore $A = QR$ is a full QR factorization of A.

69

The two methods can thus be summarized as follows:

Gram–Schmidt: triangular orthogonalization,
Householder: orthogonal triangularization.

Triangularizing by Introducing Zeros

At the heart of the Householder method is an idea originally proposed by Alston Householder in 1958. This is an ingenious way of designing the unitary matrices Q_k so that $Q_n \cdots Q_2 Q_1 A$ is upper-triangular.

The matrix Q_k is chosen to introduce zeros below the diagonal in the kth column while preserving all the zeros previously introduced. For example, in the 5×3 case, three operations Q_k are applied, as follows. In these matrices, the symbol $\times$ represents an entry that is not necessarily zero, and boldfacing indicates an entry that has just been changed. Blank entries are zero.

$$
\begin{bmatrix} \times & \times & \times \\ \times & \times & \times \\ \times & \times & \times \\ \times & \times & \times \\ \times & \times & \times \end{bmatrix} \xrightarrow{Q_1} \begin{bmatrix} \mathbf{\times} & \mathbf{\times} & \mathbf{\times} \\ \mathbf{0} & \mathbf{\times} & \mathbf{\times} \\ \mathbf{0} & \mathbf{\times} & \mathbf{\times} \\ \mathbf{0} & \mathbf{\times} & \mathbf{\times} \\ \mathbf{0} & \mathbf{\times} & \mathbf{\times} \end{bmatrix} \xrightarrow{Q_2} \begin{bmatrix} \times & \times & \times \\ & \mathbf{\times} & \mathbf{\times} \\ & \mathbf{0} & \mathbf{\times} \\ & \mathbf{0} & \mathbf{\times} \\ & \mathbf{0} & \mathbf{\times} \end{bmatrix} \xrightarrow{Q_3} \begin{bmatrix} \times & \times & \times \\ & \times & \times \\ & & \mathbf{\times} \\ & & \mathbf{0} \\ & & \mathbf{0} \end{bmatrix} \quad (10.1)
$$

$$
A \qquad\qquad Q_1 A \qquad\qquad Q_2 Q_1 A \qquad\qquad Q_3 Q_2 Q_1 A
$$

First, Q_1 operates on rows $1, \ldots, 5$, introducing zeros in positions $(2,1)$, $(3,1)$, $(4,1)$, and $(5,1)$. Next, Q_2 operates on rows $2, \ldots, 5$, introducing zeros in positions $(3,2)$, $(4,2)$, and $(5,2)$ but not destroying the zeros introduced by Q_1. Finally, Q_3 operates on rows $3, \ldots, 5$, introducing zeros in positions $(4,3)$ and $(5,3)$ without destroying any of the zeros introduced earlier.

In general, Q_k operates on rows $k, \ldots, m$. At the beginning of step k, there is a block of zeros in the first $k-1$ columns of these rows. The application of Q_k forms linear combinations of these rows, and the linear combinations of the zero entries remain zero. After n steps, all the entries below the diagonal have been eliminated and $Q_n \cdots Q_2 Q_1 A = R$ is upper-triangular.

Householder Reflectors

How can we construct unitary matrices Q_k to introduce zeros as indicated in (10.1)? The standard approach is as follows. Each Q_k is chosen to be a unitary matrix of the form

$$
Q_k = \begin{bmatrix} I & 0 \\ 0 & F \end{bmatrix}, \qquad (10.2)
$$

where I is the $(k-1) \times (k-1)$ identity and F is an $(m-k+1) \times (m-k+1)$ unitary matrix. Multiplication by F must introduce zeros into the

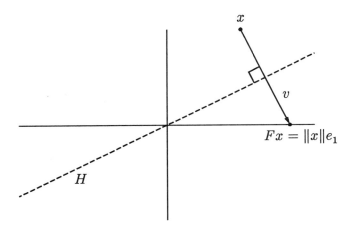

Figure 10.1. *A Householder reflection.*

kth column. The Householder algorithm chooses F to be a particular matrix called a *Householder reflector*.

Suppose, at the beginning of step k, the entries $k, \ldots, m$ of the kth column are given by the vector $x \in \mathbb{C}^{m-k+1}$. To introduce the correct zeros into the kth column, the Householder reflector F should effect the following map:

$$
x = \begin{bmatrix} \times \\ \times \\ \times \\ \vdots \\ \times \end{bmatrix} \quad \overset{F}{\longrightarrow} \quad Fx = \begin{bmatrix} \|x\| \\ 0 \\ 0 \\ \vdots \\ 0 \end{bmatrix} = \|x\|e_1. \tag{10.3}
$$

(We shall modify this idea by a $\pm$ sign in a moment.) The idea for accomplishing this is indicated in Figure 10.1. The reflector F will reflect the space $\mathbb{C}^{m-k+1}$ across the hyperplane H orthogonal to $v = \|x\|e_1 - x$. A *hyperplane* is the higher-dimensional generalization of a two-dimensional plane in three-space—a three-dimensional subspace of a four-dimensional space, a four-dimensional subspace of a five-dimensional space, and so on. In general, a hyperplane can be characterized as the set of points orthogonal to a fixed nonzero vector. In Figure 10.1, that vector is $v = \|x\|e_1 - x$, and one can think of the dashed line as a depiction of H viewed "edge on."

When the reflector is applied, every point on one side of the hyperplane H is mapped to its mirror image on the other side. In particular, x is mapped to $\|x\|e_1$. The formula for this reflection can be derived as follows. In (6.11) we have seen that for any $y \in \mathbb{C}^m$, the vector

$$
Py = \left(I - \frac{vv^*}{v^*v}\right)y = y - v\left(\frac{v^*y}{v^*v}\right)
$$

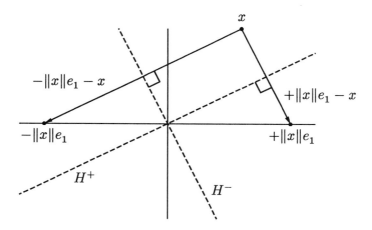

Figure 10.2. *Two possible reflections. For numerical stability, it is important to choose the one that moves x the larger distance.*

is the orthogonal projection of y onto the space H. To reflect y across H, we must not stop at this point; we must go exactly twice as far in the same direction. The reflection Fy should therefore be

$$Fy = \left(I - 2\frac{vv^*}{v^*v}\right)y = y - 2v\left(\frac{v^*y}{v^*v}\right).$$

Hence the matrix F is

$$F = I - 2\frac{vv^*}{v^*v}. \tag{10.4}$$

Note that the projector P (rank $m-1$) and the reflector F (full rank, unitary) differ only in the presence of a factor of 2.

The Better of Two Reflectors

In (10.3) and in Figure 10.1 we have simplified matters, for in fact, there are many Householder reflections that will introduce the zeros needed. The vector x can be reflected to $z\|x\|e_1$, where z is any scalar with $|z| = 1$. In the complex case, there is a circle of possible reflections, and even in the real case, there are two alternatives, represented by reflections across two different hyperplanes, H^+ and H^-, as illustrated in Figure 10.2.

Mathematically, either choice of sign is satisfactory. However, this is a case where the goal of numerical stability—insensitivity to rounding errors— dictates that one choice should be taken rather than the other. For numerical stability, it is desirable to reflect x to the vector $z\|x\|e_1$ that is not too close to x itself. To achieve this, we can choose $z = -\text{sign}(x_1)$, where x_1 denotes the first component of x, so that the reflection vector becomes $v = -\text{sign}(x_1)\|x\|e_1 - x$, or, upon clearing the factors -1,

$$v = \text{sign}(x_1)\|x\|e_1 + x. \tag{10.5}$$

To make this a complete prescription, we may arbitrarily impose the convention that $\text{sign}(x_1) = 1$ if $x_1 = 0$.

It is not hard to see why the choice of sign makes a difference for stability. Suppose that in Figure 10.2, the angle between H^+ and the e_1 axis is very small. Then the vector $v = \|x\|e_1 - x$ is much smaller than x or $\|x\|e_1$. Thus the calculation of v represents a subtraction of nearby quantities and will tend to suffer from cancellation errors. If we pick the sign as in (10.5), we avoid such effects by ensuring that $\|v\|$ is never smaller than $\|x\|$.

The Algorithm

We now formulate the whole Householder algorithm. To do this, it will be helpful to utilize a new (MATLAB-style) notation. If A is a matrix, we define $A_{i:i',j:j'}$ to be the $(i'-i+1)\times(j'-j+1)$ submatrix of A with upper-left corner a_{ij} and lower-right corner $a_{i',j'}$. In the special case where the submatrix reduces to a subvector of a single row or column, we write $A_{i,j:j'}$ or $A_{i:i',j}$, respectively.

The following algorithm computes the factor R of a QR factorization of an $m \times n$ matrix A with $m \geq n$, leaving the result in place of A. Along the way, n reflection vectors $v_1, \ldots, v_n$ are stored for later use.

Algorithm 10.1. Householder QR Factorization

for $k = 1$ **to** n

$\quad x = A_{k:m,k}$

$\quad v_k = \text{sign}(x_1)\|x\|_2 e_1 + x$

$\quad v_k = v_k/\|v_k\|_2$

$\quad A_{k:m,k:n} = A_{k:m,k:n} - 2v_k(v_k^* A_{k:m,k:n})$

Applying or Forming Q

Upon the completion of Algorithm 10.1, A has been reduced to upper-triangular form; this is the matrix R in the QR factorization $A = QR$. The unitary matrix Q has not, however, been constructed, nor has its n-column submatrix $\hat{Q}$ corresponding to a reduced QR factorization. There is a reason for this. Constructing Q or $\hat{Q}$ takes additional work, and in many applications, we can avoid this by working directly with the formula

$$Q^* = Q_n \cdots Q_2 Q_1 \tag{10.6}$$

or its conjugate

$$Q = Q_1 Q_2 \cdots Q_n. \tag{10.7}$$

(No asterisks have been forgotten here; recall that each Q_j is hermitian.)

For example, in Lecture 7 we saw that a square system of equations $Ax = b$ can be solved via QR factorization of A. The only way in which Q was used in this process was in the computation of the product Q^*b. By (10.6), we can calculate Q^*b by a sequence of n operations applied to b, the same operations that were applied to A to make it triangular. The algorithm is as follows.

Algorithm 10.2. Implicit Calculation of a Product Q^*b

for $k = 1$ to n
$$b_{k:m} = b_{k:m} - 2v_k(v_k^* b_{k:m})$$

Similarly, the computation of a product Qx can be achieved by the same process executed in reverse order.

Algorithm 10.3. Implicit Calculation of a Product Qx

for $k = n$ downto 1
$$x_{k:m} = x_{k:m} - 2v_k(v_k^* x_{k:m})$$

The work involved in either of these algorithms is of order $O(mn)$, not $O(mn^2)$ as in Algorithm 10.1 (see below).

Sometimes, of course, one may wish to construct the matrix Q explicitly. This can be achieved in various ways. We can construct QI via Algorithm 10.3 by computing its columns $Qe_1, Qe_2, \ldots, Qe_m$. Alternatively, we can construct Q^*I via Algorithm 10.2 and then conjugate the result. A variant of this idea is to conjugate each step rather than the final product, that is, to construct IQ by computing its rows $e_1^*Q, e_2^*Q, \ldots, e_m^*Q$ as suggested by (10.7). Of these various ideas, the best is the first one, based on Algorithm 10.3. The reason is that it begins with operations involving Q_n, Q_{n-1}, and so on that modify only a small part of the vector they are applied to; if advantage is taken of this sparsity property, a speed-up is achieved.

If only $\hat{Q}$ rather than Q is needed, it is enough to compute the columns $Qe_1, Qe_2, \ldots, Qe_n$.

Operation Count

The work involved in Algorithm 10.1 is dominated by the innermost loop,

$$A_{k:m,j} - 2v_k(v_k^* A_{k:m,k}). \tag{10.8}$$

If the vector length is $l = m - k + 1$, this calculation requires $4l - 1 \sim 4l$ scalar operations: l for the subtraction, l for the scalar multiplication, and $2l - 1$ for the dot product. This is ~ 4 flops for each entry operated on.

We may add up these four flops per entry by geometric reasoning, as in Lecture 8. Each successive step of the outer loop operates on fewer rows, because during step k, rows $1, \ldots, k-1$ are not changed. Furthermore, each step operates on fewer columns, because columns $1, \ldots, k-1$ of the rows operated on are zero and are skipped. Thus the work done by one outer step can be represented by a single layer of the following solid:

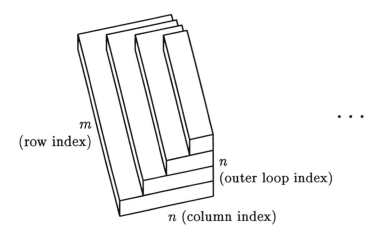

The total number of operations corresponds to four times the volume of the solid. To determine the volume pictorially we may divide the solid into two pieces:

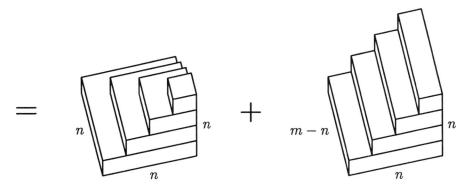

The solid on the left has the shape of a ziggurat and converges to a pyramid as $n \to \infty$, with volume $\frac{1}{3}n^3$. The solid on the right has the shape of a staircase and converges to a prism as $m, n \to \infty$, with volume $\frac{1}{2}(m-n)n^2$. Combined, the volume is $\sim \frac{1}{2}mn^2 - \frac{1}{6}n^3$. Multiplying by four flops per unit volume, we find

$$\text{Work for Householder orthogonalization:} \quad \sim 2mn^2 - \frac{2}{3}n^3 \text{ flops.} \quad (10.9)$$

Exercises

10.1. Determine the (a) eigenvalues, (b) determinant, and (c) singular values of a Householder reflector. For the eigenvalues, give a geometric argument as well as an algebraic proof.

10.2. (a) Write a MATLAB function [W,R] = house(A) that computes an implicit representation of a full QR factorization $A = QR$ of an $m \times n$ matrix A with $m \geq n$ using Householder reflections. The output variables are a lower-triangular matrix $W \in \mathbb{C}^{m \times n}$ whose columns are the vectors v_k defining the successive Householder reflections, and a triangular matrix $R \in \mathbb{C}^{n,n}$.

(b) Write a MATLAB function Q = formQ(W) that takes the matrix W produced by house as input and generates a corresponding $m \times m$ orthogonal matrix Q.

10.3. Let Z be the matrix

$$Z = \begin{bmatrix} 1 & 2 & 3 \\ 4 & 5 & 6 \\ 7 & 8 & 7 \\ 4 & 2 & 3 \\ 4 & 2 & 2 \end{bmatrix}.$$

Compute three reduced QR factorizations of Z in MATLAB: by the Gram–Schmidt routine mgs of Exercise 8.2, by the Householder routines house and formQ of Exercise 10.2, and by MATLAB's built-in command [Q,R] = qr(Z,0). Compare these three and comment on any differences you see.

10.4. Consider the 2×2 orthogonal matrices

$$F = \begin{bmatrix} -c & s \\ s & c \end{bmatrix}, \qquad J = \begin{bmatrix} c & s \\ -s & c \end{bmatrix}, \tag{10.10}$$

where $s = \sin\theta$ and $c = \cos\theta$ for some θ. The first matrix has $\det F = -1$ and is a reflector—the special case of a Householder reflector in dimension 2. The second has $\det J = 1$ and effects a rotation instead of a reflection. Such a matrix is called a *Givens rotation*.

(a) Describe exactly what geometric effects left-multiplications by F and J have on the plane $\mathbb{R}^2$. (J rotates the plane by the angle θ, for example, but is the rotation clockwise or counterclockwise?)

(b) Describe an algorithm for QR factorization that is analogous to Algorithm 10.1 but based on Givens rotations instead of Householder reflections.

(c) Show that your algorithm involves six flops per entry operated on rather than four, so that the asymptotic operation count is 50% greater than (10.9).

Lecture 11. Least Squares Problems

Least squares data-fitting has been an indispensable tool since its invention by Gauss and Legendre around 1800, with ramifications extending throughout the mathematical sciences. In the language of linear algebra, the problem here is the solution of an overdetermined system of equations $Ax = b$—rectangular, with more rows than columns. The least squares idea is to "solve" such a system by minimizing the 2-norm of the residual $b - Ax$.

The Problem

Consider a linear system of equations having n unknowns but $m > n$ equations. Symbolically, we wish to find a vector $x \in \mathbb{C}^n$ that satisfies $Ax = b$, where $A \in \mathbb{C}^{m \times n}$ and $b \in \mathbb{C}^m$. In general, such a problem has no solution. A suitable vector x exists only if b lies in range(A), and since b is an m-vector, whereas range(A) is of dimension at most n, this is true only for exceptional choices of b. We say that a rectangular system of equations with $m > n$ is *overdetermined*. The vector known as the *residual*,

$$r = b - Ax \in \mathbb{C}^m, \tag{11.1}$$

can perhaps be made quite small by a suitable choice of x, but in general it cannot be made equal to zero.

What can it mean to solve a problem that has no solution? In the case of an overdetermined system of equations, there is a natural answer to this question. Since the residual r cannot be made to be zero, let us instead make it as small as possible. Measuring the smallness of r entails choosing a norm. If we choose the 2-norm, the problem takes the following form:

$$\text{Given } A \in \mathbb{C}^{m \times n}, \ m \geq n, \ b \in \mathbb{C}^m,$$
$$\text{find } x \in \mathbb{C}^n \text{ such that } \|b - Ax\|_2 \text{ is minimized.}$$

(11.2)

This is our formulation of the general (linear) *least squares problem*. The choice of the 2-norm can be defended by various geometric and statistical arguments, and, as we shall see, it certainly leads to simple algorithms— ultimately because the derivative of a quadratic function, which must be set to zero for minimization, is linear.

The 2-norm corresponds to Euclidean distance, so there is a simple geometric interpretation of (11.2). We seek a vector $x \in \mathbb{C}^n$ such that the vector $Ax \in \mathbb{C}^m$ is the closest point in range(A) to b.

Example: Polynomial Data-Fitting

As an example, let us compare polynomial interpolation, which leads to a square system of equations, and least squares polynomial data-fitting, where the system is rectangular.

Example 11.1. Polynomial Interpolation. Suppose we are given m distinct points $x_1, \ldots, x_m \in \mathbb{C}$ and data $y_1, \ldots, y_m \in \mathbb{C}$ at these points. Then there exists a unique *polynomial interpolant* to these data in these points, that is, a polynomial of degree at most $m - 1$,

$$p(x) = c_0 + c_1 x + \cdots + c_{m-1} x^{m-1},$$

(11.3)

with the property that at each x_i, $p(x_i) = y_i$. The relationship of the data $\{x_i\}, \{y_i\}$ to the coefficients $\{c_i\}$ can be expressed by the square Vandermonde system seen already in Example 1.1:

$$\begin{bmatrix} 1 & x_1 & x_1^2 & & x_1^{m-1} \\ 1 & x_2 & x_2^2 & \cdots & x_2^{m-1} \\ 1 & x_3 & x_3^2 & & x_3^{m-1} \\ \vdots & & \vdots & & \vdots \\ 1 & x_m & x_m^2 & \cdots & x_m^{m-1} \end{bmatrix} \begin{bmatrix} c_0 \\ c_1 \\ c_2 \\ \vdots \\ c_{m-1} \end{bmatrix} = \begin{bmatrix} y_1 \\ y_2 \\ y_3 \\ \vdots \\ y_m \end{bmatrix}.$$

(11.4)

To determine the coefficients $\{c_i\}$ for a given set of data, we can solve this system of equations, which is guaranteed to be nonsingular as long as the points $\{x_i\}$ are distinct (Exercise 37.3).

Figure 11.1 presents an example of this process of polynomial interpolation. We have eleven data points in the form of a discrete square wave, represented by crosses, and the curve $p(x)$ passes through them, as it must. However, the fit is not at all pleasing. Near the ends of the interval, $p(x)$ exhibits large oscillations that are clearly an artifact of the interpolation process, not a reasonable reflection of the data.

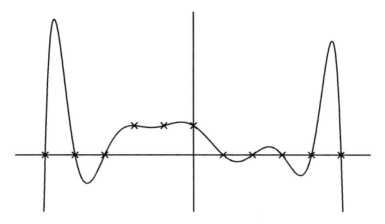

Figure 11.1. *Degree* 10 *polynomial interpolant to eleven data points. The axis scales are not given, as these have no effect on the picture.*

This unsatisfactory behavior is typical of polynomial interpolation. The fits it produces are often bad, and they tend to get worse rather than better if more data are utilized. Even if the fit is good, the interpolation process may be ill-conditioned, i.e., sensitive to perturbations of the data (next lecture). To avoid these problems, one can utilize a nonuniform set of interpolation points such as Chebyshev points in the interval $[-1, 1]$. In applications, however, it will not always be possible to choose the interpolation points at will. □

Example 11.2. Polynomial Least Squares Fitting. Without changing the data points, we can do better by reducing the degree of the polynomial. Given $x_1, \ldots, x_m$ and $y_1, \ldots, y_m$ again, consider now a degree $n-1$ polynomial

$$p(x) \; = \; c_0 + c_1 x + \cdots + c_{n-1}x^{n-1} \tag{11.5}$$

for some $n < m$. Such a polynomial is a least squares fit to the data if it minimizes the sum of the squares of the deviation from the data,

$$\sum_{i=1}^{m} |p(x_i) - y_i|^2. \tag{11.6}$$

This sum of squares is equal to the square of the norm of the residual, $\|r\|_2^2$, for the rectangular Vandermonde system

$$\begin{bmatrix} 1 & x_1 & & x_1^{n-1} \\ 1 & x_2 & \cdots & x_2^{n-1} \\ 1 & x_3 & & x_3^{n-1} \\ \vdots & & & \vdots \\ 1 & x_m & \cdots & x_m^{n-1} \end{bmatrix} \begin{bmatrix} c_0 \\ c_1 \\ \vdots \\ c_{n-1} \end{bmatrix} \approx \begin{bmatrix} y_1 \\ y_2 \\ y_3 \\ \vdots \\ y_m \end{bmatrix}. \tag{11.7}$$

Figure 11.2 illustrates what we get if we fit the same eleven data points from the last example with a polynomial of degree 7. The new polynomial does not interpolate the data, but it captures their overall behavior much better than the polynomial of Example 11.1. Though one cannot see this in the figure, it is also less sensitive to perturbations.   □

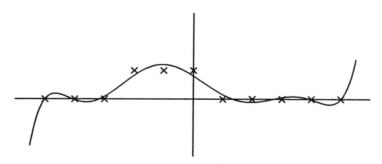

Figure 11.2. *Degree 7 polynomial least squares fit to the same eleven data points.*

Orthogonal Projection and the Normal Equations

How was Figure 11.2 computed? How are least squares problems solved in general? The key to deriving algorithms is orthogonal projection.

The idea is illustrated in Figure 11.3. Our goal is to find the closest point Ax in range(A) to b, so that the norm of the residual $r = b - Ax$ is minimized. It is clear geometrically that this will occur provided $Ax = Pb$, where $P \in \mathbb{C}^{m \times m}$ is the orthogonal projector (Lecture 6) that maps $\mathbb{C}^m$ onto range(A). In other words, *the residual $r = b - Ax$ must be orthogonal to* range(A). We formulate this condition as the following theorem.

Theorem 11.1. *Let $A \in \mathbb{C}^{m \times n}$ ($m \geq n$) and $b \in \mathbb{C}^m$ be given. A vector $x \in \mathbb{C}^n$ minimizes the residual norm $\|r\|_2 = \|b - Ax\|_2$, thereby solving the least squares problem (11.2), if and only if $r \perp$ range(A), that is,*

$$A^* r = 0, \tag{11.8}$$

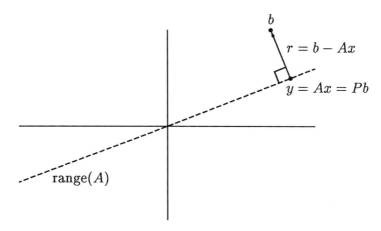

Figure 11.3. *Formulation of the least squares problem (11.2) in terms of orthogonal projection.*

or equivalently,

$$A^*Ax = A^*b, \tag{11.9}$$

or again equivalently,

$$Pb = Ax, \tag{11.10}$$

where $P \in \mathbb{C}^{m \times m}$ is the orthogonal projector onto range(A). *The $n \times n$ system of equations* (11.9), *known as the* normal equations, *is nonsingular if and only if A has full rank. Consequently the solution x is unique if and only if A has full rank.*

Proof. The equivalence of (11.8) and (11.10) follows from the properties of orthogonal projectors discussed in Lecture 6, and the equivalence of (11.8) and (11.9) follows from the definition of r. To show that $y = Pb$ is the unique point in range(A) that minimizes $\|b - y\|_2$, suppose $z \neq y$ is another point in range(A). Since $z - y$ is orthogonal to $b - y$, the Pythagorean theorem (Exercise 2.2) gives $\|b - z\|_2^2 = \|b - y\|_2^2 + \|y - z\|_2^2 > \|b - y\|_2^2$, as required. Finally, we note that if A^*A is singular, then $A^*Ax = 0$ for some nonzero x, implying $x^*A^*Ax = 0$ (see Exercise 6.3). Thus $Ax = 0$, which implies that A is rank-deficient. Conversely, if A is rank-deficient, then $Ax = 0$ for some nonzero x, implying $A^*Ax = 0$ also, so A^*A is singular. By (11.9), this characterization of nonsingular matrices A^*A implies the statement about the uniqueness of x. $\square$

Pseudoinverse

We have just seen that if A has full rank, then the solution x to the least squares problem (11.2) is unique and is given by $x = (A^*A)^{-1}A^*b$. The matrix

$(A^*A)^{-1}A^*$ is known as the *pseudoinverse* of A, denoted by A^+:

$$A^+ = (A^*A)^{-1}A^* \in \mathbb{C}^{n,m}. \qquad (11.11)$$

This matrix maps vectors $b \in \mathbb{C}^m$ to vectors $x \in \mathbb{C}^n$, which explains why it has dimensions $n \times m$—more columns than rows.

We can summarize the full-rank linear least squares problem (11.2) as follows. The problem is to compute one or both of the vectors

$$x = A^+b, \qquad y = Pb, \qquad (11.12)$$

where A^+ is the pseudoinverse of A and P is the orthogonal projector onto range(A). We now describe the three leading algorithms for doing this.

Normal Equations

The classical way to solve least squares problems is to solve the normal equations (11.9). If A has full rank, this is a square, hermitian positive definite system of equations of dimension n. The standard method of solving such a system is by *Cholesky factorization*, discussed in Lecture 23. This method constructs a factorization $A^*A = R^*R$, where R is upper-triangular, reducing (11.9) to the equations

$$R^*Rx = A^*b. \qquad (11.13)$$

Here is the algorithm.

Algorithm 11.1. Least Squares via Normal Equations

  1. Form the matrix A^*A and the vector A^*b.
  2. Compute the Cholesky factorization $A^*A = R^*R$.
  3. Solve the lower-triangular system $R^*w = A^*b$ for w.
  4. Solve the upper-triangular system $Rx = w$ for x.

The steps that dominate the work for this computation are the first two (for steps 3 and 4, see Lecture 17). Because of symmetry, the computation of A^*A requires only mn^2 flops, half what the cost would be if A and A^* were arbitrary matrices of the same dimensions. Cholesky factorization, which also exploits symmetry, requires $n^3/3$ flops. All together, solving least squares problems by the normal equations involves the following total operation count:

$$\text{Work for Algorithm 11.1:} \quad \sim mn^2 + \frac{1}{3}n^3 \text{ flops.} \qquad (11.14)$$

QR Factorization

The "modern classical" method for solving least squares problems, popular since the 1960s, is based upon reduced QR factorization. By Gram–Schmidt orthogonalization or, more usually, Householder triangularization, one constructs a factorization $A = \hat{Q}\hat{R}$. The orthogonal projector P can then be written $P = \hat{Q}\hat{Q}^*$ (6.6), so we have

$$y = Pb = \hat{Q}\hat{Q}^*b. \tag{11.15}$$

Since $y \in \text{range}(A)$, the system $Ax = y$ has an exact solution. Combining the QR factorization and (11.15) gives

$$\hat{Q}\hat{R}x = \hat{Q}\hat{Q}^*b, \tag{11.16}$$

and left-multiplication by $\hat{Q}^*$ results in

$$\hat{R}x = \hat{Q}^*b. \tag{11.17}$$

(Multiplying by $\hat{R}^{-1}$ now gives the formula $A^+ = \hat{R}^{-1}\hat{Q}$ for the pseudoinverse.) Equation (11.17) is an upper-triangular system, nonsingular if A has full rank, and it is readily solved by back substitution (Lecture 17).

Algorithm 11.2. Least Squares via QR Factorization

    1. Compute the reduced QR factorization $A = \hat{Q}\hat{R}$.

    2. Compute the vector $\hat{Q}^*b$.

    3. Solve the upper-triangular system $\hat{R}x = \hat{Q}^*b$ for x.

Notice that (11.17) can also be derived from the normal equations. If $A^*Ax = A^*b$, then $\hat{R}^*\hat{Q}^*\hat{Q}\hat{R}x = \hat{R}^*\hat{Q}^*b$, which implies $\hat{R}x = \hat{Q}^*b$.

The work for Algorithm 11.2 is dominated by the cost of the QR factorization. If Householder reflections are used for this step, we have from (10.9)

$$\text{Work for Algorithm 11.2:} \quad \sim 2mn^2 - \frac{2}{3}n^3 \text{ flops.} \tag{11.18}$$

SVD

In Lecture 31 we shall describe an algorithm for computing the reduced singular value decomposition $A = \hat{U}\hat{\Sigma}V^*$. This suggests another method for solving least squares problems. Now P is represented in the form $P = \hat{U}\hat{U}^*$, giving

$$y = Pb = \hat{U}\hat{U}^*b, \tag{11.19}$$

and the analogues of (11.16) and (11.17) are

$$\hat{U}\hat{\Sigma}V^*x = \hat{U}\hat{U}^*b \tag{11.20}$$

and
$$\hat{\Sigma}V^*x = \hat{U}^*b. \tag{11.21}$$

(Multiplying by $V\hat{\Sigma}^{-1}$ gives $A^+ = V\hat{\Sigma}^{-1}\hat{U}^*$.) The algorithm looks like this.

Algorithm 11.3. Least Squares via SVD

   1. Compute the reduced SVD $A = \hat{U}\hat{\Sigma}V^*$.

   2. Compute the vector $\hat{U}^*b$.

   3. Solve the diagonal system $\hat{\Sigma}w = \hat{U}^*b$ for w.

   4. Set $x = Vw$.

Note that whereas QR factorization reduces the least squares problem to a triangular system of equations, the SVD reduces it to a diagonal system of equations, which is of course trivially solved. If A has full rank, the diagonal system is nonsingular.

As before, (11.21) can be derived from the normal equations. If $A^*Ax = A^*b$, then $V\hat{\Sigma}^*\hat{U}^*\hat{U}\hat{\Sigma}V^*x = V\hat{\Sigma}^*\hat{U}^*b$, implying $\hat{\Sigma}V^*x = \hat{U}^*b$.

The operation count for Algorithm 11.3 is dominated by the computation of the SVD. As we shall see in Lecture 31, for $m \gg n$ this cost is approximately the same as for QR factorization, but for $m \approx n$ the SVD is more expensive. A typical estimate is

$$\text{Work for Algorithm 11.3:} \quad \sim 2mn^2 + 11n^3 \text{ flops,} \tag{11.22}$$

but see Lecture 31 for qualifications of this result.

Comparison of Algorithms

Each of the methods we have described is advantageous in certain situations. When speed is the only consideration, Algorithm 11.1 may be the best. However, solving the normal equations is not always stable in the presence of rounding errors, and thus for many years, numerical analysts have recommended Algorithm 11.2 instead as the standard method for least squares problems. This is indeed a natural and elegant algorithm, and we recommend it for "daily use." If A is close to rank-deficient, however, it turns out that Algorithm 11.2 itself has less-than-ideal stability properties, and in such cases there are good reasons to turn to Algorithm 11.3, based on the SVD.

What are these stability considerations that make one algorithm better than another in some circumstances yet not in others? It is time now to undertake a systematic discussion of such matters. We shall return to the study of algorithms for least squares problems in Lectures 18 and 19.

Exercises

11.1. Suppose the $m \times n$ matrix A has the form

$$A = \begin{bmatrix} A_1 \\ A_2 \end{bmatrix},$$

where A_1 is a nonsingular matrix of dimension $n \times n$ and A_2 is an arbitrary matrix of dimension $(m - n) \times n$. Prove that $\|A^+\|_2 \le \|A_1^{-1}\|_2$.

11.2. (a) How closely, as measured in the L^2 norm on the interval $[1, 2]$, can the function $f(x) = x^{-1}$ be fitted by a linear combination of the functions e^x, $\sin x$, and $\Gamma(x)$? ($\Gamma(x)$ is the gamma function, a built-in function in MATLAB.) Write a program that determines the answer to at least two digits of relative accuracy using a discretization of $[1, 2]$ and a discrete least squares problem. Write down your estimate of the answer and also of the coefficients of the optimal linear combination, and produce a plot of the optimal approximation.

(b) Now repeat, but with $[1, 2]$ replaced by $[0, 1]$. You may find the following fact helpful: if $g(x) = 1/\Gamma(x)$, then $g'(0) = 1$.

11.3. Take $m = 50$, $n = 12$. Using MATLAB's linspace, define t to be the m-vector corresponding to linearly spaced grid points from 0 to 1. Using MATLAB's vander and fliplr, define A to be the $m \times n$ matrix associated with least squares fitting on this grid by a polynomial of degree $n - 1$. Take b to be the function $\cos(4t)$ evaluated on the grid. Now, calculate and print (to sixteen-digit precision) the least squares coefficient vector x by six methods:

(a) Formation and solution of the normal equations, using MATLAB's \,

(b) QR factorization computed by mgs (modified Gram–Schmidt, Exercise 8.2),

(c) QR factorization computed by house (Householder triangularization, Exercise 10.2),

(d) QR factorization computed by MATLAB's qr (also Householder triangularization),

(e) x = A\b in MATLAB (also based on QR factorization),

(f) SVD, using MATLAB's svd.

(g) The calculations above will produce six lists of twelve coefficients. In each list, shade with red pen the digits that appear to be wrong (affected by rounding error). Comment on what differences you observe. Do the normal equations exhibit instability? You do not have to explain your observations.

Part III

Conditioning and Stability

Lecture 12. Conditioning and Condition Numbers

In this third part of the book we turn to a systematic discussion of two fundamental issues of numerical analysis that until now we have only skirted. *Conditioning* pertains to the perturbation behavior of a mathematical problem. *Stability* pertains to the perturbation behavior of an algorithm used to solve that problem on a computer.

Condition of a Problem

In the abstract, we can view a *problem* as a function $f : X \to Y$ from a normed vector space X of data to a normed vector space Y of solutions. This function f is usually nonlinear (even in linear algebra), but most of the time it is at least continuous.

Typically we shall be concerned with the behavior of a problem f at a particular data point $x \in X$ (the behavior may vary greatly from one point to another). The combination of a problem f with prescribed data x might be called a *problem instance*, but it is more usual, though occasionally confusing, to use the term *problem* for this notion too.

A *well-conditioned* problem (instance) is one with the property that all small perturbations of x lead to only small changes in $f(x)$. An *ill-conditioned* problem is one with the property that some small perturbation of x leads to a large change in $f(x)$.

The meaning of "small" and "large" in these statements depends on the application. In particular, sometimes it is most appropriate to measure perturbations on an absolute scale, and sometimes it is most appropriate to measure them relative to the norm of the object being perturbed.

Absolute Condition Number

Let δx denote a small perturbation of x, and write $\delta f = f(x + \delta x) - f(x)$. The *absolute condition number* $\hat{\kappa} = \hat{\kappa}(x)$ of the problem f at x is defined as

$$\hat{\kappa} = \lim_{\delta \to 0} \sup_{\|\delta x\| \leq \delta} \frac{\|\delta f\|}{\|\delta x\|}. \qquad (12.1)$$

For most problems, the limit of the supremum in this formula can be interpreted as a supremum over all infinitesimal perturbations δx, and in the interest of readability, we shall generally write the formula simply as

$$\hat{\kappa} = \sup_{\delta x} \frac{\|\delta f\|}{\|\delta x\|}, \qquad (12.2)$$

with the understanding that δx and δf are infinitesimal.

If f is differentiable, we can evaluate the condition number by means of the derivative of f. Let $J(x)$ be the matrix whose i, j entry is the partial derivative $\partial f_i / \partial x_j$ evaluated at x, known as the *Jacobian* of f at x. The definition of the derivative gives us, to first order, $\delta f \approx J(x)\delta x$, with equality in the limit $\|\delta x\| \to 0$. The absolute condition number becomes

$$\hat{\kappa} = \|J(x)\|, \qquad (12.3)$$

where $\|J(x)\|$ represents the norm of $J(x)$ induced by the norms on X and Y.

Relative Condition Number

When we are concerned with relative changes, we need the notion of relative condition. The *relative condition number* $\kappa = \kappa(x)$ is defined by

$$\kappa = \lim_{\delta \to 0} \sup_{\|\delta x\| \leq \delta} \left(\frac{\|\delta f\|}{\|f(x)\|} \bigg/ \frac{\|\delta x\|}{\|x\|} \right), \qquad (12.4)$$

or, again assuming δx and δf are infinitesimal,

$$\kappa = \sup_{\delta x} \left(\frac{\|\delta f\|}{\|f(x)\|} \bigg/ \frac{\|\delta x\|}{\|x\|} \right). \qquad (12.5)$$

If f is differentiable, we can express this quantity in terms of the Jacobian:

$$\kappa = \frac{\|J(x)\|}{\|f(x)\|/\|x\|}. \qquad (12.6)$$

Both absolute and relative condition numbers have their uses, but the latter are more important in numerical analysis. This is ultimately because the floating point arithmetic used by computers introduces relative errors rather than absolute ones; see the next lecture. A problem is *well-conditioned* if κ is small (e.g., 1, 10, 10^2), and *ill-conditioned* if κ is large (e.g., 10^6, 10^{16}).

Examples

Example 12.1. Consider the trivial problem of obtaining the scalar $x/2$ from $x \in \mathbb{C}$. The Jacobian of the function $f : x \mapsto x/2$ is just the derivative $J = f' = 1/2$, so by (12.6),

$$\kappa = \frac{\|J\|}{\|f(x)\|/\|x\|} = \frac{1/2}{(x/2)/x} = 1.$$

This problem is well-conditioned by any standard.   □

Example 12.2. Consider the problem of computing $\sqrt{x}$ for $x > 0$. The Jacobian of $f : x \mapsto \sqrt{x}$ is the derivative $J = f' = 1/(2\sqrt{x})$, so we have

$$\kappa = \frac{\|J\|}{\|f(x)\|/\|x\|} = \frac{1/(2\sqrt{x})}{\sqrt{x}/x} = \frac{1}{2}.$$

Again, this is a well-conditioned problem.   □

Example 12.3. Consider the problem of obtaining the scalar $f(x) = x_1 - x_2$ from the vector $x = (x_1, x_2)^* \in \mathbb{C}^2$. For simplicity, we use the ∞-norm on the data space $\mathbb{C}^2$. The Jacobian of f is

$$J = \left[\frac{\partial f}{\partial x_1} \ \frac{\partial f}{\partial x_2} \right] = \left[1 \ -1 \right],$$

with $\|J\|_\infty = 2$. The condition number is thus

$$\kappa = \frac{\|J\|_\infty}{\|f(x)\|/\|x\|} = \frac{2}{|x_1 - x_2|/\max\{|x_1|, |x_2|\}}.$$

This quantity is large if $|x_1 - x_2| \approx 0$, so the problem is ill-conditioned when $x_1 \approx x_2$, matching our intuition of the hazards of "cancellation error."   □

Example 12.4. Consider the computation of $f(x) = \tan x$ for x near 10^{100}. In this problem, minuscule relative perturbations in x can result in arbitrarily large changes in $\tan x$. The result: $\tan(10^{100})$ is effectively uncomputable on most computers. The same minuscule perturbations result in arbitrary changes in the derivative of $\tan x$, so there is little point in trying to calculate the Jacobian other than to observe that it is not small. For a story whose punch line depends on precisely this ill-conditioning of $\tan(10^{100})$, see "Lucky Numbers" in Richard Feynman's *Surely You're Joking, Mr. Feynman.*   □

Example 12.5. The determination of the roots of a polynomial, given the coefficients, is a classic example of an ill-conditioned problem. Consider $x^2 - 2x + 1 = (x - 1)^2$, with a double root at $x = 1$. A small perturbation in the coefficients may lead to a larger change in the roots; for example, $x^2 - 2x + 0.9999 = (x - 0.99)(x - 1.01)$. In fact, the roots can change in proportion to the square root of the change in the coefficients, so in this case the Jacobian is infinite (the problem is not differentiable), and $\kappa = \infty$.

Polynomial rootfinding is typically ill-conditioned even in cases that do not involve multiple roots. If the ith coefficient a_i of a polynomial $p(x)$ is perturbed by an infinitesimal quantity δa_i, the perturbation of the jth root x_j is $\delta x_j = (\delta a_i) x_j^i / p'(x_j)$, where p' denotes the derivative of p. The condition number of x_j with respect to perturbations of the single coefficient a_i is therefore

$$\kappa = \frac{|\delta x_j|}{|x_j|} \bigg/ \frac{|\delta a_i|}{|a_i|} = \frac{|a_i x_j^{i-1}|}{|p'(x_j)|}. \tag{12.7}$$

This number is often very large. Consider the "Wilkinson polynomial"

$$p(x) = \prod_{i=1}^{20}(x - i) = a_0 + a_1 x + \cdots + a_{19}x^{19} + x^{20}. \tag{12.8}$$

The most sensitive root of this polynomial is $x = 15$, and it is most sensitive to changes in the coefficient $a_{15} \approx 1.67 \times 10^9$. The condition number is

$$\kappa \approx \frac{1.67 \times 10^9 \cdot 15^{14}}{5! \, 14!} \approx 5.1 \times 10^{13}.$$

Figure 12.1 illustrates the ill-conditioning graphically.   □

Example 12.6. The problem of computing the eigenvalues of a nonsymmetric matrix is also often ill-conditioned. One can see this by comparing the two matrices

$$\begin{bmatrix} 1 & 1000 \\ 0 & 1 \end{bmatrix}, \qquad \begin{bmatrix} 1 & 1000 \\ 0.001 & 1 \end{bmatrix},$$

whose eigenvalues are $\{1, 1\}$ and $\{0, 2\}$, respectively. On the other hand, if a matrix A is symmetric (more generally, if it is normal), then its eigenvalues are well-conditioned. It can be shown that if λ and $\lambda + \delta\lambda$ are corresponding eigenvalues of A and $A + \delta A$, then $|\delta\lambda| \le \|\delta A\|_2$, with equality if δA is a multiple of the identity (Exercise 26.3). Thus the absolute condition number of the symmetric eigenvalue problem is $\hat{\kappa} = 1$, if perturbations are measured in the 2-norm, and the relative condition number is $\kappa = \|A\|_2 / |\lambda|$.   □

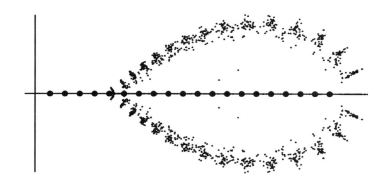

Figure 12.1. *Wilkinson's classic example of ill-conditioning. The large dots are the roots of the unperturbed polynomial* (12.8). *The small dots are the superimposed roots in the complex plane of* 100 *randomly perturbed polynomials with coefficients defined by* $\tilde{a}_k = a_k(1 + 10^{-10}r_k)$, *where* r_k *is a number from the normal distribution of mean* 0 *and variance* 1.

Condition of Matrix-Vector Multiplication

Now we come to one of the condition numbers of fundamental importance in numerical linear algebra.

Fix $A \in \mathbb{C}^{m \times n}$ and consider the problem of computing Ax from input x; that is, we are going to determine a condition number corresponding to perturbations of x but not A. Working directly from the definition of κ, with $\| \cdot \|$ denoting an arbitrary vector norm and the corresponding induced matrix norm, we find

$$\kappa = \sup_{\delta x} \left(\frac{\|A(x + \delta x) - Ax\|}{\|Ax\|} \Big/ \frac{\|\delta x\|}{\|x\|} \right) = \sup_{\delta x} \frac{\|A\delta x\|}{\|\delta x\|} \Big/ \frac{\|Ax\|}{\|x\|},$$

that is,

$$\kappa = \|A\| \frac{\|x\|}{\|Ax\|} \tag{12.9}$$

(a special case of (12.6)). This is an exact formula for κ, dependent on both A and x.

Suppose in the above calculation that A happens to be square and non-singular. Then we can use the fact that $\|x\|/\|Ax\| \leq \|A^{-1}\|$ to loosen (12.9) to a bound independent of x:

$$\kappa \leq \|A\| \|A^{-1}\|. \tag{12.10}$$

Or, one might write

$$\kappa = \alpha \|A\| \|A^{-1}\| \tag{12.11}$$

with

$$\alpha = \frac{\|x\|}{\|Ax\|} \Big/ \|A^{-1}\|. \tag{12.12}$$

For certain choices of x, we have $\alpha = 1$, and consequently $\kappa = \|A\|\|A^{-1}\|$. If $\|\cdot\| = \|\cdot\|_2$, this will occur whenever x is a multiple of a minimal right singular vector of A.

In fact, A need not have been square. If $A \in \mathbb{C}^{m \times n}$ with $m \geq n$ has full rank, equations (12.10)–(12.12) hold with A^{-1} replaced by the pseudoinverse A^+ defined in (11.11).

What about the inverse problem: given A, compute $A^{-1}b$ from input b? Mathematically, this is identical to the problem just considered, except that A has been replaced by A^{-1}. Thus we have already proved the following theorem.

Theorem 12.1. *Let $A \in \mathbb{C}^{m \times m}$ be nonsingular and consider the equation $Ax = b$. The problem of computing b, given x, has condition number*

$$\kappa = \|A\|\frac{\|x\|}{\|b\|} \leq \|A\|\|A^{-1}\| \tag{12.13}$$

with respect to perturbations of x. The problem of computing x, given b, has condition number

$$\kappa = \|A^{-1}\|\frac{\|b\|}{\|x\|} \leq \|A\|\|A^{-1}\| \tag{12.14}$$

with respect to perturbations of b. If $\|\cdot\| = \|\cdot\|_2$, then equality holds in (12.13) if x is a multiple of a right singular vector of A corresponding to the minimal singular value σ_m, and equality holds in (12.14) if b is a multiple of a left singular vector of A corresponding to the maximal singular value σ_1.

Condition Number of a Matrix

The product $\|A\|\|A^{-1}\|$ comes up so often that it has its own name: it is the *condition number of A* (relative to the norm $\|\cdot\|$), denoted by $\kappa(A)$:

$$\kappa(A) = \|A\|\|A^{-1}\|. \tag{12.15}$$

Thus, in this case the term "condition number" is attached to a matrix, not a problem. If $\kappa(A)$ is small, A is said to be *well-conditioned;* if $\kappa(A)$ is large, A is *ill-conditioned.* If A is singular, it is customary to write $\kappa(A) = \infty$.

Note that if $\|\cdot\| = \|\cdot\|_2$, then $\|A\| = \sigma_1$ and $\|A^{-1}\| = 1/\sigma_m$. Thus

$$\kappa(A) = \frac{\sigma_1}{\sigma_m} \tag{12.16}$$

in the 2-norm, and it is this formula that is generally used for computing 2-norm condition numbers of matrices. The ratio σ_1/σ_m can be interpreted as the eccentricity of the hyperellipse that is the image of the unit sphere of $\mathbb{C}^m$ under A (Figure 4.1).

For a rectangular matrix $A \in \mathbb{C}^{m \times n}$ of full rank, $m \geq n$, the condition number is defined in terms of the pseudoinverse: $\kappa(A) = \|A\|\|A^+\|$. Since A^+ is motivated by least squares problems, this definition is most useful in the case $\|\cdot\| = \|\cdot\|_2$, where we have

$$\kappa(A) = \frac{\sigma_1}{\sigma_n}. \tag{12.17}$$

Condition of a System of Equations

In Theorem 12.1, we held A fixed and perturbed x or b. What happens if we perturb A? Specifically, let us hold b fixed and consider the behavior of the problem $A \mapsto x = A^{-1}b$ when A is perturbed by infinitesimal δA. Then x must change by infinitesimal δx, where

$$(A + \delta A)(x + \delta x) = b.$$

Using the equality $Ax = b$ and dropping the doubly infinitesimal term $(\delta A)(\delta x)$, we obtain $(\delta A)x + A(\delta x) = 0$, that is, $\delta x = -A^{-1}(\delta A)x$. This equation implies $\|\delta x\| \leq \|A^{-1}\|\|\delta A\|\|x\|$, or equivalently,

$$\frac{\|\delta x\|}{\|x\|} \Big/ \frac{\|\delta A\|}{\|A\|} \leq \|A^{-1}\|\|A\| = \kappa(A).$$

Equality in this bound will hold whenever δA is such that

$$\|A^{-1}(\delta A)x\| = \|A^{-1}\|\|\delta A\|\|x\|,$$

and it can be shown by the use of dual norms (Exercise 3.6) that for any A and b and norm $\|\cdot\|$, such perturbations δA exist. This leads us to the following result.

Theorem 12.2. *Let b be fixed and consider the problem of computing $x = A^{-1}b$, where A is square and nonsingular. The condition number of this problem with respect to perturbations in A is*

$$\kappa = \|A\|\|A^{-1}\| = \kappa(A). \tag{12.18}$$

Theorems 12.1 and 12.2 are of fundamental importance in numerical linear algebra, for they determine how accurately one can solve systems of equations. If a problem $Ax = b$ contains an ill-conditioned matrix A, one must always expect to "lose $\log_{10} \kappa(A)$ digits" in computing the solution, except under very special circumstances. We shall return to this phenomenon later, and analogous results for least squares problems will be discussed in Lecture 18.

Exercises

12.1. Suppose A is a 202×202 matrix with $\|A\|_2 = 100$ and $\|A\|_F = 101$. Give the sharpest possible lower bound on the 2-norm condition number $\kappa(A)$.

12.2. In Example 11.1 we remarked that polynomial interpolation in equispaced points is ill-conditioned. To illustrate this phenomenon, let $x_1, \ldots, x_n$ and $y_1, \ldots, y_m$ be n and m equispaced points from -1 to 1, respectively.

(a) Derive a formula for the $m \times n$ matrix A that maps an n-vector of data at $\{x_j\}$ to an m-vector of sampled values $\{p(y_j)\}$, where p is the degree $n-1$ polynomial interpolant of the data (see Example 1.1).

(b) Write a program to calculate A and plot $\|A\|_\infty$ on a semilog scale for $n = 1, 2, \ldots, 30$, $m = 2n - 1$. In the continuous limit $m \to \infty$, the numbers $\|A\|_\infty$ are known as the *Lebesgue constants* for equispaced interpolation, which are asymptotic to $2^n/(e(n-1)\log n)$ as $n \to \infty$.

(c) For $n = 1, 2, \ldots, 30$ and $m = 2n-1$, what is the ∞-norm condition number κ of the problem of interpolating the constant function 1? Use (12.6).

(d) How close is your result for $n = 11$ to the bound implicit in Figure 11.1?

12.3. The goal of this problem is to explore some properties of random matrices. Your job is to be a laboratory scientist, performing experiments that lead to conjectures and more refined experiments. Do not try to prove anything. Do produce well-designed plots, which are worth a thousand numbers.

Define a *random matrix* to be an $m \times m$ matrix whose entries are independent samples from the real normal distribution with mean zero and standard deviation $m^{-1/2}$. (In MATLAB, A = randn(m,m)/sqrt(m).) The factor $\sqrt{m}$ is introduced to make the limiting behavior clean as $m \to \infty$.

(a) What do the eigenvalues of a random matrix look like? What happens, say, if you take 100 random matrices and superimpose all their eigenvalues in a single plot? If you do this for $m = 8, 16, 32, 64, \ldots$, what pattern is suggested? How does the spectral radius $\rho(A)$ (Exercise 3.2) behave as $m \to \infty$?

(b) What about norms? How does the 2-norm of a random matrix behave as $m \to \infty$? Of course, we must have $\rho(A) \leq \|A\|$ (Exercise 3.2). Does this inequality appear to approach an equality as $m \to \infty$?

(c) What about condition numbers—or more simply, the smallest singular value $\sigma_{\min}$? Even for fixed m this question is interesting. What proportions of random matrices in $\mathbb{R}^{m \times m}$ seem to have $\sigma_{\min} \leq 2^{-1}, 4^{-1}, 8^{-1}, \ldots$? In other words, what does the tail of the probability distribution of smallest singular values look like? How does the scale of all this change with m?

(d) How do the answers to (a)–(c) change if we consider random triangular instead of full matrices, i.e., upper-triangular matrices whose entries are samples from the same distribution as above?

Lecture 13. Floating Point Arithmetic

It did not take long after the invention of computers for consensus to emerge on the right way to represent real numbers on a digital machine. The secret is floating point arithmetic, the hardware analogue of scientific notation. Before we can begin to study the accuracy of the algorithms of numerical linear algebra, we must examine this topic.

Limitations of Digital Representations

Since digital computers use a finite number of bits to represent a real number, they can represent only a finite subset of the real numbers (or the complex numbers, which we discuss at the end of this lecture). This limitation presents two difficulties. First, the represented numbers cannot be arbitrarily large or small. Second, there must be gaps between them.

Modern computers represent numbers sufficiently large and small that the first constraint rarely poses difficulties. For example, the widely used IEEE double precision arithmetic permits numbers as large as 1.79×10^{308} and as small as 2.23×10^{-308}, a range great enough for most of the problems considered in this book. In other words, *overflow* and *underflow* are usually not a serious hazard (but watch out if you are asked to evaluate a determinant!).

By contrast, the problem of gaps between represented numbers is a concern throughout scientific computing. For example, in IEEE double precision arithmetic, the interval $[1, 2]$ is represented by the discrete subset

$$1, \ 1 + 2^{-52}, \ 1 + 2 \times 2^{-52}, \ 1 + 3 \times 2^{-52}, \ \ldots, \ 2. \qquad (13.1)$$

The interval $[2, 4]$ is represented by the same numbers multiplied by 2,

$$2, \quad 2 + 2^{-51}, \quad 2 + 2 \times 2^{-51}, \quad 2 + 3 \times 2^{-51}, \quad \ldots, \quad 4,$$

and in general, the interval $[2^j, 2^{j+1}]$ is represented by (13.1) times 2^j. Thus in IEEE double precision arithmetic, the gaps between adjacent numbers are in a relative sense never larger than $2^{-52} \approx 2.22 \times 10^{-16}$. This may seem negligible, and so it is for most purposes if one uses stable algorithms (see the next lecture). But it is surprising how many carelessly constructed algorithms turn out to be unstable!

Floating Point Numbers

IEEE arithmetic is an example of an arithmetic system based on a *floating point* representation of the real numbers. This is the universal practice on general purpose computers nowadays. In a floating point number system, the position of the decimal (or binary) point is stored separately from the digits, and the gaps between adjacent represented numbers scale in proportion to the size of the numbers. This is distinguished from a *fixed point* representation, where the gaps are all of the same size.

Specifically, let us consider an idealized floating point number system defined as follows. The system consists of a discrete subset $\mathbf{F}$ of the real numbers $\mathbb{R}$ determined by an integer $\beta \geq 2$ known as the *base* or *radix* (typically 2) and an integer $t \geq 1$ known as the *precision* (24 and 53 for IEEE single and double precision, respectively). The elements of $\mathbf{F}$ are the number 0 together with all numbers of the form

$$x = \pm(m/\beta^t)\beta^e, \tag{13.2}$$

where m is an integer in the range $1 \leq m \leq \beta^t$ and e is an arbitrary integer. Equivalently, we can restrict the range to $\beta^{t-1} \leq m \leq \beta^t - 1$ and thereby make the choice of m unique. The quantity $\pm(m/\beta^t)$ is then known as the *fraction* or *mantissa* of x, and e is the *exponent*.

Our floating point number system is idealized in that it ignores over- and underflow. As a result, $\mathbf{F}$ is a countably infinite set, and it is self-similar: $\mathbf{F} = \beta\mathbf{F}$.

Machine Epsilon

The resolution of $\mathbf{F}$ is traditionally summarized by a number known as *machine epsilon*. Provisionally, let us define this number by

$$\epsilon_{\text{machine}} = \tfrac{1}{2}\beta^{1-t}. \tag{13.3}$$

(We shall modify the definition after (13.7).) This number is half the distance between 1 and the next larger floating point number. In a relative sense, this is as large as the gaps between floating point numbers get. That is, $\epsilon_{\text{machine}}$ has the following property:

$$\text{For all } x \in \mathbb{R}, \text{ there exists } x' \in \mathbf{F} \text{ such that } |x - x'| \leq \epsilon_{\text{machine}}|x|. \quad (13.4)$$

For the values of β and t common on various computers, $\epsilon_{\text{machine}}$ usually lies between 10^{-6} and 10^{-35}. In IEEE single and double precision arithmetic, $\epsilon_{\text{machine}}$ is specified to be $2^{-24} \approx 5.96 \times 10^{-8}$ and $2^{-53} \approx 1.11 \times 10^{-16}$, respectively.

Let fl $: \mathbb{R} \to \mathbf{F}$ be a function giving the closest floating point approximation to a real number, its *rounded* equivalent in the floating point system. (For our purposes, ties can be broken arbitrarily, though the treatment of ties so as to avoid statistical bias is an interesting matter in itself.) The inequality (13.4) can be stated in terms of fl:

$$\begin{array}{c} \text{For all } x \in \mathbb{R}, \text{ there exists } \epsilon \text{ with } |\epsilon| \leq \epsilon_{\text{machine}} \\ \text{such that } \text{fl}(x) = x(1 + \epsilon). \end{array} \quad (13.5)$$

That is, the difference between a real number and its closest floating point approximation is always smaller than $\epsilon_{\text{machine}}$ in relative terms.

Floating Point Arithmetic

It is not enough to represent real numbers, of course; one must compute with them. On a computer, all mathematical computations are reduced to certain elementary arithmetic operations, of which the classical set is $+$, $-$, $\times$, and $\div$. Mathematically, these symbols represent operations on $\mathbb{R}$. On a computer, they have analogues that are operations on $\mathbf{F}$. It is common practice to denote these floating point operations by $\oplus$, $\ominus$, $\otimes$, and $\oslash$.

A computer might be built on the following design principle. Let x and y be arbitrary floating point numbers, that is, $x, y \in \mathbf{F}$. Let $*$ be one of the operations $+$, $-$, $\times$, or $\div$, and let $\circledast$ be its floating point analogue. Then $x \circledast y$ must be given exactly by

$$x \circledast y = \text{fl}(x * y). \quad (13.6)$$

If this property holds, then from (13.5) and (13.6) we conclude that the computer has a simple and powerful property.

Fundamental Axiom of Floating Point Arithmetic

For all $x, y \in \mathbf{F}$, there exists ϵ with $|\epsilon| \leq \epsilon_{\text{machine}}$ such that

$$x \circledast y = (x * y)(1 + \epsilon). \quad (13.7)$$

In words, *every operation of floating point arithmetic is exact up to a relative error of size at most $\epsilon_{\text{machine}}$.*

Machine Epsilon, Again

The rounding error analysis in this book is based on (13.5) and (13.7), not on the other details of floating point arithmetic described above. This means that we can be generous in allowing for hardware implementations that may not perform floating point computations as perfectly as indicated by (13.6). For such a machine, (13.5) and (13.7) may still be satisfied if $\epsilon_{\text{machine}}$ is replaced by a somewhat larger value. For example, on a computer in which intermediate quantities are truncated rather than rounded, (13.7) may hold with $\epsilon_{\text{machine}}$ replaced by $2\epsilon_{\text{machine}}$.

The simplest way to allow for such complications is to retain (13.5) and (13.7) as written, but to modify the definition of $\epsilon_{\text{machine}}$. From now on, let us assume that $\epsilon_{\text{machine}}$ is defined not by (13.3), but as the smallest number for which (13.5) and (13.7) hold. For most computers, including all those implementing IEEE arithmetic, this change in the definition of $\epsilon_{\text{machine}}$ makes no significant change in its value.

Occasionally an unexpectedly large value of $\epsilon_{\text{machine}}$ may be needed to make (13.7) hold. In late 1994 the Intel Pentium^{TM} microprocessor acquired notoriety when it was discovered that because of a bug in a table used in implementing the double precision IEEE standard, its effective precision was eleven orders of magnitude too coarse, $\epsilon_{\text{machine}} \approx 6.1 \times 10^{-5}$. (The bug was soon corrected.) In fact, there are machines for which (13.7) holds only with $\epsilon_{\text{machine}} = 1$. For example, floating point subtraction on Cray computers produced up to the mid-1990s had this property, because the operation of subtraction was implemented without a "guard digit." Such computers are not useless, but they demand a different style of error analysis from the one in this book.

Fortunately, the benefits of the axiom (13.7), and of the adoption of uniform standards of computer arithmetic, have become widely accepted by computer manufacturers in recent years, and the number of machines on the market that fail to satisfy (13.7) with a small value of $\epsilon_{\text{machine}}$ is dwindling. Indeed, IEEE arithmetic itself is rapidly becoming the standard for computers of all sizes, including, as of 1996, all IBM-compatible personal computers and all workstations manufactured by SUN, DEC, Hewlett-Packard, and IBM.

Complex Floating Point Arithmetic

Floating point complex numbers are generally represented as pairs of floating point real numbers, and the elementary operations upon them are computed by reduction to real and imaginary parts. The result is that the axiom (13.7) is valid for complex as well as real floating point numbers, except that for $\otimes$ and $\ominus$, $\epsilon_{\text{machine}}$ must be enlarged from (13.3) by factors on the order of $2^{3/2}$ and $2^{5/2}$, respectively. Once $\epsilon_{\text{machine}}$ is adjusted in this manner, rounding error analysis for complex numbers can proceed just as for real numbers.

Exercises

13.1. Between an adjacent pair of nonzero IEEE single precision real numbers, how many IEEE double precision numbers are there?

13.2. The floating point system $\mathbf{F}$ defined by (13.2) includes many integers, but not all of them.

(a) Give an exact formula for the smallest positive integer n that does not belong to $\mathbf{F}$.

(b) In particular, what are the values of n for IEEE single and double precision arithmetic?

(c) Figure out a way to verify this result for your own computer. Specifically, design and run a program that produces evidence that $n-3$, $n-2$, and $n-1$ belong to $\mathbf{F}$ but n does not. What about $n+1$, $n+2$, and $n+3$?

13.3. Consider the polynomial $p(x) = (x-2)^9 = x^9 - 18x^8 + 144x^7 - 672x^6 + 2016x^5 - 4032x^4 + 5376x^3 - 4608x^2 + 2304x - 512$.

(a) Plot $p(x)$ for $x = -1.920, -1.919, -1.918, \ldots, 2.080$, evaluating p via its coefficients $1, -18, 144, \ldots$.

(b) Produce the same plot again, now evaluating p via the expression $(x-2)^9$.

13.4. The polynomial $p(x) = x^5 - 2x^4 - 3x^3 + 3x^2 - 2x - 1$ has three real zeros. Applying Newton's method to p with initial guess $x_0 = 0$ produces a series of estimates $x_1, x_2, x_3 \ldots$ that converge rapidly to a zero $x_* \approx -0.315$.

(a) Compute $x_1, \ldots, x_6$ in floating point arithmetic with $\epsilon_{\text{machine}} \approx 10^{-16}$. How many digits do you estimate are correct in each of these numbers?

(b) Compute $x_1, \ldots, x_6$ again *exactly* with the aid of a symbolic algebra system such as MAPLE or MATHEMATICA. Each x_j is a rational number. How many digits are there in the numerator and the denominator for each j?

Lecture 14. Stability

It would be a fine thing if numerical algorithms could provide exact solutions to numerical problems. Since the problems are continuous while digital computers are discrete, however, this is generally not possible. The notion of stability is the standard way of characterizing what *is* possible—numerical analysts' idea of what it means to get the "right answer," even if it is not exact.

Algorithms

In Lecture 12, we defined a mathematical *problem* as a function $f : X \rightarrow Y$ from a vector space X of data to a vector space Y of solutions.

An *algorithm* can be viewed as another map $\tilde{f} : X \rightarrow Y$ between the same two spaces. We make this definition precise as follows. Let a problem f, a computer whose floating point system satisfies (13.7) (but not necessarily (13.6)), an algorithm for f (in the loose sense of the term), and an implementation of this algorithm in the form of a computer program be fixed. Given data $x \in X$, let this data be rounded to floating point in a matter satisfying (13.5) and then supplied as input to the computer program. Now, run the program. The result is a collection of floating point numbers that belong to the vector space Y (since the algorithm was designed to solve f). Let this computed result be called $\tilde{f}(x)$.

The situation couldn't be uglier! As a minimum, $\tilde{f}(x)$ will be affected by rounding errors. Depending on the circumstances, it may also be affected by all kinds of other complications such as convergence tolerances or even the other jobs running on the computer, in cases where the assignment of computations to processors is not determined until runtime. Thus the "function" $\tilde{f}(x)$ may even take different values from one run to the next; it may be multivalued. (In fact, the problem f should really be allowed to be multivalued too; this permits handling of cases where a nonunique solution is acceptable, e.g., either of the two square roots of a complex number.) Yet despite all these complications, we shall find that we can make surprisingly clean statements about $\tilde{f}(x)$, and hence about the accuracy of the algorithms of numerical linear algebra, based only on the fundamental axioms (13.5) and (13.7).

The tilde ($\tilde{}$) notation is very convenient. Just as $\tilde{f}$ is the computed analogue of f, other computed quantities in this book will frequently be marked by tildes. For example, the computed solution to a system of equations $Ax = b$ may be denoted by $\tilde{x}$.

Accuracy

Except in trivial cases, $\tilde{f}$ cannot be continuous. Nevertheless, a good algorithm should approximate the associated problem f. To make this idea quantitative, we may consider the *absolute error* of a computation, $\|\tilde{f}(x) - f(x)\|$, or the *relative error*,

$$\frac{\|\tilde{f}(x) - f(x)\|}{\|f(x)\|}. \tag{14.1}$$

In this book we mainly utilize relative quantities, and thus (14.1) will be our standard error measure.

If $\tilde{f}$ is a good algorithm, one might naturally expect the relative error to be small, of order $\epsilon_{\text{machine}}$. One might say that an algorithm $\tilde{f}$ for a problem f is *accurate* if for each $x \in X$,

$$\frac{\|\tilde{f}(x) - f(x)\|}{\|f(x)\|} = O(\epsilon_{\text{machine}}). \tag{14.2}$$

Loosely speaking, the symbol $O(\epsilon_{\text{machine}})$ in (14.3) means "on the order of machine epsilon." However, $O(\epsilon_{\text{machine}})$ also has a precise meaning, which we shall discuss in a moment. That discussion will also clarify how a formula like (14.2) is to be interpreted if the denominator is zero.

Stability

If the problem f is ill-conditioned, however, the goal of accuracy as defined by (14.2) is unreasonably ambitious. Rounding of the input data is unavoidable on a digital computer, and even if all the subsequent computations could be carried out perfectly, this perturbation alone might lead to a significant change in the result. Instead of aiming for accuracy in all cases, the most it is appropriate to aim for in general is *stability*. We say that an algorithm $\tilde{f}$ for a problem f is *stable* if for each $x \in X$,

$$\frac{\|\tilde{f}(x) - f(\tilde{x})\|}{\|f(\tilde{x})\|} = O(\epsilon_{\text{machine}}) \tag{14.3}$$

for some $\tilde{x}$ with

$$\frac{\|\tilde{x} - x\|}{\|x\|} = O(\epsilon_{\text{machine}}). \tag{14.4}$$

In words,

> A stable algorithm gives nearly the right answer
> to nearly the right question.

The motivation for this definition will become clear in the next lecture and in applications throughout the remainder of this book.

We caution the reader that whereas the definitions of stability given here are useful in many parts of numerical linear algebra, the condition $O(\epsilon_{\text{machine}})$ is probably too strict to be appropriate for all numerical problems in other areas such as differential equations.

Backward Stability

Many algorithms of numerical linear algebra satisfy a condition that is both stronger and simpler than stability. We say that an algorithm $\tilde{f}$ for a problem f is *backward stable* if for each $x \in X$,

$$\tilde{f}(x) = f(\tilde{x}) \quad \text{for some } \tilde{x} \text{ with} \quad \frac{\|\tilde{x} - x\|}{\|x\|} = O(\epsilon_{\text{machine}}). \tag{14.5}$$

This is a tightening of the definition of stability in that the $O(\epsilon_{\text{machine}})$ in (14.3) has been replaced by zero. In words,

> A backward stable algorithm gives exactly the right answer
> to nearly the right question.

Examples are given in the next lecture.

The Meaning of $O(\epsilon_{\text{machine}})$

We now explain the precise meaning of "$O(\epsilon_{\text{machine}})$" in (14.2)–(14.5).

The notation

$$\varphi(t) = O(\psi(t)) \tag{14.6}$$

is a standard one in mathematics, with a precise definition. This equation asserts that there exists some positive constant C such that, for all t sufficiently close to an understood limit (e.g., $t \to 0$ or $t \to \infty$),

$$|\varphi(t)| \leq C\psi(t). \tag{14.7}$$

For example, the statement $\sin^2 t = O(t^2)$ as $t \to 0$ asserts that there exists a constant C such that, for all sufficiently small t, $|\sin^2 t| \le Ct^2$.

Also standard in mathematics are statements of the form

$$\varphi(s,t) = O(\psi(t)) \quad \text{uniformly in } s, \tag{14.8}$$

where φ is a function that depends not only on t but also on another variable s. The word "uniformly" indicates that there exists a single constant C as in (14.7) that holds for all choices of s. Thus, for example,

$$(\sin^2 t)(\sin^2 s) = O(t^2)$$

holds uniformly as $t \to 0$, but the uniformity is lost if we replace $\sin^2 s$ by s^2.

In this book, our use of the "O" symbol follows these standard definitions. Specifically, we often state results along the lines of

$$\|\text{computed quantity}\| = O(\epsilon_{\text{machine}}). \tag{14.9}$$

Here is what (14.9) means. First, "$\|$computed quantity$\|$" represents the norm of some number or collection of numbers determined by an algorithm $\tilde{f}$ for a problem f, depending both on the data $x \in X$ for f and on $\epsilon_{\text{machine}}$. An example is the relative error (14.1). Second, the implicit limit process is $\epsilon_{\text{machine}} \to 0$ (i.e., $\epsilon_{\text{machine}}$ is the variable corresponding to t in (14.8)). Third, the "O" applies uniformly for all data $x \in X$ (i.e., x is the variable corresponding to s). We shall rarely mention the uniformity with respect to $x \in X$, but it is always implicit.

In any particular machine arithmetic, the number $\epsilon_{\text{machine}}$ is a fixed quantity. In speaking of the limit $\epsilon_{\text{machine}} \to 0$, we are considering an idealization of a computer, or perhaps one should say, of a family of computers. Equation (14.9) means that if we were to run the algorithm in question on computers satisfying (13.5) and (13.7) for a sequence of values of $\epsilon_{\text{machine}}$ decreasing to zero, then $\|$computed quantity$\|$ would be guaranteed to decrease in proportion to $\epsilon_{\text{machine}}$ or faster. These ideal computers are required to satisfy (13.5) and (13.7) but nothing else.

Dependence on m and n, not A and b

It cannot hurt to discuss the meaning of $O(\epsilon_{\text{machine}})$ in (14.2)–(14.5) a bit further. The uniformity of the constant implicit in the "O" can be illustrated by the following example. Suppose we are considering an algorithm for solving a nonsingular $m \times m$ system of equations $Ax = b$ for x, and we assert that the computed result $\tilde{x}$ for this algorithm satisfies

$$\frac{\|\tilde{x} - x\|}{\|x\|} = O(\kappa(A)\,\epsilon_{\text{machine}}). \tag{14.10}$$

This assertion means that the bound

$$\frac{\|\tilde{x} - x\|}{\|x\|} \leq C\kappa(A)\,\epsilon_{\text{machine}} \qquad (14.11)$$

holds for a single constant C, independently of the matrix A or the right-hand side b, for all sufficiently small $\epsilon_{\text{machine}}$.

If the denominator in a formula like (14.11) is zero, its meaning is defined by the following convention. When we write (14.11), what we really mean is

$$\|\tilde{x} - x\| \leq C\kappa(A)\,\epsilon_{\text{machine}}\|x\|. \qquad (14.12)$$

There is no difference if $\|x\| \neq 0$, but if $\|x\| = 0$, (14.12) makes it clear that the precise meaning of (14.10) is that $\|\tilde{x} - x\| = 0$ for all sufficiently small $\epsilon_{\text{machine}}$.

Though the constant C of (14.11) or (14.12) does not depend on A or b, it does depend, in general, on the dimension m. Formally speaking, this is a consequence of our definition of a problem in Lecture 12. If the dimensions such as m or n that define a problem f change, then the vector spaces X and Y must change too, and thus we have a new problem, f'. As a practical matter, too, the effects of rounding errors on the algorithms of numerical linear algebra generally grow with m and n. However, this growth is usually slow enough that it is not serious. The dependence on m or n is typically linear, quadratic, or cubic in the worst case (the exponent depends on the choice of norm as well as the choice of algorithm), and the errors for most data are much smaller than in the worst case, thanks to statistical cancellation.

In principle, a statement like (14.9) might conceal a dimension-dependent factor such as 2^m that would make the bound useless in practice. However, there is only one place in this book where such a thing happens—in the discussion of Gaussian elimination with partial pivoting—and we shall give the reader ample warning at that point to avoid misunderstanding. As a rule, when the expression $O(\epsilon_{\text{machine}})$ is printed in this book, the chances are that in an actual calculation on an actual machine, the quantity in question will be at most 100 or perhaps 1000 times as large as $\epsilon_{\text{machine}}$.

Independence of Norm

Our definitions involving $O(\epsilon_{\text{machine}})$ have the convenient property that, provided X and Y are finite-dimensional, they are norm-independent.

Theorem 14.1. *For problems f and algorithms $\tilde{f}$ defined on finite-dimensional spaces X and Y, the properties of accuracy, stability, and backward stability all hold or fail to hold independently of the choice of norms in X and Y.*

Proof. It is well known (and easily proved) that in a finite-dimensional vector space, all norms are equivalent in the sense that if $\|\cdot\|$ and $\|\cdot\|'$ are two norms on the same space, then there exist positive constants C_1 and C_2 such that $C_1\|x\| \leq \|x\|' \leq C_2\|x\|$ for all x in that space. It follows that a change of norm may affect the size of the constant C implicit in a statement involving $O(\epsilon_{\mathrm{machine}})$, but not the existence of such a constant.   □

Exercises

14.1. True or False?

(a) $\sin x = O(1)$ as $x \to \infty$.

(b) $\sin x = O(1)$ as $x \to 0$.

(c) $\log x = O(x^{1/100})$ as $x \to \infty$.

(d) $n! = O((n/e)^n)$ as $n \to \infty$.

(e) $A = O(V^{2/3})$ as $V \to \infty$, where A and V are the surface area and volume of a sphere measured in square miles and cubic microns, respectively.

(f) $\mathrm{fl}(\pi) - \pi = O(\epsilon_{\mathrm{machine}})$. (We do not mention that the limit is $\epsilon_{\mathrm{machine}} \to 0$, since that is implicit for all expressions $O(\epsilon_{\mathrm{machine}})$ in this book.)

(g) $\mathrm{fl}(n\pi) - n\pi = O(\epsilon_{\mathrm{machine}})$, uniformly for all integers n. (Here $n\pi$ represents the exact mathematical quantity, not the result of a floating point calculation.)

14.2. (a) Show that $(1 + O(\epsilon_{\mathrm{machine}}))(1 + O(\epsilon_{\mathrm{machine}})) = 1 + O(\epsilon_{\mathrm{machine}})$. The precise meaning of this statement is that if f is a function satisfying $f(\epsilon_{\mathrm{machine}}) = (1 + O(\epsilon_{\mathrm{machine}}))(1 + O(\epsilon_{\mathrm{machine}}))$ as $\epsilon_{\mathrm{machine}} \to 0$, then f also satisfies $f(\epsilon_{\mathrm{machine}}) = 1 + O(\epsilon_{\mathrm{machine}})$ as $\epsilon_{\mathrm{machine}} \to 0$.

(b) Show that $(1 + O(\epsilon_{\mathrm{machine}}))^{-1} = 1 + O(\epsilon_{\mathrm{machine}})$.

Lecture 15. More on Stability

連続はけ

Stability

和 conditioning

We continue the discussion of stability by considering examples of stable and unstable algorithms. Then we discuss a fundamental idea linking conditioning and stability, whose power has been proved in innumerable applications since the 1950s: backward error analysis.

Stability of Floating Point Arithmetic

The four simplest computational problems are $+$, $-$, $\times$, and $\div$. There is not much to say about choice of algorithms! Of course, we shall normally use the floating point operations $\oplus$, $\ominus$, $\otimes$, and $\oslash$ provided with the computer. As it happens, the axioms (13.5) and (13.7) imply that these four canonical examples of algorithms are all backward stable.

Let us show this for subtraction, since this is the elementary operation one might expect to be at greatest risk of instability. As in Example 12.3, the data space X is the set of 2-vectors, $\mathbb{C}^2$, and the solution space Y is the set of scalars, $\mathbb{C}$. By Theorem 14.1, we need not specify the norms in these spaces.

For data $x = (x_1, x_2)^* \in X$, the problem of subtraction corresponds to the function $f(x_1, x_2) = x_1 - x_2$, and the algorithm we are considering can be written

$$\tilde{f}(x_1, x_2) = \mathrm{fl}(x_1) \ominus \mathrm{fl}(x_2).$$

This equation means that we first round x_1 and x_2 to floating point values, then apply the operation $\ominus$. Now by (13.5), we have

$$\mathrm{fl}(x_1) = x_1(1 + \epsilon_1), \qquad \mathrm{fl}(x_2) = x_2(1 + \epsilon_2)$$

108

for some $|\epsilon_1|, |\epsilon_2| \le \epsilon_{\text{machine}}$. By (13.7), we have

$$\text{fl}(x_1) \ominus \text{fl}(x_2) = (\text{fl}(x_1) - \text{fl}(x_2))(1 + \epsilon_3)$$

for some $|\epsilon_3| \le \epsilon_{\text{machine}}$. Combining these equations gives

$$\begin{aligned}
\text{fl}(x_1) \ominus \text{fl}(x_2) &= [x_1(1 + \epsilon_1) - x_2(1 + \epsilon_2)](1 + \epsilon_3) \\
&= x_1(1 + \epsilon_1)(1 + \epsilon_3) - x_2(1 + \epsilon_2)(1 + \epsilon_3) \\
&= x_1(1 + \epsilon_4) - x_2(1 + \epsilon_5)
\end{aligned}$$

for some $|\epsilon_4|, |\epsilon_5| \le 2\epsilon_{\text{machine}} + O(\epsilon_{\text{machine}}^2)$ (see Exercise 14.2). In other words, the computed result $\tilde{f}(x) = \text{fl}(x_1) \ominus \text{fl}(x_2)$ is exactly equal to the difference $\tilde{x}_1 - \tilde{x}_2$, where $\tilde{x}_1$ and $\tilde{x}_2$ satisfy

$$\frac{|\tilde{x}_1 - x_1|}{|x_1|} = O(\epsilon_{\text{machine}}), \qquad \frac{|\tilde{x}_2 - x_2|}{|x_2|} = O(\epsilon_{\text{machine}}),$$

and any $C > 2$ will suffice for the constants implicit in the "O" symbols. For any choice of norm $\|\cdot\|$ in the space $\mathbb{C}^2$, this implies (14.5).

Further Examples

Example 15.1. Inner Product. Suppose we are given vectors $x, y \in \mathbb{C}^m$ and wish to compute the inner product $\alpha = x^* y$. The obvious algorithm is to compute the pairwise products $\bar{x}_i y_i$ with $\otimes$ and add them with $\oplus$ to obtain a computed result $\tilde{\alpha}$. It can be shown that this algorithm is backward stable; this is done implicitly in Lecture 17. $\qquad\square$

Example 15.2. Outer Product. On the other hand, suppose we wish to compute the rank-one outer product $A = xy^*$ for vectors $x \in \mathbb{C}^m$, $y \in \mathbb{C}^n$. The obvious algorithm is to compute the mn products $x_i \bar{y}_j$ with $\otimes$ and collect them into a matrix $\tilde{A}$. This algorithm is stable, but it is not backward stable. The explanation is that the matrix $\tilde{A}$ will be most unlikely to have rank exactly 1, and thus it cannot generally be written in the form $(x + \delta x)(y + \delta y)^*$. As a rule, for problems where the dimension of the solution space Y is greater than that of the problem space X, backward stability is rare. $\qquad\square$

Example 15.3. Suppose we use $\oplus$ to compute $x + 1$, given $x \in \mathbb{C}$: $\tilde{f}(x) = \text{fl}(x) \oplus 1$. This algorithm is stable but not backward stable. The reason is that for $x \approx 0$, the addition $\oplus$ will introduce absolute errors of size $O(\epsilon_{\text{machine}})$. Relative to the size of x, these are unbounded, so they cannot be interpreted as caused by small relative perturbations in the data. This example indicates that backward stability is a rather special property, a reasonable goal in some contexts but not others. Note that if the problem had been to compute $x + y$ for data x and y, then the algorithm would have been backward stable. $\qquad\square$

Example 15.4. What is it reasonable to expect of a computer program or calculator that computes $\sin x$ or $\cos x$? Again the answer is stability, not backward stability. For $\cos x$, this follows from the fact that $\cos 0 \neq 0$, as in the previous example. For both $\sin x$ and $\cos x$, backward stability is also ruled out by the fact that the function has derivative equal to zero at certain points. For example, suppose we evaluate $f(x) = \sin x$ on a computer for $x = \pi/2 - \delta$, $0 < \delta \ll 1$. Suppose we are lucky enough to get as a computed result the exactly correct answer, rounded to the floating point system: $\tilde{f}(x) = \mathrm{fl}(\sin x)$. Since $f'(x) = \cos x \approx \delta$, we have $\tilde{f}(x) = f(\tilde{x})$ for some $\tilde{x}$ with $\tilde{x} - x \approx (\tilde{f}(x) - f(x))/\delta = O(\epsilon_{\mathrm{machine}}/\delta)$. Since δ can be arbitrarily small, this backward error is not of magnitude $O(\epsilon_{\mathrm{machine}})$. $\square$

An Unstable Algorithm

These are toy examples. Here is a more substantial one: the use of the characteristic polynomial to find the eigenvalues of a matrix.

Since z is an eigenvalue of A if and only if $p(z) = 0$, where $p(z)$ is the characteristic polynomial $\det(zI - A)$, the roots of p are the eigenvalues of A (Lecture 24). This suggests a method for computing the eigenvalues:

1. Find the coefficients of the characteristic polynomial,
2. Find its roots.

This algorithm is not only backward unstable but unstable, and it should not be used. Even in cases where extracting eigenvalues is a well-conditioned problem, it may produce answers that have relative errors vastly larger than $\epsilon_{\mathrm{machine}}$.

The instability is revealed in the rootfinding of the second step. As we saw in Example 12.5, the problem of finding the roots of a polynomial, given the coefficients, is generally ill-conditioned. It follows that small errors in the coefficients of the characteristic polynomial will tend to be amplified when finding roots, even if the rootfinding is done to perfect accuracy.

For example, suppose $A = I$, the 2×2 identity matrix. The eigenvalues of A are insensitive to perturbations of the entries, and a stable algorithm should be able to compute them with errors $O(\epsilon_{\mathrm{machine}})$. However, the algorithm described above produces errors on the order of $\sqrt{\epsilon_{\mathrm{machine}}}$. To explain this, we note that the characteristic polynomial is $x^2 - 2x + 1$, just as in Example 12.5. When the coefficients of this polynomial are computed, they can be expected to have errors on the order of $\epsilon_{\mathrm{machine}}$, and these can cause the roots to change by order $\sqrt{\epsilon_{\mathrm{machine}}}$. For example, if $\epsilon_{\mathrm{machine}} = 10^{-16}$, the roots of the computed characteristic polynomial can be perturbed from the actual eigenvalues by approximately 10^{-8}, a loss of eight digits of accuracy.

Before you try this computation for yourself, we must be a little more honest. If you use the algorithm just described to compute the eigenvalues of the 2×2 identity matrix, you will probably find that there are no errors at all, because the coefficients and roots of $x^2 - 2x + 1$ are small integers that will be represented exactly on your computer. However, if the experiment is done on a slightly perturbed matrix, such as

$$A = \begin{bmatrix} 1 + 10^{-14} & 0 \\ 0 & 1 \end{bmatrix},$$

the computed eigenvalues will differ from the actual ones by the expected order $\sqrt{\epsilon_{\text{machine}}}$. Try it!

Accuracy of a Backward Stable Algorithm

Suppose we have a backward stable algorithm $\tilde{f}$ for a problem $f : X \to Y$. Will the results it delivers be accurate? The answer depends on the condition number $\kappa = \kappa(x)$ of f. If $\kappa(x)$ is small, the results will be accurate in the relative sense, but if it is large, the accuracy will suffer proportionately.

Theorem 15.1. *Suppose a backward stable algorithm is applied to solve a problem $f : X \to Y$ with condition number κ on a computer satisfying the axioms (13.5) and (13.7). Then the relative errors satisfy*

$$\frac{\|\tilde{f}(x) - f(x)\|}{\|f(x)\|} = O(\kappa(x)\,\epsilon_{\text{machine}}). \tag{15.1}$$

Proof. By the definition (14.5) of backward stability, we have $\tilde{f}(x) = f(\tilde{x})$ for some $\tilde{x} \in X$ satisfying

$$\frac{\|\tilde{x} - x\|}{\|x\|} = O(\epsilon_{\text{machine}}).$$

By the definition (12.5) of $\kappa(x)$, this implies

$$\frac{\|\tilde{f}(x) - f(x)\|}{\|f(x)\|} \leq (\kappa(x) + o(1)) \frac{\|\tilde{x} - x\|}{\|x\|}, \tag{15.2}$$

where $o(1)$ denotes a quantity that converges to zero as $\epsilon_{\text{machine}} \to 0$. Combining these bounds gives (15.1). $\qquad\square$

Backward Error Analysis

The process we have just carried out in proving Theorem 15.1 is known as *backward error analysis*. We obtained an accuracy estimate by two steps. One step was to investigate the condition of the problem. The other was to investigate the stability of the algorithm. Our conclusion was that if the algorithm is stable, then the final accuracy reflects that condition number.

Mathematically, this is straightforward, but it is certainly not the first idea an unprepared person would think of if called upon to analyze a numerical algorithm. The first idea would be *forward error analysis*. Here, the rounding errors introduced at each step of the calculation are estimated, and somehow, a total is maintained of how they may compound from step to step.

Experience has shown that for most of the algorithms of numerical linear algebra, forward error analysis is harder to carry out than backward error analysis. With the benefit of hindsight, it is not hard to explain why this is so. Suppose a tried-and-true algorithm is used, say, to solve $Ax = b$ on a computer. It is an established fact (see Lecture 22) that the results obtained will be consistently less accurate when A is ill-conditioned. Now, how could a forward error analysis capture this phenomenon? The condition number of A is so global a property as to be more or less invisible at the level of the individual floating point operations involved in solving $Ax = b$. (We dramatize this by an example in the next lecture.) Yet one way or another, the forward analysis will have to detect that condition number if it is to end up with a correct result.

In short, it is an established fact that the best algorithms for most problems do no better, in general, than to compute exact solutions for slightly perturbed data. Backward error analysis is a method of reasoning fitted neatly to this backward reality.

Exercises

15.1. Each of the following problems describes an algorithm implemented on a computer satisfying the axioms (13.5) and (13.7). For each one, state whether the algorithm is *backward stable*, *stable but not backward stable*, or *unstable*, and prove it or at least give a reasonably convincing argument. Be sure to follow the definitions as given in the text.

(a) Data: $x \in \mathbb{C}$. Solution: $2x$, computed as $x \oplus x$.

(b) Data: $x \in \mathbb{C}$. Solution: x^2, computed as $x \otimes x$.

(c) Data: $x \in \mathbb{C} \setminus \{0\}$. Solution: 1, computed as $x \oslash x$. (A machine satisfying (13.6) will give exactly the right answer, but our definitions are based on the weaker condition (13.7).)

(d) Data: $x \in \mathbb{C}$. Solution: 0, computed as $x \ominus x$. (Again, a real machine may do better than our definitions based on (13.7).)

(e) Data: none. Solution: e, computed by summing $\sum_{k=0}^{\infty} 1/k!$ from left to right using $\otimes$ and $\oplus$, stopping when a summand is reached of magnitude $< \epsilon_{\text{machine}}$.

(f) Data: none. Solution: e, computed by the same algorithm as above except with the series summed from right to left.

(g) Data: none. Solution: π, computed by doing an exhaustive search to find the smallest floating point number x in the interval $[3, 4]$ such that $s(x) \otimes s(x') \leq 0$. Here $s(x)$ is an algorithm that calculates $\sin(x)$ stably in the given interval, and x' denotes the next floating point number after x in the floating point system.

15.2. Consider an algorithm for the problem of computing the (full) SVD of a matrix. The data for this problem is a matrix A, and the solution is three matrices U (unitary), Σ (diagonal), and V (unitary) such that $A = U\Sigma V^*$. (We are speaking here of explicit matrices U and V, not implicit representations as products of reflectors.)

(a) Explain what it would mean for this algorithm to be backward stable.

(b) In fact, for a simple reason, this algorithm cannot be backward stable. Explain.

(c) Fortunately, the standard algorithms for computing the SVD (Lecture 31) are stable. Explain what stability means for such an algorithm.

Lecture 16. Stability of Householder Triangularization

In this lecture we see backward error analysis in action. First we observe in a MATLAB experiment the remarkable phenomenon of backward stability of Householder triangularization. We then consider how the triangularization step can be combined with other backward stable pieces to obtain a stable algorithm for solving $Ax = b$.

Experiment

Householder factorization is a backward stable algorithm for computing QR factorizations. We can illustrate this by a MATLAB experiment carried out in IEEE double precision arithmetic, $\epsilon_{\text{machine}} \approx 1.11 \times 10^{-16}$.

`R = triu(randn(50));`	Set R to a 50×50 upper-triangular matrix with normal random entries.
`[Q,X] = qr(randn(50));`	Set Q to a random orthogonal matrix by orthogonalizing a random matrix.
`A = Q*R;`	Set A to the product QR, up to rounding errors.
`[Q2,R2] = qr(A);`	Compute QR factorization $A \approx Q_2 R_2$ by Householder triangularization.

The purpose of these four lines of MATLAB is to construct a matrix with a known QR factorization, $A = QR$, which can then be compared with the QR factorization $A = Q_2 R_2$ computed by Householder triangularization. Actually, because of rounding errors, the QR factors of the computed matrix A are not exactly Q and R. However, for the purposes of this experiment they are close enough. The results about to be presented would not be significantly different if A were exactly equal to QR (which we could achieve, in effect, by calculating $A = QR$ in higher precision arithmetic on the computer).

For Q_2 and R_2, as it happens, are very far from exact:

```
norm(Q2-Q)                How accurate is Q₂?
    ans = 0.00889
norm(R2-R)/norm(R)        How accurate is R₂?
    ans = 0.00071
```

These errors are huge! Our calculations have been done with sixteen digits of accuracy, yet the final results are accurate to only two or three digits. The individual rounding errors have been amplified by factors on the order of 10^{13}. (Note that the computed Q_2 is close enough to Q to indicate that changes in the signs of the columns cannot be responsible for any of the errors. If you try this experiment and get entirely different results, it may be that you need to multiply the columns of Q and rows of R by appropriate factors ± 1.)

We seem to have lost twelve digits of accuracy. But now, an astonishing thing happens when we multiply these inaccurate matrices Q_2 and R_2:

```
norm(A-Q2*R2)/norm(A)     How accurate is Q₂R₂?
    ans = 1.432e-15
```

The product $Q_2 R_2$ is accurate to a full fifteen digits! The errors in Q_2 and R_2 must be "diabolically correlated," as Wilkinson used to say. To one part in 10^{12}, they cancel out in the product $Q_2 R_2$.

To highlight how special this accuracy of $Q_2 R_2$ is, let us construct another pair of matrices Q_3 and R_3 that are equally accurate approximations to Q and R, and multiply them.

```
Q3 = Q+1e-4*randn(50);    Set Q₃ to a random perturbation of Q
                          that is closer to Q than Q₂ is.
R3 = R+1e-4*randn(50);    Set R₃ to a random perturbation of R
                          that is closer to R than R₂ is.
norm(A-Q3*R3)/norm(A)     How accurate is Q₃R₃?
    ans = 0.00088
```

This time, the error in the product is huge. Q_2 is no better than Q_3, and R_2 is no better than R_3, but $Q_2 R_2$ is twelve orders of magnitude better than $Q_3 R_3$.

In this experiment, we did not take the trouble to make R_3 upper-triangular or Q_3 orthogonal, but there would have been little difference had we done so.

The errors in Q_2 and R_2 are *forward errors*. In general, a large forward error can be the result of an ill-conditioned problem or an unstable algorithm (Theorem 15.1). In our experiment, they are due to the former. As a rule, the sequences of column spaces of a random triangular matrix are exceedingly ill-conditioned as a function of the entries of the matrix.

The error in $Q_2 R_2$ is the *backward error* or *residual*. The smallness of this error suggests that Householder triangularization is backward stable.

Theorem

In fact, Householder triangularization is backward stable for all matrices A and all computers satisfying (13.5) and (13.7). We shall now state a theorem to this effect. As with most stability results in this book, we shall not give a proof.

Our result will take the form

$$\tilde{Q}\tilde{R} = A + \delta A, \qquad (16.1)$$

where δA is small. In words, the computed Q times the computed R equals a small perturbation of the given matrix A. However, a subtlety arises in the way we shall use these symbols. By $\tilde{R}$, we mean just what one would expect: the upper-triangular matrix that is constructed by Householder triangularization in floating point arithmetic. By $\tilde{Q}$, however, we mean a certain matrix that is *exactly unitary*. Recall that Q is equal to the product $Q = Q_1 Q_2 \cdots Q_n$ (10.7), where Q_k is the Householder reflector defined by the vector v_k (10.4) determined at the kth step of Algorithm 10.1. In the floating point computation, we obtain instead a sequence of vectors $\tilde{v}_k$. Let $\tilde{Q}_k$ denote the *exactly unitary* reflector defined—mathematically, not on the computer—by the floating point vector $\tilde{v}_k$. Now define

$$\tilde{Q} = \tilde{Q}_1 \tilde{Q}_2 \cdots \tilde{Q}_n. \qquad (16.2)$$

This exactly unitary matrix $\tilde{Q}$ will take the role of our "computed Q." This definition may seem odd at first, but it is the natural one. In applications, as discussed in Lecture 10, the matrix Q is generally not formed explicitly anyway, so it would not be useful to define a "computed Q" of the more obvious kind. It is the vectors $\tilde{v}_k$ that are formed explicitly, and these are what enter into (16.2).

Here is the theorem that explains our MATLAB experiment.

Theorem 16.1. *Let the QR factorization $A = QR$ of a matrix $A \in \mathbb{C}^{m \times n}$ be computed by Householder triangularization (Algorithm 10.1) on a computer satisfying the axioms (13.5) and (13.7), and let the computed factors $\tilde{Q}$ and $\tilde{R}$*

be defined as indicated above. Then we have

$$\tilde{Q}\tilde{R} = A + \delta A, \qquad \frac{\|\delta A\|}{\|A\|} = O(\epsilon_{\text{machine}}) \qquad (16.3)$$

for some $\delta A \in \mathbb{C}^{m \times n}$.

As always in this book, the expression $O(\epsilon_{\text{machine}})$ in (16.3) has the precise meaning discussed in Lecture 14. The bound holds as $\epsilon_{\text{machine}} \to 0$, uniformly for all matrices A of any fixed dimensions m and n, but not uniformly with respect to m and n. Because all norms on a finite-dimensional vector space are equivalent, we need not specify the norm (Theorem 14.1).

Analyzing an Algorithm to Solve $Ax = b$

We have seen that Householder triangularization is backward stable but not always accurate in the forward sense. (The same is true of most of the matrix factorizations of numerical linear algebra.) Now, QR factorization is generally not an end in itself, but a means to other ends such as solution of a system of equations, a least squares problem, or an eigenvalue problem. Is its backward stability enough to make it a satisfactory piece of a larger algorithm? That is, is accuracy of the product QR enough for applications, or do we need accuracy of Q and R individually?

The happy answer is that accuracy of QR is indeed enough for most purposes. We can show this by surprisingly simple arguments.

The example we shall consider is the use of Householder triangularization to solve a nonsingular $m \times m$ linear system $Ax = b$. This idea was discussed at the end of Lecture 7. Here is a more complete statement of that algorithm.

Algorithm 16.1. Solving $Ax = b$ by QR Factorization

$QR = A$ Factor A into QR by Algorithm 10.1, with Q represented as the product of reflectors.

$y = Q^*b$ Construct Q^*b by Algorithm 10.2.

$x = R^{-1}y$ Solve the triangular system $Rx = y$ by back substitution (Algorithm 17.1).

This algorithm is backward stable, and proving this is straightforward, given that each of the three steps is itself backward stable. Here, we shall state backward stability results for the three steps, without proof, and then give the details of how they can be combined.

The first step of Algorithm 16.1 is QR factorization of A, leading to computed matrices $\tilde{R}$ and $\tilde{Q}$. The backward stability of this process has already been expressed by (16.3).

The second step is computation of $\tilde{Q}^*b$ by Algorithm 10.2. (Note that for the purposes of this argument we do not write Q^*b, for at this point of the computation, the first step is complete, and the matrix it has produced is not Q but $\tilde{Q}$.) When $\tilde{Q}^*b$ is computed by Algorithm 10.2, rounding errors will be made, so the result will not be exactly $\tilde{Q}^*b$. Instead it will be some vector $\tilde{y}$. It can be shown that this vector satisfies the following backward stability estimate:

$$(\tilde{Q} + \delta Q)\tilde{y} = b, \qquad \|\delta Q\| = O(\epsilon_{\text{machine}}). \qquad (16.4)$$

Like (16.3), this equality is exact. In words, the result of applying the Householder reflectors in floating point arithmetic is exactly equivalent to multiplying b by a slightly perturbed matrix, $(\tilde{Q} + \delta Q)^{-1}$.

The final step of Algorithm 16.1 is back substitution to compute $\tilde{R}^{-1}\tilde{y}$. (Again, it is $\tilde{R}$ and $\tilde{y}$ that are available at this stage; R and y have nothing to do with it.) In this step new rounding errors will be introduced, but once more, the computation is backward stable. This time the estimate takes the form

$$(\tilde{R} + \delta R)\tilde{x} = \tilde{y}, \qquad \frac{\|\delta R\|}{\|\tilde{R}\|} = O(\epsilon_{\text{machine}}). \qquad (16.5)$$

As always, the equality on the left is exact. It asserts that the floating point result $\tilde{x}$ is the exact solution of a slight perturbation of the system $\tilde{R}x = \tilde{y}$.

In the next lecture we shall derive (16.5) in full detail—more interesting than it sounds! The result is stated as Theorem 17.1.

Now, taking (16.3)–(16.5) as given, here is the theorem that results. This is typical of backward stability theorems that can be proved for many of the algorithms of numerical linear algebra.

Theorem 16.2. *Algorithm 16.1 is backward stable, satisfying*

$$(A + \Delta A)\tilde{x} = b, \qquad \frac{\|\Delta A\|}{\|A\|} = O(\epsilon_{\text{machine}}) \qquad (16.6)$$

for some $\Delta A \in \mathbb{C}^{m \times m}$.

Proof. Composing (16.4) and (16.5), we have

$$b = (\tilde{Q} + \delta Q)(\tilde{R} + \delta R)\tilde{x} = [\tilde{Q}\tilde{R} + (\delta Q)\tilde{R} + \tilde{Q}(\delta R) + (\delta Q)(\delta R)]\tilde{x}.$$

Therefore, by (16.3),

$$b = [A + \delta A + (\delta Q)\tilde{R} + \tilde{Q}(\delta R) + (\delta Q)(\delta R)]\tilde{x}.$$

This equation has the form

$$b = (A + \Delta A)\tilde{x},$$

where ΔA is a sum of four terms. To establish (16.6), we must show that each of these terms is small relative to A.

Since $\tilde{Q}\tilde{R} = A + \delta A$ and $\tilde{Q}$ is unitary, we have

$$\frac{\|\tilde{R}\|}{\|A\|} \leq \|\tilde{Q}^*\| \frac{\|A + \delta A\|}{\|A\|} = O(1)$$

as $\epsilon_{\text{machine}} \to 0$, by (16.3). (It is $1 + O(\epsilon_{\text{machine}})$ if $\|\cdot\| = \|\cdot\|_2$, but we have made no assumptions about $\|\cdot\|$.) This gives us

$$\frac{\|(\delta Q)\tilde{R}\|}{\|A\|} \leq \|\delta Q\| \frac{\|\tilde{R}\|}{\|A\|} = O(\epsilon_{\text{machine}})$$

by (16.4). Similarly,

$$\frac{\|\tilde{Q}(\delta R)\|}{\|A\|} \leq \|\tilde{Q}\| \frac{\|\delta R\|}{\|\tilde{R}\|} \frac{\|\tilde{R}\|}{\|A\|} = O(\epsilon_{\text{machine}})$$

by (16.5). Finally,

$$\frac{\|(\delta Q)(\delta R)\|}{\|A\|} \leq \|\delta Q\| \frac{\|\delta R\|}{\|A\|} = O(\epsilon_{\text{machine}}^2).$$

The total perturbation ΔA thus satisfies

$$\frac{\|\Delta A\|}{\|A\|} \leq \frac{\|\delta A\|}{\|A\|} + \frac{\|(\delta Q)\tilde{R}\|}{\|A\|} + \frac{\|\tilde{Q}(\delta R)\|}{\|A\|} + \frac{\|(\delta Q)(\delta R)\|}{\|A\|} = O(\epsilon_{\text{machine}}),$$

as claimed. □

Combining Theorems 12.2, 15.1, and 16.2 gives the following result about accuracy of solutions of $Ax = b$.

Theorem 16.3. *The solution $\tilde{x}$ computed by Algorithm 16.1 satisfies*

$$\frac{\|\tilde{x} - x\|}{\|x\|} = O(\kappa(A)\epsilon_{\text{machine}}). \tag{16.7}$$

Exercises

16.1. (a) Let unitary matrices $Q_1, \ldots, Q_k \in \mathbb{C}^{m,m}$ be fixed and consider the problem of computing, for $A \in \mathbb{C}^{m \times n}$, the product $B = Q_k \cdots Q_1 A$. Let the computation be carried out from right to left by straightforward floating point operations on a computer satisfying (13.5) and (13.7). Show that this algorithm is backward stable. (Here A is thought of as data that can be perturbed; the matrices Q_j are fixed and not to be perturbed.)

(b) Give an example to show that this result no longer holds if the unitary matrices Q_j are replaced by arbitrary matrices $X_j \in \mathbb{C}^{m \times m}$.

16.2. The idea of this exercise is to carry out an experiment analogous to the one described in this lecture, but for the SVD instead of QR factorization.

(a) Write a MATLAB program that constructs a 50×50 matrix A=U*S*V', where U and V are random orthogonal matrices and S is a diagonal matrix whose diagonal entries are random uniformly distributed numbers in $[0, 1]$, sorted into nonincreasing order. Have your program compute [U2,S2,V2] = svd(A) and the norms of U-U2, V-V2, S-S2, and A-U2*S2*V2'. Do this for five matrices A and comment on the results. (Hint: Plots of diag(U2'*U) and diag(V2'*V) may be informative.)

(b) Fix the signs in your computed SVD so that the difficulties of (a) go away. Run the program again for five random matrices and comment on the various norms. Do they have a connection with cond(A)?

(c) Replace the diagonal entries of S by their sixth powers and repeat (b). Do you see significant differences between the results of this exercise and those of the experiment for QR factorization?

Lecture 17. Stability of Back Substitution

One of the easiest problems of numerical linear algebra is the solution of a triangular system of equations. The standard algorithm is successive substitution, called back substitution when the system is upper-triangular. Here we show in full detail that this algorithm is backward stable, obtaining quantitative bounds on the effects of rounding errors, with no "$O(\epsilon_{\text{machine}})$".

Triangular Systems

We have seen that a general system of equations $Ax = b$ can be reduced to an upper-triangular system $Rx = y$ by QR factorization. Lower- and upper-triangular systems also arise in Gaussian elimination, in Cholesky factorization, and in numerous other computations of numerical linear algebra. These systems are easily solved by a process of successive substitution, called *forward substitution* if the system is lower-triangular and *back substitution* if it is upper-triangular. Although the two cases are mathematically identical, for definiteness, we treat back substitution in this lecture.

Suppose we wish to solve $Rx = b$, that is,

$$
\begin{bmatrix}
r_{11} & r_{12} & \cdots & r_{1m} \\
 & r_{22} & & \vdots \\
 & & \ddots & \vdots \\
 & & & r_{mm}
\end{bmatrix}
\begin{bmatrix}
x_1 \\
x_2 \\
\vdots \\
x_m
\end{bmatrix}
=
\begin{bmatrix}
b_1 \\
b_2 \\
\vdots \\
b_m
\end{bmatrix},
\qquad (17.1)
$$

121

where $b \in \mathbb{C}^m$ and $R \in \mathbb{C}^{m \times m}$, nonsingular and upper-triangular, are given, and $x \in \mathbb{C}^m$ is unknown. We can do this by solving for the components of x one after another, beginning with x_m and finishing with x_1. For later convenience we write the algorithm as a sequence of formulas rather than a loop.

Algorithm 17.1. Back Substitution

$$x_m = b_m / r_{mm}$$

$$x_{m-1} = \left(b_{m-1} - x_m r_{m-1,m} \right) \big/ r_{m-1,m-1}$$

$$x_{m-2} = \left(b_{m-2} - x_{m-1} r_{m-2,m-1} - x_m r_{m-2,m} \right) \big/ r_{m-2,m-2}$$

$$\vdots$$

$$x_j = \left(b_j - \sum_{k=j+1}^{m} x_k r_{jk} \right) \big/ r_{jj}$$

The structure is triangular, with a subtraction and a multiplication at each position. The operation count is accordingly twice the area of an $m \times m$ triangle:

$$\text{Work for back substitution:} \quad \sim m^2 \text{ flops.} \tag{17.2}$$

Backward Stability Theorem

In the last lecture, back substitution appeared as one of three steps in the solution of $Ax = b$ by QR factorization. In (16.3)–(16.5) we asserted that each of these steps is backward stable, but we did not prove these claims. In this lecture we shall fill one of these gaps by deriving a bound that implies (16.5). Our argument is an example of how proofs of backward stability are organized. This will be the only case in this book in which we give all the details of such a proof.

Before we can prove that Algorithm 17.1 is backward stable, however, we must pin down one detail of the algorithm that is not specified by the formulas as written. Let us decide, arbitrarily, that in the expressions in parentheses above, the subtractions will be carried out from left to right. (Other orders are also stable; only the details of the estimates are different.) Now we can state our theorem.

Theorem 17.1. *Let Algorithm 17.1 be applied to a problem (17.1) consisting of floating point numbers on a computer satisfying (13.7). This algorithm is backward stable in the sense that the computed solution $\tilde{x} \in \mathbb{C}^m$ satisfies*

$$(R + \delta R)\tilde{x} = b \tag{17.3}$$

for some upper-triangular $\delta R \in \mathbb{C}^{m \times m}$ with

$$\frac{\|\delta R\|}{\|R\|} = O(\epsilon_{\text{machine}}). \tag{17.4}$$

Specifically, for each i, j,

$$\frac{|\delta r_{ij}|}{|r_{ij}|} \leq m\epsilon_{\text{machine}} + O(\epsilon_{\text{machine}}^2). \tag{17.5}$$

In (17.5) and throughout this lecture, we continue to use the convention of (14.12) that if the denominator is zero, the numerator is implicitly asserted to be zero also (for all sufficiently small $\epsilon_{\text{machine}}$).

To keep the ideas clear and interesting, our proof will be most leisurely.

$m = 1$

According to (17.3), our task is to express every floating point error as a perturbation of the input. Let us begin with the simplest case, where R is of dimension 1×1. Back substitution in this case consists of a single step,

$$\tilde{x}_1 = b_1 \oslash r_{11}.$$

(Recall from Lecture 13 that $\oslash$, $\otimes$, $\oplus$, and $\ominus$ denote floating point operations.) The axiom (13.7) for $\oslash$ guarantees that the computed solution is close to correct:

$$\tilde{x}_1 = \frac{b_1}{r_{11}}(1 + \epsilon_1), \quad |\epsilon_1| \leq \epsilon_{\text{machine}}.$$

However, we would like to express the error as if it resulted from a perturbation in R. To this end, we set $\epsilon_1' = -\epsilon_1/(1 + \epsilon_1)$, whereupon the formula becomes

$$\tilde{x}_1 = \frac{b_1}{r_{11}(1 + \epsilon_1')}, \quad |\epsilon_1'| \leq \epsilon_{\text{machine}} + O(\epsilon_{\text{machine}}^2). \tag{17.6}$$

Note that ϵ_1' is equal to $-\epsilon_1$ plus a term of order $\epsilon_1{}^2$. We can freely move small relative perturbations from numerators to denominators or vice versa, and the result changes by terms of order $\epsilon_{\text{machine}}^2$ (Exercise 14.2(b)).

In (17.6), the equality is exact; the division is mathematical, not floating point. The formula states that 1×1 back substitution is backward stable, for $\tilde{x}_1$ is exactly the correct solution to a perturbed problem, namely

$$(r_{11} + \delta r_{11})\tilde{x}_1 = b_1,$$

with $\delta r_{11} = \epsilon_1' r_{11}$; hence

$$\frac{|\delta r_{11}|}{|r_{11}|} \leq \epsilon_{\text{machine}} + O(\epsilon_{\text{machine}}^2).$$

$m = 2$

The 2×2 case is slightly less trivial. Suppose we have an upper-triangular matrix $R \in \mathbb{C}^{2 \times 2}$ and a vector $b \in \mathbb{C}^2$. The computation of $\tilde{x} \in \mathbb{C}^2$ proceeds in two steps. The first is the same as in the 1×1 case:

$$\tilde{x}_2 = b_2 \oslash r_{22} = \frac{b_2}{r_{22}(1 + \epsilon_1)}, \quad |\epsilon_1| \leq \epsilon_{\text{machine}} + O(\epsilon_{\text{machine}}^2). \qquad (17.7)$$

The second step is defined by the formula

$$\tilde{x}_1 = (b_1 \ominus (\tilde{x}_2 \otimes r_{12})) \oslash r_{11}.$$

To establish backward stability, we must express the errors in these three floating point operations as perturbations in the entries r_{ij}.

The multiplication is easy; we use the axiom (13.7) directly, interpreting the floating point multiplication as a perturbation in r_{12}:

$$\tilde{x}_1 = (b_1 \ominus \tilde{x}_2 r_{12}(1 + \epsilon_2)) \oslash r_{11}, \quad |\epsilon_2| \leq \epsilon_{\text{machine}}.$$

The division and subtraction are more subtle. First, we write the formula with exact mathematics according to (13.7):

$$\tilde{x}_1 = (b_1 - \tilde{x}_2 r_{12}(1 + \epsilon_2))(1 + \epsilon_3) \oslash r_{11} \qquad (17.8)$$

$$= \frac{(b_1 - \tilde{x}_2 r_{12}(1 + \epsilon_2))(1 + \epsilon_3)}{r_{11}}(1 + \epsilon_4). \qquad (17.9)$$

Here (13.7) guarantees $|\epsilon_3|, |\epsilon_4| \leq \epsilon_{\text{machine}}$. Now we shift the ϵ_3 and ϵ_4 terms from the numerator to the denominator, as before. This gives

$$\tilde{x}_1 = \frac{b_1 - \tilde{x}_2 r_{12}(1 + \epsilon_2)}{r_{11}(1 + \epsilon_3')(1 + \epsilon_4')},$$

with $|\epsilon_3'|, |\epsilon_4'| \leq \epsilon_{\text{machine}} + O(\epsilon_{\text{machine}}^2)$, or equivalently,

$$\tilde{x}_1 = \frac{b_1 - \tilde{x}_2 r_{12}(1 + \epsilon_2)}{r_{11}(1 + 2\epsilon_5)}, \qquad (17.10)$$

with $|\epsilon_5| \leq \epsilon_{\text{machine}} + O(\epsilon_{\text{machine}}^2)$. This formula states that $\tilde{x}_1$ would be exactly correct if r_{22}, r_{12}, and r_{11} were perturbed by factors $(1 + \epsilon_1)$, $(1 + \epsilon_2)$, and $(1 + 2\epsilon_5)$, respectively. These perturbations can be summarized by the equation

$$(R + \delta R)\tilde{x} = b,$$

where the entries δr_{ij} of δR satisfy

$$\begin{bmatrix} |\delta r_{11}|/|r_{11}| & |\delta r_{12}|/|r_{12}| \\ & |\delta r_{22}|/|r_{22}| \end{bmatrix} = \begin{bmatrix} 2|\epsilon_5| & |\epsilon_2| \\ & |\epsilon_1| \end{bmatrix} \leq \begin{bmatrix} 2 & 1 \\ & 1 \end{bmatrix} \epsilon_{\text{machine}} + O(\epsilon_{\text{machine}}^2).$$

(The "$\leq$" here and in similar results below is to be interpreted entrywise.) This formula guarantees $\|\delta R\|/\|R\| = O(\epsilon_{\text{machine}})$ in any matrix norm and thus that 2×2 back substitution is backward stable.

$m = 3$

The analysis for a 3×3 matrix includes all the reasoning necessary for the general case. The first two steps are the same as before:

$$\tilde{x}_3 = b_3 \oslash r_{33} = \frac{b_3}{r_{33}(1 + \epsilon_1)}, \tag{17.11}$$

$$\tilde{x}_2 = (b_2 \ominus (\tilde{x}_3 \otimes r_{23})) \oslash r_{22} = \frac{b_2 - \tilde{x}_3 r_{23}(1 + \epsilon_2)}{r_{22}(1 + 2\epsilon_3)}, \tag{17.12}$$

where

$$\begin{bmatrix} 2|\epsilon_3| & |\epsilon_2| \\ & |\epsilon_1| \end{bmatrix} \leq \begin{bmatrix} 2 & 1 \\ & 1 \end{bmatrix} \epsilon_{\text{machine}} + O(\epsilon^2_{\text{machine}}).$$

The third step involves the computation

$$\tilde{x}_1 = [(b_1 \ominus (\tilde{x}_2 \otimes r_{12})) \ominus (\tilde{x}_3 \otimes r_{13})] \oslash r_{11}. \tag{17.13}$$

We convert the two $\otimes$ operations in (17.13) to mathematical multiplication by introducing perturbations ϵ_4 and ϵ_5:

$$\tilde{x}_1 = [(b_1 \ominus \tilde{x}_2 r_{12}(1 + \epsilon_4)) \ominus \tilde{x}_3 r_{13}(1 + \epsilon_5)] \oslash r_{11}.$$

We convert the $\ominus$ operations to mathematical subtractions via perturbations ϵ_6 and ϵ_7:

$$\tilde{x}_1 = [(b_1 - \tilde{x}_2 r_{12}(1 + \epsilon_4))(1 + \epsilon_6) - \tilde{x}_3 r_{13}(1 + \epsilon_5)](1 + \epsilon_7) \oslash r_{11}.$$

Finally, the $\oslash$ is eliminated using ϵ_8; let us immediately replace this by ϵ'_8 with $|\epsilon_8| \leq \epsilon_{\text{machine}} + O(\epsilon^2_{\text{machine}})$ and put the result in the denominator:

$$\tilde{x}_1 = \frac{[(b_1 - \tilde{x}_2 r_{12}(1 + \epsilon_4))(1 + \epsilon_6) - \tilde{x}_3 r_{13}(1 + \epsilon_5)](1 + \epsilon_7)}{r_{11}(1 + \epsilon'_8)}.$$

Now, the expression above has everything as we need it except the terms involving ϵ_6 and ϵ_7, which originated from operations $\ominus$. If these are distributed, they will affect the number b_1, whereas our aim is to perturb only the entries r_{ij}. The term involving ϵ_7 is easily dispatched: we change ϵ_7 to ϵ'_7 and move it to the denominator as usual. The term involving ϵ_6 requires a new trick. We move it to the denominator too, but to keep the equality valid, we compensate by putting the new factor $(1 + \epsilon'_6)$ into the r_{13} term as well. Thus

$$\tilde{x}_1 = \frac{b_1 - \tilde{x}_2 r_{12}(1 + \epsilon_4) - \tilde{x}_3 r_{13}(1 + \epsilon_5)(1 + \epsilon'_6)}{r_{11}(1 + \epsilon'_6)(1 + \epsilon'_7)(1 + \epsilon'_8)}.$$

Now r_{13} has two perturbations of size at most $\epsilon_{\text{machine}}$, and r_{11} has three. In this formula, all the errors in the computation have been expressed as perturbations in the entries of R.

The result can be summarized as

$$(R + \delta R)\tilde{x} = b,$$

where the entries δr_{ij} satisfy

$$\begin{bmatrix} |\delta r_{11}|/|r_{11}| & |\delta r_{12}|/|r_{12}| & |\delta r_{13}|/|r_{13}| \\ & |\delta r_{22}|/|r_{22}| & |\delta r_{23}|/|r_{23}| \\ & & |\delta r_{33}|/|r_{33}| \end{bmatrix} \leq \begin{bmatrix} 3 & 1 & 2 \\ & 2 & 1 \\ & & 1 \end{bmatrix} \epsilon_{\text{machine}} + O(\epsilon_{\text{machine}}^2).$$

General m

The analysis in higher-dimensional cases is similar. For example, in the 5×5 case we obtain the componentwise bound

$$\frac{|\delta R|}{|R|} \leq \begin{bmatrix} 5 & 1 & 2 & 3 & 4 \\ & 4 & 1 & 2 & 3 \\ & & 3 & 1 & 2 \\ & & & 2 & 1 \\ & & & & 1 \end{bmatrix} \epsilon_{\text{machine}} + O(\epsilon_{\text{machine}}^2). \tag{17.14}$$

The entries of the matrix in this formula are obtained from three components. The multiplications $\tilde{x}_k r_{jk}$ introduce $\epsilon_{\text{machine}}$ perturbations in the pattern

$$\otimes : \begin{bmatrix} 0 & 1 & 1 & 1 & 1 \\ & 0 & 1 & 1 & 1 \\ & & 0 & 1 & 1 \\ & & & 0 & 1 \\ & & & & 0 \end{bmatrix}. \tag{17.15}$$

The divisions by r_{kk} introduce perturbations in the pattern

$$\oslash : \begin{bmatrix} 1 & & & & \\ & 1 & & & \\ & & 1 & & \\ & & & 1 & \\ & & & & 1 \end{bmatrix}. \tag{17.16}$$

Finally, the subtractions also occur in the pattern (17.15), and, due to the decision to compute from left to right, each one introduces a perturbation on the diagonal and at each position to the right. This adds up to the pattern

$$\ominus : \begin{bmatrix} 4 & 0 & 1 & 2 & 3 \\ & 3 & 0 & 1 & 2 \\ & & 2 & 0 & 1 \\ & & & 1 & 0 \\ & & & & 0 \end{bmatrix}. \tag{17.17}$$

Adding (17.15), (17.16), and (17.17) produces the result in (17.14). This completes the proof of Theorem 17.1.

Remarks

The analysis leading to (17.14) is typical of backward error analysis for all kinds of floating point computations. The only raw ingredient is the floating point axiom (13.7), ensuring that each operation $\oplus$, $\ominus$, $\otimes$, or $\oslash$ (sometimes also floating point $\sqrt{\ }$) introduces a small relative error. One utilizes this axiom repeatedly and carefully, interpreting each error in the course of the calculation as an error in the initial data. Perturbations of order $\epsilon_{\text{machine}}$ are composed additively and moved freely between numerators and denominators, since the difference is of order $\epsilon_{\text{machine}}^2$.

More than one error bound can be derived for a given algorithm. In the present case, we could have perturbed b_j as well as r_{ij}, avoiding the need for the trickery represented in (17.17). On the other hand, a final result in which only R is perturbed is appealingly clean.

Equation (17.5) is a *componentwise* backward error bound, meaning that each entry r_{ij} is perturbed by a quantity that is small relative to itself, not just relative to the norm of R. For example, if $r_{ij} = 0$, this entry undergoes no perturbation at all: δR has the same sparsity pattern as R. Some algorithms of numerical linear algebra satisfy componentwise backward error estimates, and some do not; in the latter case we must settle for a weaker *normwise* estimate such as (17.4). In the early decades of numerical analysis after the Second World War, most error estimates were obtained in the normwise form, whereas in more recent years there has been a shift toward componentwise analysis, since such results are sharper and algorithms that satisfy componentwise bounds are less sensitive to scaling of variables. However, we shall not pursue these matters in this book.

We close with a comment about the relationship between quantitative bounds like (17.5) or (17.14) and those such as (17.4) that we express in the $O(\epsilon_{\text{machine}})$ notation (Exercise 17.1). Why do we not dispense with crude statements like (17.4)? The reason is that quantitative bounds must include factors such as $\sqrt{m}$ or m, and these are norm-dependent, unmemorable, and often pessimistic in practice anyway because of statistical cancellation. We prefer to avoid such complications by expressing most of our results in terms of $O(\epsilon_{\text{machine}})$—which is, most assuredly, the form in which numerical analysts remember them.

Exercises

17.1. For any particular choice of norm $\|\cdot\|$, the bound (17.5) implies a more quantitative normwise bound than (17.4). Derive such bounds for the norms (a) $\|\cdot\|_1$, (b) $\|\cdot\|_2$, (c) $\|\cdot\|_\infty$.

17.2. A triangular system (17.1) is solved by back substitution. Exactly what does Theorem 17.1 imply about the error $\|\tilde{x} - x\|$?

17.3. Let $L \in \mathbb{C}^{m \times m}$ be a unit lower-triangular matrix (i.e., with diagonal entries equal to 1). For convenience, write L in the form

$$L = \begin{bmatrix} 1 & & & & \\ -\ell_{2,1} & 1 & & & \\ -\ell_{3,1} & -\ell_{3,2} & 1 & & \\ \vdots & \vdots & & \ddots & \\ -\ell_{m,1} & -\ell_{m,2} & -\ell_{m,3} & \cdots & 1 \end{bmatrix},$$

and define $M = L^{-1}$.

(a) Derive a formula for m_{ij} (which may involve other entries of M). Which entries of L does m_{ij} depend on?

(b) Suppose the subdiagonal entries of L are independent random numbers ± 1 with equal probability. Fix k and define $\mu_1 = m_{kk}$, $\mu_2 = m_{k+1,k}$, $\mu_3 = m_{k+2,k}, \ldots$. Write down a system of recurrence relations with random coefficients for the numbers μ_j.

(c) Experiments show that random triangular matrices with entries ± 1 are exponentially ill-conditioned in the sense that if κ_m denotes the 2-norm condition number of a matrix of this kind of dimension m, then $\lim_{m \to \infty} (\kappa_m)^{1/m} = C$ for some constant $1 < C < 1.5$. (The limit process can be made precise in various ways, but we shall not go into the technicalities; think of it as holding "with probability 1.") Perform numerical experiments involving random matrices of various dimensions to estimate C to 10% accuracy or better.

(d) Larger scale experiments become feasible if the random matrices of (c) are replaced by the random sequences $\mu_1, \mu_2, \mu_3, \ldots$ of (b). Explain (without proof) why the constant C can also be obtained by considering these sequences, and carry out numerical experiments to estimate C to 1% accuracy or better.

Lecture 18. Conditioning of Least Squares Problems

The conditioning of least squares problems is a subtle topic, combining the conditioning of square systems of equations with the geometry of orthogonal projection. It is important because it has nontrivial implications for the stability of least squares algorithms.

Four Conditioning Problems

In this lecture we return to the linear least squares problem (11.2), illustrated again in Figure 18.1. We assume the matrix defining the problem is of full rank, and write throughout this lecture $\| \cdot \| = \| \cdot \|_2$:

$$
\begin{aligned}
&\text{Given } A \in \mathbb{C}^{m \times n} \text{ of full rank, } m \geq n, \, b \in \mathbb{C}^m, \\
&\text{find } x \in \mathbb{C}^n \text{ such that } \|b - Ax\| \text{ is minimized.}
\end{aligned}
\tag{18.1}
$$

The solution x and the corresponding point $y = Ax$ that is closest to b in range(A) are given by

$$
x = A^+ b, \qquad y = Pb,
\tag{18.2}
$$

where $A^+ \in \mathbb{C}^{n \times m}$ is the pseudoinverse (11.11) of A and $P = AA^+ \in \mathbb{C}^{m \times m}$ is the orthogonal projector onto range(A). We consider the conditioning of (18.1) with respect to perturbations. Just as the last lecture represented the most detailed analysis of stability in this book, the present lecture represents our most detailed analysis of conditioning. We pick the least squares problem

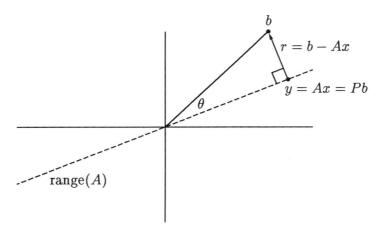

Figure 18.1. *The least squares problem (repetition of Figure 11.3).*

because the details are interesting and because they have an important practical consequence, to be discussed in the next lecture: the instability of the normal equations as a general purpose least squares algorithm.

Conditioning pertains to the sensitivity of solutions to perturbations in data. For (18.1), we shall investigate two choices of each. The data for the problem are the $m \times n$ matrix A and the m-vector b. The solution is either the coefficient vector x or the corresponding point $y = Ax$. Thus

$$\text{Data: } A, b, \qquad \text{Solution: } x, y.$$

Together, these two pairs of choices define four conditioning questions that we shall consider, all of which have application in certain contexts.

Theorem

The centerpiece of this lecture is Theorem 18.1, which provides answers to these questions. The results are expressed in terms of three dimensionless parameters that appear repeatedly in the analysis of least squares problems. The first is the condition number of A. For a square matrix, this is $\kappa(A) = \|A\|\,\|A^{-1}\|$, and in the rectangular case, the definition generalizes to (12.17),

$$\kappa(A) = \|A\|\,\|A^{+}\| = \frac{\sigma_1}{\sigma_n}. \tag{18.3}$$

The second is the angle θ marked in Figure 18.1, a measure of the closeness of the fit:

$$\theta = \cos^{-1}\frac{\|y\|}{\|b\|}. \tag{18.4}$$

The third is a measure of how much $\|y\|$ falls short of its maximum possible value, given $\|A\|$ and $\|x\|$:

$$\eta = \frac{\|A\|\,\|x\|}{\|y\|} = \frac{\|A\|\,\|x\|}{\|Ax\|}. \tag{18.5}$$

These parameters lie in the ranges

$$1 \le \kappa(A) < \infty, \qquad 0 \le \theta \le \pi/2, \qquad 1 \le \eta \le \kappa(A). \tag{18.6}$$

Theorem 18.1. *Let $b \in \mathbb{C}^m$ and $A \in \mathbb{C}^{m\times n}$ of full rank be fixed. The least squares problem (18.1) has the following 2-norm relative condition numbers (12.5) describing the sensitivities of y and x to perturbations in b and A:*

	y	x
b	$\dfrac{1}{\cos\theta}$	$\dfrac{\kappa(A)}{\eta\cos\theta}$
A	$\dfrac{\kappa(A)}{\cos\theta}$	$\kappa(A) + \dfrac{\kappa(A)^2\tan\theta}{\eta}$

The results in the first row are exact, being attained for certain perturbations δb, and the results in the second row are upper bounds.

In the special case $m = n$, (18.1) reduces to a square, nonsingular system of equations, with $\theta = 0$. In this case, the numbers in the second column of the theorem reduce to $\kappa(A)/\eta$ and $\kappa(A)$, which are the results (12.14) and (12.18) derived earlier, and the number in the lower-left position can be replaced by 0 (Exercise 18.4).

Transformation to a Diagonal Matrix

As a first step in the proof of Theorem 18.1, we note that the least squares problem becomes easier to analyze if we transform to a convenient choice of bases. Let A have an SVD of the form $A = U\Sigma V^*$, where Σ is an $m\times n$ diagonal matrix with positive diagonal entries. Since perturbations are measured in the 2-norm, their sizes are unaffected by a unitary change of basis, so the perturbation behavior of A is the same as that of Σ. Therefore, without loss of generality, we can deal with Σ directly. For the remainder of the discussion, we assume $A = \Sigma$ and write

$$A = \begin{bmatrix} \sigma_1 & & & \\ & \sigma_2 & & \\ & & \ddots & \\ & & & \sigma_n \\ & & & \\ & & & \end{bmatrix} = \begin{bmatrix} A_1 \\ 0 \end{bmatrix}. \tag{18.7}$$

Here A_1 is $n \times n$ and diagonal; the rest of A is zero.

The orthogonal projection of b onto range(A) is now a triviality. Write

$$b = \begin{bmatrix} b_1 \\ b_2 \end{bmatrix},$$

where b_1 contains the first n entries of b. Then the projection $y = Pb$ is

$$y = \begin{bmatrix} b_1 \\ 0 \end{bmatrix}.$$

To find the corresponding x we can write $Ax = y$ as

$$\begin{bmatrix} A_1 \\ 0 \end{bmatrix} x = \begin{bmatrix} b_1 \\ 0 \end{bmatrix},$$

which implies

$$x = A_1^{-1}b_1. \tag{18.8}$$

From these formulas it is evident that the orthogonal projector and pseudo-inverse are the block 2×2 and 1×2 matrices (see Exercise 11.1)

$$P = \begin{bmatrix} I & 0 \\ 0 & 0 \end{bmatrix}, \qquad A^+ = \begin{bmatrix} A_1^{-1} & 0 \end{bmatrix}. \tag{18.9}$$

Sensitivity of y to Perturbations in b

We begin with the simplest of our four conditioning results. By (18.2), the relationship between b and y is just the linear equation $y = Pb$. The Jacobian of this mapping is P itself, with $\|P\| = 1$ by (18.9). By (12.6) and (18.4), the condition number of y with respect to perturbations in b is accordingly

$$\kappa_{b \mapsto y} = \frac{\|P\|}{\|y\|/\|b\|} = \frac{1}{\cos\theta}.$$

This establishes the upper-left result of Theorem 18.1. The condition number is realized (that is, the supremum in (12.5) is attained) for perturbations δb with $\|P(\delta b)\| = \|\delta b\|$, which occurs when δb is zero except in the first n entries.

Sensitivity of x to Perturbations in b

The relationship between b and x is also linear, $x = A^+b$, with Jacobian A^+. By (12.6), (18.4), and (18.5), the condition number of x with respect to perturbations in b is consequently

$$\kappa_{b \mapsto x} = \frac{\|A^+\|}{\|x\|/\|b\|} = \|A^+\| \frac{\|b\|}{\|y\|} \frac{\|y\|}{\|x\|} = \|A^+\| \frac{1}{\cos\theta} \frac{\|A\|}{\eta} = \frac{\kappa(A)}{\eta \cos\theta}.$$

This establishes the upper-right result of Theorem 18.1. Here, the condition number is realized by perturbations δb satisfying $\|A^+(\delta b)\| = \|A^+\| \|\delta b\| = \|\delta b\|/\sigma_n$, which occurs when δb is zero except in the nth entry (or perhaps also in other entries, if A has more than one singular value equal to σ_n).

Tilting the Range of A

The analysis of perturbations in A is a nonlinear problem and more subtle. We could proceed by calculating Jacobians algebraically, but instead, we shall take a geometric view. Our starting point is the observation that perturbations in A affect the least squares problem in two ways: they distort the mapping of $\mathbb{C}^n$ onto range(A), and they alter range(A) itself. Let us consider this latter effect for a moment.

We can visualize slight changes in range(A) as small "tiltings" of this space. The question is, what is the maximum angle of tilt $\delta\alpha$ that can be imparted by a small perturbation δA? The answer can be determined as follows. The image under A of the unit n-sphere is a hyperellipse that lies flat in range(A). To change range(A) as efficiently as possible, we grasp a point $p = Av$ on the hyperellipse (hence $\|v\| = 1$) and nudge it in a direction δp orthogonal to range(A). A matrix perturbation that achieves this most efficiently is $\delta A = (\delta p)v^*$, which gives $(\delta A)v = \delta p$ with $\|\delta A\| = \|\delta p\|$ (Example 3.6). Now it is clear that to obtain the maximum tilt with a given $\|\delta p\|$, we should take p to be as close to the origin as possible. That is, we want $p = \sigma_n u_n$, where σ_n is the smallest singular value of A and u_n is the corresponding left singular vector. With A in the diagonal form (18.7), p is equal to the last column of A, v^* is the n-vector $(0, 0, \ldots, 0, 1)$, and δA is a perturbation of the entries of A below the diagonal in this column. Such a perturbation tilts range(A) by the angle $\delta\alpha$ given by $\tan(\delta\alpha) = \|\delta p\|/\sigma_n$. Since $\|\delta p\| = \|\delta A\|$ and $\delta\alpha \leq \tan(\delta\alpha)$, we have

$$\delta\alpha \leq \frac{\|\delta A\|}{\sigma_n} = \frac{\|\delta A\|}{\|A\|}\kappa(A), \tag{18.10}$$

with equality attained for choices δA of the kind just described, provided they are infinitesimal (so that $\delta\alpha = \tan(\delta\alpha)$).

Sensitivity of y to Perturbations in A

Now we are prepared to derive the second row of the table in Theorem 18.1. We begin with its left-hand entry. Since y is the orthogonal projection of b onto range(A), it is determined by b and range(A) alone. Therefore, to analyze the sensitivity of y to perturbations in A, we can simply study the effect on y of tilting range(A) by some angle $\delta\alpha$.

An elegant geometrical property becomes apparent when we imagine fixing b and watching y vary as range(A) is tilted (Figure 18.2). No matter how range(A) is tilted, the vector $y \in$ range(A) must always be orthogonal to $y - b$. That is, the line b–y must lie at right angles to the line 0–y. In other words, as range(A) is adjusted, y moves along the sphere of radius $\|b\|/2$ centered at the point $b/2$.

Tilting range(A) in the plane 0–b–y by an angle $\delta\alpha$ changes the angle 2θ at the central point $b/2$ by $2\delta\alpha$. Thus the corresponding perturbation δy is the

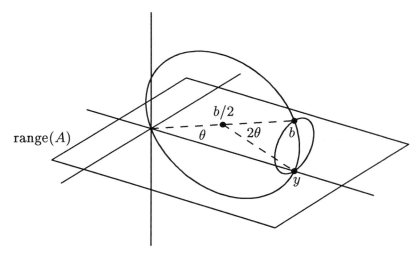

Figure 18.2. *Two circles on the sphere along which y moves as* range(A) *varies. The large circle, of radius* $\|b\|/2$, *corresponds to tilting* range(A) *in the plane* 0–b–y, *and the small circle, of radius* $(\|b\|/2)\sin\theta$, *corresponds to tilting it in an orthogonal direction. However* range(A) *is tilted, y remains on the sphere of radius* $\|b\|/2$ *centered at* $b/2$.

base of an isosceles triangle with central angle $2\delta\alpha$ and edge length $\|b\|/2$. This implies $\|\delta y\| = \|b\|\sin(\delta\alpha)$. Tilting range($A$) in any other direction results in a similar geometry in a different plane and perturbations smaller by a factor as small as $\sin\theta$. Thus for arbitrary perturbations by an angle $\delta\alpha$ we have

$$\|\delta y\| \;\le\; \|b\|\sin(\delta\alpha) \;\le\; \|b\|\delta\alpha. \qquad (18.11)$$

By (18.4) and (18.10), this gives us $\|\delta y\| \le \|\delta A\|\kappa(A)\|y\|/\|A\|\cos\theta$, that is,

$$\frac{\|\delta y\|}{\|y\|} \;\Big/\; \frac{\|\delta A\|}{\|A\|} \;\le\; \frac{\kappa(A)}{\cos\theta}. \qquad (18.12)$$

This establishes the lower-left result of Theorem 18.1.

Sensitivity of x to Perturbations in A

We are now ready to analyze the most interesting relationship of Theorem 18.1: the sensitivity of x to perturbations in A.

A perturbation δA splits naturally into two parts: one part δA_1 in the first n rows of A, and another part δA_2 in the remaining $m - n$ rows:

$$\delta A \;=\; \begin{bmatrix} \delta A_1 \\ \delta A_2 \end{bmatrix} \;=\; \begin{bmatrix} \delta A_1 \\ 0 \end{bmatrix} + \begin{bmatrix} 0 \\ \delta A_2 \end{bmatrix}.$$

First, let us consider the effect of perturbations δA_1. Such a perturbation changes the mapping of A in its range, but not range(A) itself or y. It perturbs A_1 by δA_1 in the square system (18.8) without changing b_1. The condition number for such perturbations is given by (12.18), which here takes the form

$$\frac{\|\delta x\|}{\|x\|} \Big/ \frac{\|\delta A_1\|}{\|A\|} \leq \kappa(A_1) = \kappa(A). \tag{18.13}$$

Next we consider the effect of (infinitesimal) perturbations δA_2. Such a perturbation tilts range(A) without changing the mapping of A within this space. The point y and thus the vector b_1 are perturbed, but not A_1. This corresponds to perturbing b_1 in (18.8) without changing A_1. The condition number for such perturbations is given by (12.14), which takes the form

$$\frac{\|\delta x\|}{\|x\|} \Big/ \frac{\|\delta b_1\|}{\|b_1\|} \leq \frac{\kappa(A_1)}{\eta(A_1; x)} = \frac{\kappa(A)}{\eta}. \tag{18.14}$$

To finish the argument we need to relate δb_1 to δA_2. Now the vector b_1 is y expressed in the coordinates of range(A). Therefore, the only changes in y that are realized as changes in b_1 are those that lie parallel to range(A); orthogonal changes have no effect. In particular, if range(A) is tilted by an angle $\delta\alpha$ in the plane 0–b–y, the resulting perturbation δy lies not parallel to range(A) but at an angle $\pi/2 - \theta$. Consequently, the change in b_1 satisfies $\|\delta b_1\| = \sin\theta\|\delta y\|$. By (18.11), we therefore have

$$\|\delta b_1\| \leq (\|b\|\delta\alpha)\sin\theta. \tag{18.15}$$

Curiously, if range(A) is tilted in a direction orthogonal to the plane 0–b–y, we obtain the same bound, but for a different reason. Now δy is parallel to range(A), but it is a factor of $\sin\theta$ smaller, as discussed above in connection with Figure 18.2. Thus we have $\|\delta y\| \leq (\|b\|\delta\alpha)\sin\theta$, and since $\|\delta b_1\| \leq \|\delta y\|$, we again arrive at (18.15).

All the pieces are now in place. Since $\|b_1\| = \|b\|\cos\theta$, we can rewrite (18.15) as

$$\frac{\|\delta b_1\|}{\|b_1\|} \leq (\delta\alpha)\tan\theta. \tag{18.16}$$

Relating $\delta\alpha$ to $\|\delta A_2\|$ by (18.10) and combining (18.14) with (18.16), we obtain

$$\frac{\|\delta x\|}{\|x\|} \Big/ \frac{\|\delta A_2\|}{\|A\|} \leq \frac{\kappa(A)^2\tan\theta}{\eta}.$$

Adding this to (18.13) establishes the lower-right result of Theorem 18.1.

Exercises

18.1. Consider the example

$$A = \begin{bmatrix} 1 & 1 \\ 1 & 1.0001 \\ 1 & 1.0001 \end{bmatrix}, \qquad b = \begin{bmatrix} 2 \\ 0.0001 \\ 4.0001 \end{bmatrix}.$$

(a) What are the matrices A^+ and P for this example? Give exact answers.

(b) Find the exact solutions x and $y = Ax$ to the least squares problem $Ax \approx b$.

(c) What are $\kappa(A)$, θ, and η? From here on, numerical answers are acceptable.

(d) What are the four condition numbers of Theorem 18.1?

(e) Give examples of perturbations δb and δA that approximately attain these four condition numbers.

18.2. Social scientists depend on the technique of *regression*, in which a vector of observations of some quantity is approximated in the least squares sense by a linear combination of other vectors. The coefficients of the fit are then interpreted as representing, say, the effects on annual income of IQ, years of education, parents' years of education, and parents' income.

One might think that the more variables one included in such a model, the more information one would obtain, but this is not always true. Explain this phenomenon from the point of view of conditioning, making specific reference to the results of Theorem 18.1.

18.3. Suppose you look across Lake Cayuga at a light from a house on the other side. If the lake surface is rippled, the reflected light appears as a long vertical streak. The same effect appears with taillights of the car ahead of you on a rainy road, or even with reflections of hallway lights on a shiny waxed floor. It is a real effect, not an optical illusion, and the explanation is a matter of geometry.

(a) Derive a quantitative theory explaining this phenomenon. Specifically, suppose you and the house across the lake are each fifty meters above the surface, and the lake is one kilometer wide. What is the length-to-width ratio of the streak as it appears in your visual field?

(b) Describe a connection between this problem and one of the geometrical arguments of this lecture.

18.4. Explain why, as remarked after Theorem 18.1, the condition number of y with respect to perturbations in A becomes 0 in the case $m = n$.

Lecture 19. Stability of Least Squares Algorithms

Least squares problems can be solved by various methods, as described in Lecture 11, including the normal equations, Householder triangularization, Gram–Schmidt orthogonalization, and the SVD. Here we compare these methods and show that the use of the normal equations is in general unstable.

Example

To illustrate the behavior of our algorithms, we shall apply them to a numerical example with $m = 100$, $n = 15$. Here is the MATLAB setup:

```
m = 100; n = 15;
t = (0:m-1)'/(m-1);              Set t to a discretization of [0, 1].
A = []; for i=1:n,               Construct Vandermonde matrix.
   A = [A t.^(i-1)]; end
b = exp(sin(4*t));               Right-hand side.
b = b/2006.787453080206;         Normalization (see text).
```

The idea behind this example is the least squares fitting of the function $\exp(\sin(4\tau))$ on the interval $[0, 1]$ by a polynomial of degree 14. First we discretize $[0, 1]$, defining a vector t of 100 points equally spaced from 0 to 1. The matrix A is the 100×15 Vandermonde matrix whose columns are the powers $1, \tau, \ldots, \tau^{14}$ sampled at the points of t, and the right-hand side b is the function $\exp(\sin(4\tau))$ sampled at these points.

The reason for the bizarre final line of the code is as follows. For simplicity, we are going to compare just the coefficients x_{15} computed by our various algorithms. Without this final line, the correct value of x_{15} would be $2006.787453080206\ldots$ (this figure was obtained with an extended precision arithmetic package). By dividing by this number, we obtain a problem whose solution has $x_{15} = 1$, making our comparisons easier to follow.

To explain our observations, we shall need the quantities (18.3)–(18.5). One can determine these to sufficient accuracy by solving the least squares problem numerically with the aid of MATLAB's \ operator:

```
x = A\b; y = A*x;              Solve least squares problem.
kappa = cond(A)
    kappa = 2.2718e+10                      κ(A)
theta = asin(norm(b-y)/norm(b))
    theta = 3.7461e-06                       θ
eta = norm(A)*norm(x)/norm(y)
    eta = 2.1036e+05                         η
```

The result $\kappa(A) \approx 10^{10}$ indicates that the monomials $1, t, \ldots, t^{14}$ form a highly ill-conditioned basis. The result $\theta \approx 10^{-6}$ indicates that $\exp(\sin(4t))$ can be fitted very closely by a polynomial of degree 14. (The fit is so close that we computed θ with the formula $\theta = \sin^{-1}(\|b - y\|/\|b\|)$ instead of (18.4), to avoid cancellation error.) As for η, its value of about 10^5 is about midway between the extremes 1 and $\kappa(A)$ permitted by (18.6).

Inserting these numbers into the formulas of Theorem 18.1, we find that for our example problem, the condition numbers of y and x with respect to perturbations in b and A are approximately

	y	x
b	1.0	1.1×10^5
A	2.3×10^{10}	3.2×10^{10}

Householder Triangularization

As mentioned in Lecture 11, the standard algorithm for solving least squares problems is QR factorization via Householder triangularization (Algorithm 11.2). Here is what we get with a MATLAB experiment:

```
[Q,R] = qr(A,0);              Householder triang. of A.
x = R\(Q'*b);                 Solve for x.
x(15)
    ans = 1.00000031528723
```

What can we make of this result? Thanks to our normalization, the correct answer would be $x_{15} = 1$. Thus we have a relative error of about 3×10^{-7}. Since the calculation was done in IEEE double precision arithmetic with $\epsilon_{\text{machine}} \approx 10^{-16}$, this means that the rounding errors have been amplified by a factor of order 10^9. At first sight this looks bad, but a glance at the table above reminds us that the condition number of x with respect to perturbations in A is of order 10^{10}. Thus the inaccuracy in x_{15} can be entirely explained by ill-conditioning, not instability. Algorithm 11.2 appears to be backward stable.

Above, we formed $\hat{Q}$ explicitly, but as emphasized in Lectures 10 and 16, this is not necessary. It is enough to store the vectors v_k determined at the kth step of Algorithm 10.1 (equation (10.5)), which can then be utilized to compute $\hat{Q}^*b$ by Algorithm 10.2. In MATLAB, we can achieve this effect by computing a QR factorization not just of A but of the $m \times (n+1)$ "augmented" matrix $[A \ b]$. In the course of this factorization, the n Householder reflectors that make A upper-triangular are applied to b also, leaving the vector $\hat{Q}^*b$ in the first n positions of column $n + 1$. An additional $(n + 1)$st reflector is then applied to make entries $n + 2, \ldots, m$ of column $n + 1$ zero, but this does not change the first n entries of that column, which are the ones we care about. Thus:

```
[Q2,R2] = qr([A b],0);        Householder triang. of [A b].
R2 = R2(1:n,1:n);             Extract R̂ ...
Qb = R2(1:n,n+1);                 ... and Q̂*b.
x = R2\Qb;                    Solve for x.
x(15)
    ans = 1.00000031529465
```

The answer is almost the same as before. This indicates that the errors introduced in the QR factorization of A swamp those introduced in the computation of $\hat{Q}^*b$.

There is also a third way to solve the least squares problem via Householder triangularization in MATLAB. We can use the built-in operator $\backslash$, as we did already in finding $\kappa(A)$, θ, and η:

```
x = A\b;                      Solve for x.
x(15)
    ans = 0.99999994311087
```

This result is distinctly different from the others, and an order of magnitude more accurate. The reason for this is that MATLAB's $\backslash$ operator makes use of *QR factorization with column pivoting*, based on a factorization $AP = \hat{Q}\hat{R}$, where P is a permutation matrix. In this book we shall not discuss column pivoting.

From the point of view of normwise stability analysis, these three variants of QR factorization are equal. All of them, it can be proved, are backward stable.

Theorem 19.1. *Let the full-rank least squares problem (11.2) be solved by Householder triangularization (Algorithm 11.2) on a computer satisfying (13.5) and (13.7). This algorithm is backward stable in the sense that the computed solution $\tilde{x}$ has the property*

$$\|(A + \delta A)\tilde{x} - b\| = \min, \qquad \frac{\|\delta A\|}{\|A\|} = O(\epsilon_{\text{machine}}) \qquad (19.1)$$

for some $\delta A \in \mathbb{C}^{m \times n}$. This is true whether $\hat{Q}^*b$ is computed via explicit formation of $\hat{Q}$ or implicitly by Algorithm 10.2. It also holds for Householder triangularization with arbitrary column pivoting.

Gram–Schmidt Orthogonalization

Another way to solve a least squares problem is by modified Gram–Schmidt orthogonalization (Algorithm 8.1). For $m \approx n$, this takes somewhat more operations than the Householder approach, but for $m \gg n$, the flop counts for both algorithms are asymptotic to $2mn^2$.

The following MATLAB sequence implements this algorithm in the obvious fashion. The function mgs is an implementation (not shown) of Algorithm 8.1—the same as in Experiment 2 of Lecture 9.

```
[Q,R] = mgs(A);            Gram-Schmidt orthog. of A.
x = R\(Q'*b);              Solve for x.
x(15)
   ans = 1.02926594532672
```

This result is very poor. Rounding errors have been amplified by a factor on the order of 10^{14}, far greater than the condition number of the problem. In fact, this algorithm is unstable, and the reason is easily identified. As mentioned at the end of Lecture 9, Gram–Schmidt orthogonalization produces matrices $\hat{Q}$, in general, whose columns are not accurately orthonormal. Since the algorithm above depends on that orthonormality, it suffers accordingly.

The instability can be avoided by a reformulation of the algorithm. Since the Gram–Schmidt iteration delivers an accurate product $\hat{Q}\hat{R}$, even if $\hat{Q}$ does not have accurately orthogonal columns, one approach is to set up the normal equations $Rx = (\hat{Q}^*\hat{Q})^{-1}\hat{Q}^*b$ for the vector Rx, then get x by back substitution. As long as the computed $\hat{Q}$ is at least well-conditioned, this method will be free of the instabilities described below for the normal equations applied to arbitrary matrices. However, it involves unnecessary extra work and should not be used in practice.

A better method of stabilizing the Gram–Schmidt method is to make use of an augmented system of equations, just as in the second of our two House-holder experiments above:

```
[Q2,R2] = mgs([A b]);          Gram–Schmidt orthog. of [A b].
R2 = R2(1:n,1:n);              Extract R̂ ...
Qb = R2(1:n,n+1);                 ... and Q̂*b.
x = R2\Qb;                     Solve for x.
x(15)
    ans = 1.00000005653399
```

Now the result looks as good as with Householder triangularization. It can be proved that this is always the case.

Theorem 19.2. *The solution of the full-rank least squares problem* (11.2) *by Gram–Schmidt orthogonalization is also backward stable, satisfying* (19.1), *provided that $\hat{Q}^*b$ is formed implicitly as indicated in the code segment above.*

Normal Equations

A fundamentally different approach to least squares problems is the solution of the normal equations (Algorithm 11.1), typically by Cholesky factorization (Lecture 23). For $m \gg n$, this method is twice as fast as methods depending on explicit orthogonalization, requiring asymptotically only mn^2 flops (11.14). In the following experiment, the problem is solved in a single line of MATLAB by the \ operator:

```
x = (A'*A)\(A'*b);             Form and solve normal equations.
x(15)
    ans = 0.39339069870283
```

This result is terrible! It is the worst we have obtained, with not even a single digit of accuracy. The use of the normal equations is clearly an unstable method for solving least squares problems. We shall take a moment to explain this phenomenon, for the explanation is a perfect example of the interplay of ideas of conditioning and stability. Also, the normal equations are so often used that an understanding of the risks involved is important.

Suppose we have a backward stable algorithm for the full-rank problem (11.2) that delivers a solution $\tilde{x}$ satisfying $\|(A + \delta A)\tilde{x} - b\| = \min$ for some δA with $\|\delta A\|/\|A\| = O(\epsilon_{\text{machine}})$. (Allowing perturbations in b as well as A, or considering stability instead of backward stability, does not change our main points.) By Theorems 15.1 and 18.1, we have

$$\frac{\|\tilde{x} - x\|}{\|x\|} = O\left(\left(\kappa + \frac{\kappa^2 \tan\theta}{\eta}\right)\epsilon_{\text{machine}}\right), \qquad (19.2)$$

where $\kappa = \kappa(A)$. Now suppose A is ill-conditioned, i.e., $\kappa \gg 1$, and θ is bounded away from $\pi/2$. Depending on the values of the various parameters, two very different situations may arise. If $\tan\theta$ is of order 1 (that is, the least squares fit is not especially close) and $\eta \ll \kappa$, the right-hand side (19.2) is $O(\kappa^2\epsilon_{\text{machine}})$. On the other hand, if $\tan\theta$ is close to zero (a very close fit) or η is close to κ, the bound is $O(\kappa\epsilon_{\text{machine}})$. *The condition number of the least squares problem may lie anywhere in the range κ to κ^2.*

Now consider what happens when we solve (11.2) by the normal equations, $(A^*A)x = A^*b$. Cholesky factorization is a stable algorithm for this system of equations in the sense that it produces a solution $\tilde{x}$ satisfying $(A^*A + \delta H)\tilde{x} = A^*b$ for some δH with $\|\delta H\|/\|A^*A\| = O(\epsilon_{\text{machine}})$ (Theorem 23.3). However, the matrix A^*A has condition number κ^2, not κ. Thus the best we can expect from the normal equations is

$$\frac{\|\tilde{x} - x\|}{\|x\|} = O(\kappa^2\epsilon_{\text{machine}}). \tag{19.3}$$

The behavior of the normal equations is governed by κ^2, not κ.

The conclusion is now clear. If $\tan\theta$ is of order 1 and $\eta \ll \kappa$, or if κ is of order 1, then (19.2) and (19.3) are of the same order and the normal equations are stable. If κ is large and either $\tan\theta$ is close to zero or η is close to κ, however, then (19.3) is much bigger than (19.2) and the normal equations are unstable. *The normal equations are typically unstable for ill-conditioned problems involving close fits.* In our example problem, with $\kappa^2 \approx 10^{20}$, it is hardly surprising that Cholesky factorization yielded no correct digits.

According to our definitions, an algorithm is stable only if it has satisfactory behavior uniformly across all the problems under consideration. The following result is thus a natural formalization of the observations just made.

Theorem 19.3. *The solution of the full-rank least squares problem (11.2) via the normal equations (Algorithm 11.1) is unstable. Stability can be achieved, however, by restriction to a class of problems in which $\kappa(A)$ is uniformly bounded above or $(\tan\theta)/\eta$ is uniformly bounded below.*

SVD

One further algorithm for least squares problems was mentioned in Lecture 11: the use of the SVD (Algorithm 11.3). Like most computations based on the SVD, this one is stable:

```
[U,S,V] = svd(A,0);              Reduced SVD of A.
x = V*(S\(U'*b));                Solve for x.
x(15)
    ans = 0.99999998230471
```

In fact, this is the most accurate of all the results obtained in our experiments, beating Householder triangularization with column pivoting (MATLAB's \ ) by a factor of about 3. A theorem in the usual form can be proved.

Theorem 19.4. *The solution of the full-rank least squares problem* (11.2) *by the SVD (Algorithm 11.3) is backward stable, satisfying the estimate* (19.1).

Rank-Deficient Least Squares Problems

In this lecture we have identified four backward stable algorithms for linear least squares problems: Householder triangularization, Householder triangularization with column pivoting, modified Gram–Schmidt with implicit calculation of $\hat{Q}^*b$, and the SVD. From the point of view of classical normwise stability analysis of the full-rank problem (11.2), the differences among these algorithms are minor, so one might as well make use of the simplest and cheapest, Householder triangularization without pivoting.

However, there are other kinds of least squares problems where column pivoting and the SVD take on a special importance. These are problems where A has rank $< n$, possibly with $m < n$, so that the system of equations is *underdetermined*. Such problems do not have a unique solution unless one adds an additional condition, typically that x itself should have as small a norm as possible. A further complication is that the correct solution depends on the rank of A, and determining ranks numerically in the presence of rounding errors is never a trivial matter.

Thus rank-deficient least squares problems are not a challenging subclass of least squares problems, but fundamentally different. Since the definition of a solution is new, there is no reason that an algorithm that is stable for full-rank problems must be stable also in the rank-deficient case. In fact, the only fully stable algorithms for rank-deficient problems are those based on the SVD. An alternative is Householder triangularization with column pivoting, which is stable for almost all problems. We shall not give details.

Exercises

19.1. Given $A \in \mathbb{C}^{m \times n}$ of rank n and $b \in \mathbb{C}^m$, consider the block 2×2 system of equations

$$\begin{bmatrix} I & A \\ A^* & 0 \end{bmatrix} \begin{bmatrix} r \\ x \end{bmatrix} = \begin{bmatrix} b \\ 0 \end{bmatrix}, \tag{19.4}$$

where I is the $m \times m$ identity. Show that this system has a unique solution $(r, x)^T$, and that the vectors r and x are the residual and the solution of the least squares problem (18.1).

19.2. Here is a stripped-down version of one of MATLAB's built-in m-files.

```
[U,S,V] = svd(A);
S = diag(S);
tol = max(size(A))*S(1)*eps;
r = sum(S > tol);
S = diag(ones(r,1)./S(1:r));
X = V(:,1:r)*S*U(:,1:r)';
```

What does this program compute?

Part IV

Systems of Equations

Lecture 20. Gaussian Elimination

Gaussian elimination is undoubtedly familiar to the reader. It is the simplest way to solve linear systems of equations by hand, and also the standard method for solving them on computers. We first describe Gaussian elimination in its pure form, and then, in the next lecture, add the feature of row pivoting that is essential to stability.

LU Factorization

Gaussian elimination transforms a full linear system into an upper-triangular one by applying simple linear transformations on the left. In this respect it is analogous to Householder triangularization for computing QR factorizations. The difference is that the transformations applied in Gaussian elimination are not unitary.

Let $A \in \mathbb{C}^{m \times m}$ be a square matrix. (The algorithm can also be applied to rectangular matrices, but as this is rarely done in practice, we shall confine our attention to the square case.) The idea is to transform A into an $m \times m$ upper-triangular matrix U by introducing zeros below the diagonal, first in column 1, then in column 2, and so on—just as in Householder triangularization. This is done by subtracting multiples of each row from subsequent rows. This "elimination" process is equivalent to multiplying A by a sequence of lower-triangular matrices L_k on the left:

$$\underbrace{L_{m-1} \cdots L_2 L_1}_{L^{-1}} A = U. \tag{20.1}$$

147

Setting $L = L_1^{-1}L_2^{-1} \cdots L_{m-1}^{-1}$ gives $A = LU$. Thus we obtain an *LU factorization* of A,

$$A = LU, \tag{20.2}$$

where U is upper-triangular and L is lower-triangular. It turns out that L is *unit lower-triangular*, which means that all of its diagonal entries are equal to 1.

L 的 对角

项皆为 1

For example, suppose we start with a 4×4 matrix. The algorithm proceeds in three steps (compare (10.1)):

$$\begin{bmatrix} \times & \times & \times & \times \\ \times & \times & \times & \times \\ \times & \times & \times & \times \\ \times & \times & \times & \times \end{bmatrix} \xrightarrow{L_1} \begin{bmatrix} \times & \times & \times & \times \\ \mathbf{0} & \mathbf{\times} & \mathbf{\times} & \mathbf{\times} \\ \mathbf{0} & \mathbf{\times} & \mathbf{\times} & \mathbf{\times} \\ \mathbf{0} & \mathbf{\times} & \mathbf{\times} & \mathbf{\times} \end{bmatrix} \xrightarrow{L_2} \begin{bmatrix} \times & \times & \times & \times \\ & \times & \times & \times \\ & \mathbf{0} & \mathbf{\times} & \mathbf{\times} \\ & \mathbf{0} & \mathbf{\times} & \mathbf{\times} \end{bmatrix} \xrightarrow{L_3} \begin{bmatrix} \times & \times & \times & \times \\ & \times & \times & \times \\ & & \times & \times \\ & & \mathbf{0} & \mathbf{\times} \end{bmatrix}.$$
$$\quad\ A \qquad\qquad\qquad L_1 A \qquad\qquad\qquad L_2 L_1 A \qquad\qquad L_3 L_2 L_1 A$$

(As in Lecture 10, boldfacing indicates entries just operated upon, and blank entries are zero.) The kth transformation L_k introduces zeros below the diagonal in column k by subtracting multiples of row k from rows $k+1, \ldots, m$. Since the first $k-1$ entries of row k are already zero, this operation does not destroy any zeros previously introduced.

Gaussian elimination thus augments our taxonomy of algorithms for factoring a matrix:

Gram–Schmidt: $A = QR$ by triangular orthogonalization,

Householder: $A = QR$ by orthogonal triangularization,

Gaussian elimination: $A = LU$ by triangular triangularization.

Example

In discussing the details, it will help to have a numerical example on the table. Suppose we start with the 4×4 matrix

$$A = \begin{bmatrix} 2 & 1 & 1 & 0 \\ 4 & 3 & 3 & 1 \\ 8 & 7 & 9 & 5 \\ 6 & 7 & 9 & 8 \end{bmatrix}. \tag{20.3}$$

(The entries of A are anything but random; they were chosen to give a simple LU factorization.) The first step of Gaussian elimination looks like this:

$$L_1 A = \begin{bmatrix} 1 & & & \\ -2 & 1 & & \\ -4 & & 1 & \\ -3 & & & 1 \end{bmatrix} \begin{bmatrix} 2 & 1 & 1 & 0 \\ 4 & 3 & 3 & 1 \\ 8 & 7 & 9 & 5 \\ 6 & 7 & 9 & 8 \end{bmatrix} = \begin{bmatrix} 2 & 1 & 1 & 0 \\ & 1 & 1 & 1 \\ & 3 & 5 & 5 \\ & 4 & 6 & 8 \end{bmatrix}.$$

In words, we have subtracted twice the first row from the second, four times the first row from the third, and three times the first row from the fourth. The second step looks like this:

$$L_2 L_1 A = \begin{bmatrix} 1 & & & \\ & 1 & & \\ & -3 & 1 & \\ & -4 & & 1 \end{bmatrix} \begin{bmatrix} 2 & 1 & 1 & 0 \\ & 1 & 1 & 1 \\ & 3 & 5 & 5 \\ & 4 & 6 & 8 \end{bmatrix} = \begin{bmatrix} 2 & 1 & 1 & 0 \\ & 1 & 1 & 1 \\ & & 2 & 2 \\ & & 2 & 4 \end{bmatrix}.$$

This time we have subtracted three times the second row from the third and four times the second row from the fourth. Finally, in the third step we subtract the third row from the fourth:

在A的左侧不断地加上L

$$L_3 L_2 L_1 A = \begin{bmatrix} 1 & & & \\ & 1 & & \\ & & 1 & \\ & & -1 & 1 \end{bmatrix} \begin{bmatrix} 2 & 1 & 1 & 0 \\ & 1 & 1 & 1 \\ & & 2 & 2 \\ & & 2 & 4 \end{bmatrix} = \begin{bmatrix} 2 & 1 & 1 & 0 \\ & 1 & 1 & 1 \\ & & 2 & 2 \\ & & & 2 \end{bmatrix} = U.$$

Now, to exhibit the full factorization $A = LU$, we need to compute the product $L = L_1^{-1} L_2^{-1} L_3^{-1}$. Perhaps surprisingly, this turns out to be a triviality. The inverse of L_1 is just L_1 itself, but with each entry below the diagonal negated:

$$\begin{bmatrix} 1 & & & \\ -2 & 1 & & \\ -4 & & 1 & \\ -3 & & & 1 \end{bmatrix}^{-1} = \begin{bmatrix} 1 & & & \\ 2 & 1 & & \\ 4 & & 1 & \\ 3 & & & 1 \end{bmatrix}. \tag{20.4}$$

Similarly, the inverses of L_2 and L_3 are obtained by negating their subdiagonal entries. Finally, the product $L_1^{-1} L_2^{-1} L_3^{-1}$ is just the unit lower-triangular matrix with the nonzero subdiagonal entries of L_1^{-1}, L_2^{-1}, and L_3^{-1} inserted in the appropriate places. All together, we have

$$\underbrace{\begin{bmatrix} 2 & 1 & 1 & 0 \\ 4 & 3 & 3 & 1 \\ 8 & 7 & 9 & 5 \\ 6 & 7 & 9 & 8 \end{bmatrix}}_{A} = \underbrace{\begin{bmatrix} 1 & & & \\ 2 & 1 & & \\ 4 & 3 & 1 & \\ 3 & 4 & 1 & 1 \end{bmatrix}}_{L} \underbrace{\begin{bmatrix} 2 & 1 & 1 & 0 \\ & 1 & 1 & 1 \\ & & 2 & 2 \\ & & & 2 \end{bmatrix}}_{U}. \tag{20.5}$$

General Formulas and Two Strokes of Luck x_k：第k列

Here are the general formulas for an $m \times m$ matrix. Suppose x_k denotes the kth column of the matrix at the beginning of step k. Then the transformation

L_k must be chosen so that

$$x_k = \begin{bmatrix} x_{1k} \\ \vdots \\ x_{kk} \\ x_{k+1,k} \\ \vdots \\ x_{mk} \end{bmatrix} \xrightarrow{\ L_k\ } L_k x_k = \begin{bmatrix} x_{1k} \\ \vdots \\ x_{kk} \\ 0 \\ \vdots \\ 0 \end{bmatrix}.$$

To do this we wish to subtract ℓ_{jk} times row k from row j, where ℓ_{jk} is the *multiplier*

$$\ell_{jk} = \frac{x_{jk}}{x_{kk}} \qquad (k < j \leq m). \tag{20.6}$$

The matrix L_k takes the form

$$L_k = \begin{bmatrix} 1 \\ & \ddots \\ & & 1 \\ & & -\ell_{k+1,k} & 1 \\ & & \vdots & & \ddots \\ & & -\ell_{mk} & & & 1 \end{bmatrix},$$

with the nonzero subdiagonal entries situated in column k. This is analogous to (10.2) for Householder triangularization.

In the numerical example above, we noted two strokes of luck: that L_k can be inverted by negating its subdiagonal entries (20.4), and that L can be formed by collecting the entries ℓ_{jk} in the appropriate places (20.5). We can explain these bits of good fortune as follows. Let us define

$$\ell_k = \begin{bmatrix} 0 \\ \vdots \\ 0 \\ \ell_{k+1,k} \\ \vdots \\ \ell_{m,k} \end{bmatrix}.$$

Then L_k can be written $L_k = I - \ell_k e_k^*$, where e_k is, as usual, the column vector with 1 in position k and 0 elsewhere. The sparsity pattern of ℓ_k implies $e_k^* \ell_k = 0$, and therefore $(I - \ell_k e_k^*)(I + \ell_k e_k^*) = I - \ell_k e_k^* \ell_k e_k^* = I$. In other words, the inverse of L_k is $I + \ell_k e_k^*$, as in (20.4).

For the second stroke of luck we argue as follows. Consider, for example, the product $L_k^{-1} L_{k+1}^{-1}$. From the sparsity pattern of ℓ_{k+1}, we have $e_k^* \ell_{k+1} = 0$, and therefore

$$L_k^{-1} L_{k+1}^{-1} = (I + \ell_k e_k^*)(I + \ell_{k+1} e_{k+1}^*) = I + \ell_k e_k^* + \ell_{k+1} e_{k+1}^*.$$

Thus $L_k^{-1} L_{k+1}^{-1}$ is just the unit lower-triangular matrix with the entries of both L_k^{-1} and L_{k+1}^{-1} inserted in their usual places below the diagonal. When we take the product of all of these matrices to form L, we have the same convenient property everywhere below the diagonal:

$$L \; = \; L_1^{-1} L_2^{-1} \cdots L_{m-1}^{-1} \; = \; \begin{bmatrix} 1 & & & & \\ \ell_{21} & 1 & & & \\ \ell_{31} & \ell_{32} & 1 & & \\ \vdots & \vdots & \ddots & \ddots & \\ \ell_{m1} & \ell_{m2} & \cdots & \ell_{m,m-1} & 1 \end{bmatrix}. \qquad (20.7)$$

Though we did not mention it in Lecture 8, the sparsity considerations that led to (20.7) also appeared in the interpretation (8.10) of the modified Gram–Schmidt process as a succession of right-multiplications by triangular matrices R_k.

In practical Gaussian elimination, the matrices L_k are never formed and multiplied explicitly. The multipliers ℓ_{jk} are computed and stored directly into L, and the transformations L_k are then applied implicitly.

Algorithm 20.1. Gaussian Elimination without Pivoting

$U = A, \; L = I$
for $k = 1$ **to** $m - 1$
      for $j = k + 1$ **to** m
            $\ell_{jk} = u_{jk}/u_{kk}$
            $u_{j,k:m} = u_{j,k:m} - \ell_{jk} u_{k,k:m}$

(Three matrices A, L, U are not really needed; to minimize memory use on the computer, both L and U can be written into the same array as A.) See Exercise 20.4 for an alternative "outer product" formulation of Gaussian elimination, involving one explicit loop rather than two.

Operation Count

As usual, the asymptotic operation count of this algorithm can be derived geometrically. The work is dominated by the vector operation in the inner loop, $u_{j,k:m} = u_{j,k:m} - \ell_{jk} u_{k,k:m}$, which executes one scalar-vector multiplication and one vector subtraction. If $l = m - k + 1$ denotes the length of the row vectors being manipulated, the number of flops is $2l$: two flops per entry.

For each value of k, the inner loop is repeated for rows $k + 1, \ldots, m$. The work involved corresponds to one layer of the following solid:

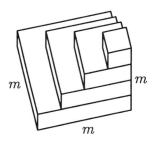

This is the same figure we displayed in Lecture 10 to represent the work done in Householder triangularization (assuming $m = n$). There, however, each unit cube represented four flops rather than two. As before, the solid converges as $m \to \infty$ to a pyramid, with volume $\frac{1}{3}m^3$. At two flops per unit of volume, this adds up to

$$\text{Work for Gaussian elimination:} \quad \sim \frac{2}{3}m^3 \text{ flops.} \qquad (20.8)$$

$LUx = b,$ $Ax = b$

Solution of $Ax = b$ by LU Factorization

If A is factored into L and U, a system of equations $Ax = b$ is reduced to the form $LUx = b$. Thus it can be solved by solving two triangular systems: first $Ly = b$ for the unknown y (forward substitution), then $Rx = y$ for the unknown x (back substitution). The first step requires $\sim \frac{2}{3}m^3$ flops, and the second and third each require $\sim m^2$ flops. The total work is $\sim \frac{2}{3}m^3$ flops, half the figure of $\sim \frac{4}{3}m^3$ flops (10.9) for a solution by Householder triangularization (Algorithm 16.1).

Why is Gaussian elimination usually used rather than QR factorization to solve square systems of equations? The factor of 2 is certainly one reason. Also important, however, may be the historical fact that the elimination idea has been known for centuries, whereas QR factorization of matrices did not come along until after the invention of computers. To supplant Gaussian elimination as the method of choice, QR factorization would have to have had a compelling advantage.

Instability of Gaussian Elimination without Pivoting

Unfortunately, Gaussian elimination as presented so far is unusable for solving general linear systems, for it is not backward stable. The instability is related to another, more obvious difficulty. For certain matrices, Gaussian elimination fails entirely, because it attempts division by zero.

For example, consider

$$A = \begin{bmatrix} 0 & 1 \\ 1 & 1 \end{bmatrix}.$$

This matrix has full rank and is well-conditioned, with $\kappa(A) = (3 + \sqrt{5})/2 \approx$ 2.618 in the 2-norm. Nevertheless, Gaussian elimination fails at the first step.

A slight perturbation of the same matrix reveals the more general problem. Suppose we apply Gaussian elimination to

$$A = \begin{bmatrix} 10^{-20} & 1 \\ 1 & 1 \end{bmatrix}. \tag{20.9}$$

Now the process does not fail. Instead, 10^{20} times the first row is subtracted from the second row, and the following factors are produced:

$$L = \begin{bmatrix} 1 & 0 \\ 10^{20} & 1 \end{bmatrix}, \qquad U = \begin{bmatrix} 10^{-20} & 1 \\ 0 & 1 - 10^{20} \end{bmatrix}.$$

However, suppose these computations are performed in floating point arithmetic with $\epsilon_{\text{machine}} \approx 10^{-16}$. The number $1 - 10^{20}$ will not be represented exactly; it will be rounded to the nearest floating point number. For simplicity, imagine that this is exactly -10^{20}. Then the floating point matrices produced by the algorithm will be

$$\tilde{L} = \begin{bmatrix} 1 & 0 \\ 10^{20} & 1 \end{bmatrix}, \qquad \tilde{U} = \begin{bmatrix} 10^{-20} & 1 \\ 0 & -10^{20} \end{bmatrix}.$$

This degree of rounding might seem tolerable at first. After all, the matrix $\tilde{U}$ is close to the correct U relative to $\|U\|$. However, the problem becomes apparent when we compute the product $\tilde{L}\tilde{U}$:

$$\tilde{L}\tilde{U} = \begin{bmatrix} 10^{-20} & 1 \\ 1 & 0 \end{bmatrix}.$$

This matrix is not at all close to A, for the 1 in the $(2,2)$ position has been replaced by 0. If we now solve the system $\tilde{L}\tilde{U}x = b$, the result will be nothing like the solution to $Ax = b$. For example, with $b = (1,0)^*$ we get $\tilde{x} = (0,1)^*$, whereas the correct solution is $x \approx (-1,1)^*$. LU 算法不是可回推的（若无枢轴化）

A careful consideration of what has occurred in this example reveals the following. Gaussian elimination has computed the LU factorization stably: $\tilde{L}$ and $\tilde{U}$ are close to the exact factors for a matrix close to A (in fact, A itself). Yet it has not solved $Ax = b$ stably. The explanation is that the LU factorization, though stable, was *not backward stable*. As a rule, if one step of an algorithm is a stable but not backward stable algorithm for solving a subproblem, the stability of the overall calculation may be in jeopardy.

In fact, for general $m \times m$ matrices A, the situation is worse than this. Gaussian elimination without pivoting is neither backward stable nor stable as a general algorithm for LU factorization. Additionally, the triangular matrices it generates have condition numbers that may be arbitrarily greater than those of A itself, leading to additional sources of instability in the forward and back substitution phases of the solution of $Ax = b$.

Exercises

20.1. Let $A \in \mathbb{C}^{m \times m}$ be nonsingular. Show that A has an LU factorization if and only if for each k with $1 \leq k \leq m$, the upper-left $k \times k$ block $A_{1:k,1:k}$ is nonsingular. (Hint: The row operations of Gaussian elimination leave the determinants $\det(A_{1:k,1:k})$ unchanged.) Prove that this LU factorization is unique.

20.2. Suppose $A \in \mathbb{C}^{m \times m}$ satisfies the condition of Exercise 20.1 and is banded with bandwidth $2p+1$, i.e., $a_{ij} = 0$ for $|i - j| > p$. What can you say about the sparsity patterns of the factors L and U of A?

20.3. Suppose an $m \times m$ matrix A is written in the block form $A = \begin{bmatrix} A_{11} & A_{12} \\ A_{21} & A_{22} \end{bmatrix}$, where A_{11} is $n \times n$ and A_{22} is $(m-n) \times (m-n)$.
Assume that A satisfies the condition of Exercise 20.1.
(a) Verify the formula

$$\begin{bmatrix} I & \\ -A_{21}A_{11}^{-1} & I \end{bmatrix} \begin{bmatrix} A_{11} & A_{12} \\ A_{21} & A_{22} \end{bmatrix} = \begin{bmatrix} A_{11} & A_{12} \\ & A_{22} - A_{21}A_{11}^{-1}A_{12} \end{bmatrix}$$

for "elimination" of the block A_{21}. The matrix $A_{22} - A_{21}A_{11}^{-1}A_{12}$ is known as the *Schur complement* of A_{11} in A.
(b) Suppose A_{21} is eliminated row by row by means of n steps of Gaussian elimination. Show that the bottom-right $(m-n) \times (m-n)$ block of the result is again $A_{22} - A_{21}A_{11}^{-1}A_{12}$.

20.4. Like most of the algorithms in this book, Gaussian elimination involves a triply nested loop. In Algorithm 20.1, there are two explicit **for** loops, and the third loop is implicit in the vectors $u_{j,k:m}$ and $u_{k,k:m}$. Rewrite this algorithm with just one explicit **for** loop indexed by k. Inside this loop, U will be updated at each step by a certain rank-one outer product. This "outer product" form of Gaussian elimination may be a better starting point than Algorithm 20.1 if one wants to optimize computer performance.

20.5. We have seen that Gaussian elimination yields a factorization $A = LU$, where L has ones on the diagonal but U does not. Describe at a high level the factorization that results if this process is varied in the following ways:
(a) Elimination by columns from left to right, rather than by rows from top to bottom, so that A is made lower-triangular.
(b) Gaussian elimination applied after a preliminary scaling of the columns of A by a diagonal matrix D. What form does a system $Ax = b$ take under this rescaling? Is it the equations or the unknowns that are rescaled by D?
(c) Gaussian elimination carried further, so that after A (assumed nonsingular) is brought to upper-triangular form, additional column operations are carried out so that this upper-triangular matrix is made diagonal.

Lecture 21. Pivoting

In the last lecture we saw that Gaussian elimination in its pure form is unstable. The instability can be controlled by permuting the order of the rows of the matrix being operated on, an operation called *pivoting*. Pivoting has been a standard feature of Gaussian elimination computations since the 1950s.

Pivots

At step k of Gaussian elimination, multiples of row k are subtracted from rows $k+1, \ldots, m$ of the working matrix X in order to introduce zeros in entry k of these rows. In this operation row k, column k, and especially the entry x_{kk} play special roles. We call x_{kk} the *pivot*. From every entry in the submatrix $X_{k+1:m,k:m}$ is subtracted the product of a number in row k and a number in column k, divided by x_{kk}:

$$
\begin{bmatrix}
\times & \times & \times & \times & \times \\
 & x_{kk} & \mathbf{\times} & \mathbf{\times} & \mathbf{\times} \\
 & \times & \times & \times & \times \\
 & \times & \times & \times & \times \\
 & \times & \times & \times & \times
\end{bmatrix}
\longrightarrow
\begin{bmatrix}
\times & \times & \times & \times & \times \\
 & x_{kk} & \times & \times & \times \\
 & 0 & \mathbf{\times} & \mathbf{\times} & \mathbf{\times} \\
 & 0 & \mathbf{\times} & \mathbf{\times} & \mathbf{\times} \\
 & 0 & \mathbf{\times} & \mathbf{\times} & \mathbf{\times}
\end{bmatrix}.
$$

However, there is no reason why the kth row and column must be chosen for the elimination. For example, we could just as easily introduce zeros in column k by adding multiples of some row i with $k < i \leq m$ to the other rows

155

$k, \ldots, m$. In this case, the entry x_{ik} would be the pivot. Here is an illustration with $k = 2$ and $i = 4$:

$$\begin{bmatrix} \times & \times & \times & \times & \times \\ & \times & \times & \times & \times \\ & \times & \times & \times & \times \\ & \boldsymbol{x_{ik}} & \boldsymbol{\times} & \boldsymbol{\times} & \boldsymbol{\times} \\ & \times & \times & \times & \times \end{bmatrix} \longrightarrow \begin{bmatrix} \times & \times & \times & \times & \times \\ & \boldsymbol{0} & \boldsymbol{\times} & \boldsymbol{\times} & \boldsymbol{\times} \\ & \boldsymbol{0} & \boldsymbol{\times} & \boldsymbol{\times} & \boldsymbol{\times} \\ & x_{ik} & \times & \times & \times \\ & \boldsymbol{0} & \boldsymbol{\times} & \boldsymbol{\times} & \boldsymbol{\times} \end{bmatrix}.$$

Similarly, we could introduce zeros in column j rather than column k. Here is an illustration with $k = 2$, $i = 4$, $j = 3$:

$$\begin{bmatrix} \times & \times & \times & \times & \times \\ & \times & \times & \times & \times \\ & \times & \times & \times & \times \\ & \boldsymbol{\times} & \boldsymbol{x_{ij}} & \boldsymbol{\times} & \boldsymbol{\times} \\ & \times & \times & \times & \times \end{bmatrix} \longrightarrow \begin{bmatrix} \times & \times & \times & \times & \times \\ & \boldsymbol{\times} & \boldsymbol{0} & \boldsymbol{\times} & \boldsymbol{\times} \\ & \boldsymbol{\times} & \boldsymbol{0} & \boldsymbol{\times} & \boldsymbol{\times} \\ & \times & x_{ij} & \times & \times \\ & \boldsymbol{\times} & \boldsymbol{0} & \boldsymbol{\times} & \boldsymbol{\times} \end{bmatrix}.$$

All in all, we are free to choose any entry of $X_{k:m,k:m}$ as the pivot, as long as it is nonzero. The possibility that an entry $x_{kk} = 0$ might arise implies that some flexibility of choice of the pivot may sometimes be necessary, even from a pure mathematical point of view. For numerical stability, however, it is desirable to pivot even when x_{kk} is nonzero if there is a larger element available. In practice, it is common to pick as pivot the largest number among a set of entries being considered as candidates.

The structure of the elimination process quickly becomes confusing if zeros are introduced in arbitrary patterns through the matrix. To see what is going on, we want to retain the triangular structure described in the last lecture, and there is an easy way to do this. We shall not think of the pivot x_{ij} as left in place, as in the illustrations above. Instead, at step k, we shall imagine that the rows and columns of the working matrix are permuted so as to move x_{ij} into the (k, k) position. Then, when the elimination is done, zeros are introduced into entries $k+1, \ldots, m$ of column k, just as in Gaussian elimination without pivoting. This interchange of rows and perhaps columns is what is usually thought of as *pivoting*.

The idea that rows and columns are interchanged is indispensable conceptually. Whether it is a good idea to interchange them physically on the computer is less clear. In some implementations, the data in computer memory are indeed swapped at each pivot step. In others, an equivalent effect is achieved by indirect addressing with permuted index vectors. Which approach is best varies from machine to machine and depends on many factors.

Partial Pivoting

If every entry of $X_{k:m,k:m}$ is considered as a possible pivot at step k, there are $O((m - k)^2)$ entries to be examined to determine the largest. Summing over

m steps, the total cost of selecting pivots becomes $O(m^3)$ operations, adding significantly to the cost of Gaussian elimination, not to mention the potential difficulties of global communication in an unpredictable pattern across all the entries of a matrix. This expensive strategy is called *complete pivoting*.

In practice, equally good pivots can be found by considering a much smaller number of entries. The standard method for doing this is *partial pivoting*. Here, only rows are interchanged. The pivot at each step is chosen as the largest of the $m - k + 1$ subdiagonal entries in column k, incurring a total cost of only $O(m - k)$ operations for selecting the pivot at each step, hence $O(m^2)$ operations overall. To bring the kth pivot into the (k, k) position, no columns need to be permuted; it is enough to swap row k with the row containing the pivot.

$$
\begin{bmatrix}
\times & \times & \times & \times & \times \\
 & \times & \times & \times & \times \\
 & \times & \times & \times & \times \\
 & \boldsymbol{x_{ik}} & \boldsymbol{\times} & \boldsymbol{\times} & \boldsymbol{\times} \\
 & \times & \times & \times & \times
\end{bmatrix}
\xrightarrow{\ P_1\ }
\begin{bmatrix}
\times & \times & \times & \times & \times \\
 & \boldsymbol{x_{ik}} & \boldsymbol{\times} & \boldsymbol{\times} & \boldsymbol{\times} \\
 & \times & \times & \times & \times \\
 & \boldsymbol{\times} & \boldsymbol{\times} & \boldsymbol{\times} & \boldsymbol{\times} \\
 & \times & \times & \times & \times
\end{bmatrix}
\xrightarrow{\ L_1\ }
\begin{bmatrix}
\times & \times & \times & \times & \times \\
 & \boldsymbol{x_{ik}} & \times & \times & \times \\
 & 0 & \boldsymbol{\times} & \boldsymbol{\times} & \boldsymbol{\times} \\
 & 0 & \boldsymbol{\times} & \boldsymbol{\times} & \boldsymbol{\times} \\
 & 0 & \boldsymbol{\times} & \boldsymbol{\times} & \boldsymbol{\times}
\end{bmatrix}.
$$

$$\text{Pivot selection} \qquad\qquad \text{Row interchange} \qquad\qquad \text{Elimination}$$

As usual in numerical linear algebra, this algorithm can be expressed as a matrix product. We saw in the last lecture that an elimination step corresponds to left-multiplication by an elementary lower-triangular matrix L_k. Partial pivoting complicates matters by applying a permutation matrix P_k on the left of the working matrix before each elimination. (A permutation matrix is a matrix with 0 everywhere except for a single 1 in each row and column. That is, it is a matrix obtained from the identity by permuting rows or columns.) After $m - 1$ steps, A becomes an upper-triangular matrix U:

$$L_{m-1}P_{m-1} \cdots L_2 P_2 L_1 P_1 A = U. \tag{21.1}$$

Example

To see what is going on, it will be helpful to return to the numerical example (20.3) of the last lecture,

$$A = \begin{bmatrix} 2 & 1 & 1 & 0 \\ 4 & 3 & 3 & 1 \\ 8 & 7 & 9 & 5 \\ 6 & 7 & 9 & 8 \end{bmatrix}. \tag{21.2}$$

With partial pivoting, the first thing we do is interchange the first and third rows (left-multiplication by P_1):

$$\begin{bmatrix} & 1 & & \\ & & 1 & \\ 1 & & & \\ & & & 1 \end{bmatrix} \begin{bmatrix} 2 & 1 & 1 & 0 \\ 4 & 3 & 3 & 1 \\ 8 & 7 & 9 & 5 \\ 6 & 7 & 9 & 8 \end{bmatrix} = \begin{bmatrix} 8 & 7 & 9 & 5 \\ 4 & 3 & 3 & 1 \\ 2 & 1 & 1 & 0 \\ 6 & 7 & 9 & 8 \end{bmatrix}.$$

The first elimination step now looks like this (left-multiplication by L_1):

$$\begin{bmatrix} 1 & & & \\ -\frac{1}{2} & 1 & & \\ -\frac{1}{4} & & 1 & \\ -\frac{3}{4} & & & 1 \end{bmatrix} \begin{bmatrix} 8 & 7 & 9 & 5 \\ 4 & 3 & 3 & 1 \\ 2 & 1 & 1 & 0 \\ 6 & 7 & 9 & 8 \end{bmatrix} = \begin{bmatrix} 8 & 7 & 9 & 5 \\ & -\frac{1}{2} & -\frac{3}{2} & -\frac{3}{2} \\ & -\frac{3}{4} & -\frac{5}{4} & -\frac{5}{4} \\ & \frac{7}{4} & \frac{9}{4} & \frac{17}{4} \end{bmatrix}.$$

Now the second and fourth rows are interchanged (multiplication by P_2):

$$\begin{bmatrix} 1 & & & \\ & & & 1 \\ & & 1 & \\ & 1 & & \end{bmatrix} \begin{bmatrix} 8 & 7 & 9 & 5 \\ & -\frac{1}{2} & -\frac{3}{2} & -\frac{3}{2} \\ & -\frac{3}{4} & -\frac{5}{4} & -\frac{5}{4} \\ & \frac{7}{4} & \frac{9}{4} & \frac{17}{4} \end{bmatrix} = \begin{bmatrix} 8 & 7 & 9 & 5 \\ & \frac{7}{4} & \frac{9}{4} & \frac{17}{4} \\ & -\frac{3}{4} & -\frac{5}{4} & -\frac{5}{4} \\ & -\frac{1}{2} & -\frac{3}{2} & -\frac{3}{2} \end{bmatrix}.$$

The second elimination step then looks like this (multiplication by L_1):

$$\begin{bmatrix} 1 & & & \\ & 1 & & \\ & \frac{3}{7} & 1 & \\ & \frac{2}{7} & & 1 \end{bmatrix} \begin{bmatrix} 8 & 7 & 9 & 5 \\ & \frac{7}{4} & \frac{9}{4} & \frac{17}{4} \\ & -\frac{3}{4} & -\frac{5}{4} & -\frac{5}{4} \\ & -\frac{1}{2} & -\frac{3}{2} & -\frac{3}{2} \end{bmatrix} = \begin{bmatrix} 8 & 7 & 9 & 5 \\ & \frac{7}{4} & \frac{9}{4} & \frac{17}{4} \\ & & -\frac{2}{7} & \frac{4}{7} \\ & & -\frac{6}{7} & -\frac{2}{7} \end{bmatrix}.$$

Now the third and fourth rows are interchanged (multiplication by P_3):

$$\begin{bmatrix} 1 & & & \\ & 1 & & \\ & & & 1 \\ & & 1 & \end{bmatrix} \begin{bmatrix} 8 & 7 & 9 & 5 \\ & \frac{7}{4} & \frac{9}{4} & \frac{17}{4} \\ & & -\frac{2}{7} & \frac{4}{7} \\ & & -\frac{6}{7} & -\frac{2}{7} \end{bmatrix} = \begin{bmatrix} 8 & 7 & 9 & 5 \\ & \frac{7}{4} & \frac{9}{4} & \frac{17}{4} \\ & & -\frac{6}{7} & -\frac{2}{7} \\ & & -\frac{2}{7} & \frac{4}{7} \end{bmatrix}.$$

The final elimination step looks like this (multiplication by L_3):

$$\begin{bmatrix} 1 & & & \\ & 1 & & \\ & & 1 & \\ & & -\frac{1}{3} & 1 \end{bmatrix} \begin{bmatrix} 8 & 7 & 9 & 5 \\ & \frac{7}{4} & \frac{9}{4} & \frac{17}{4} \\ & & -\frac{6}{7} & -\frac{2}{7} \\ & & -\frac{2}{7} & \frac{4}{7} \end{bmatrix} = \begin{bmatrix} 8 & 7 & 9 & 5 \\ & \frac{7}{4} & \frac{9}{4} & \frac{17}{4} \\ & & -\frac{6}{7} & -\frac{2}{7} \\ & & & \frac{2}{3} \end{bmatrix}.$$

$PA = LU$ Factorization and a Third Stroke of Luck

Have we just computed an LU factorization of A? Not quite, but almost. In fact, we have computed an LU factorization of PA, where P is a permutation matrix. It looks like this: 束 = 1×8 = 8   P 的意义就是让A 换行

$$
\begin{bmatrix} & & & 1 \\ & & 1 & \\ & 1 & & \\ 1 & & & \end{bmatrix}
\begin{bmatrix} 2 & 1 & 1 & 0 \\ 4 & 3 & 3 & 1 \\ 8 & 7 & 9 & 5 \\ 6 & 7 & 9 & 8 \end{bmatrix}
=
\begin{bmatrix} 1 & & & \\ \frac{3}{4} & 1 & & \\ \frac{1}{2} & -\frac{2}{7} & 1 & \\ \frac{1}{4} & -\frac{3}{7} & \frac{1}{3} & 1 \end{bmatrix}
\begin{bmatrix} 8 & 7 & 9 & 5 \\ & \frac{7}{4} & \frac{9}{4} & \frac{17}{4} \\ & & -\frac{6}{7} & -\frac{2}{7} \\ & & & \frac{2}{3} \end{bmatrix}.
$$
$$
\quad P \qquad\qquad A \qquad\qquad\qquad L \qquad\qquad\qquad U
$$

$$(21.3)$$

This formula should be compared with (20.5). The presence of integers there and fractions here is not a general distinction, but an artifact of our choice of A. The distinction that matters is that here, all the subdiagonal entries of L are ≤ 1 in magnitude, a consequence of the property $|x_{kk}| = \max_j |x_{jk}|$ in (20.6) introduced by pivoting.

It is not obvious where (21.3) comes from. Our elimination process took the form

$$L_3 P_3 L_2 P_2 L_1 P_1 A = U,$$

which doesn't look lower-triangular at all. But here, a third stroke of good fortune has come to our aid. These six elementary operations can be reordered in the form

$$L_3 P_3 L_2 P_2 L_1 P_1 = L_3' L_2' L_1' P_3 P_2 P_1, \qquad (21.4)$$

where L_k' is equal to L_k but with the subdiagonal entries permuted. To be precise, define

$$L_3' = L_3, \quad L_2' = P_3 L_2 P_3^{-1}, \quad L_1' = P_3 P_2 L_1 P_2^{-1} P_3^{-1}.$$

Since each of these definitions applies only permutations P_j with $j > k$ to L_k, it is easily verified that L_k' has the same structure as L_k. Computing the product of the matrices L_k' reveals

$$L_3' L_2' L_1' P_3 P_2 P_1 = L_3 (P_3 L_2 P_3^{-1})(P_3 P_2 L_1 P_2^{-1} P_3^{-1}) P_3 P_2 P_1 = L_3 P_3 L_1 P_2 L_2 P_1,$$

as in (21.4).

In general, for an $m \times m$ matrix, the factorization (21.1) provided by Gaussian elimination with partial pivoting can be written in the form

$$(L_{m-1}' \cdots L_2' L_1')(P_{m-1} \cdots P_2 P_1) A = U, \qquad (21.5)$$

where L_k' is defined by

$$L_k' = P_{m-1} \cdots P_{k+1} L_k P_{k+1}^{-1} \cdots P_{m-1}^{-1}. \qquad (21.6)$$

The product of the matrices L_k' is unit lower-triangular and easily invertible by negating the subdiagonal entries, just as in Gaussian elimination without pivoting. Writing $L = (L_{m-1}' \cdots L_2' L_1')^{-1}$ and $P = P_{m-1} \cdots P_2 P_1$, we have

$$PA = LU. \tag{21.7}$$

In general, any square matrix A, singular or nonsingular, has a factorization (21.7), where P is a permutation matrix, L is unit lower-triangular with lower-triangular entries ≤ 1 in magnitude, and U is upper-triangular. Partial pivoting is such a universal practice that this factorization is usually known simply as an *LU factorization* of A.

The famous formula (21.7) has a simple interpretation. Gaussian elimination with partial pivoting is equivalent to the following procedure:

1. Permute the rows of A according to P

2. Apply Gaussian elimination without pivoting to PA.

Partial pivoting is not carried out this way in practice, of course, since P is not known ahead of time.

Here is a formal statement of the algorithm.

Algorithm 21.1. Gaussian Elimination with Partial Pivoting

$U = A$, $L = I$, $P = I$
for $k = 1$ **to** $m - 1$
     Select $i \geq k$ to maximize $|u_{ik}|$
     $u_{k,k:m} \leftrightarrow u_{i,k:m}$ (interchange two rows)
     $\ell_{k,1:k-1} \leftrightarrow \ell_{i,1:k-1}$
     $p_{k,:} \leftrightarrow p_{i,:}$
     for $j = k + 1$ **to** m
         $\ell_{jk} = u_{jk}/u_{kk}$
         $u_{j,k:m} = u_{j,k:m} - \ell_{jk} u_{k,k:m}$

To leading order, this algorithm requires the same number of floating point operations (20.8) as Gaussian elimination without pivoting, namely, $\frac{2}{3}m^3$. As with Algorithm 20.1, the use of computer memory can be minimized if desired by overwriting U and L into the same array used to store A.

In practice, of course, P is not represented explicitly as a matrix. The rows are swapped at each step, or an equivalent effect is achieved via a permutation vector, as indicated earlier.

Complete Pivoting

In complete pivoting, the selection of pivots takes a significant amount of time. In practice this is rarely done, because the improvement in stability is marginal. However, we shall outline how the algebra changes in this case.

In matrix form, complete pivoting precedes each elimination step with a permutation P_k of the rows applied on the left and also a permutation Q_k of the columns applied on the right:

$$L_{m-1}P_{m-1}\cdots L_2 P_2 L_1 P_1 A Q_1 Q_2 \cdots Q_{m-1} = U. \tag{21.8}$$

Once again, this is not quite an LU factorization of A, but it is close. If the L'_k are defined as in (21.6) (the column permutations are not involved), then

$$(L'_{m-1}\cdots L'_2 L'_1)(P_{m-1}\cdots P_2 P_1)A(Q_1 Q_2 \cdots Q_{m-1}) = U. \tag{21.9}$$

Setting $L = (L'_{m-1}\cdots L'_2 L'_1)^{-1}$, $P = P_{m-1}\cdots P_2 P_1$, and $Q = Q_1 Q_2 \cdots Q_{m-1}$, we obtain

$$PAQ = LU. \tag{21.10}$$

Exercises

21.1. Let A be the 4×4 matrix (20.3) considered in this lecture and the previous one.

(a) Determine $\det A$ from (20.5).

(b) Determine $\det A$ from (21.3).

(c) Describe how Gaussian elimination with partial pivoting can be used to find the determinant of a general square matrix.

21.2. Suppose $A \in \mathbb{C}^{m \times m}$ is banded with bandwidth $2p+1$, as in Exercise 20.2, and a factorization $PA = LU$ is computed by Gaussian elimination with partial pivoting. What can you say about the sparsity patterns of L and U?

21.3. Consider Gaussian elimination carried out with pivoting by columns instead of rows, leading to a factorization $AQ = LU$, where Q is a permutation matrix.

(a) Show that if A is nonsingular, such a factorization always exists.

(b) Show that if A is singular, such a factorization does not always exist.

21.4. Gaussian elimination can be used to compute the inverse A^{-1} of a nonsingular matrix $A \in \mathbb{C}^{m \times m}$, though it is rarely really necessary to do so.

(a) Describe an algorithm for computing A^{-1} by solving m systems of equations, and show that its asymptotic operation count is $8m^3/3$ flops.

(b) Describe a variant of your algorithm, taking advantage of sparsity, that reduces the operation count to $2m^3$ flops.

(c) Suppose one wishes to solve n systems of equations $Ax_j = b_j$, or equivalently, a block system $AX = B$ with $B \in \mathbb{C}^{m \times n}$. What is the asymptotic operation count (a function of m and n) for doing this (i) directly from the LU factorization and (ii) with a preliminary computation of A^{-1}?

21.5. Suppose $A \in \mathbb{C}^{m \times m}$ is hermitian, or in the real case, symmetric (but not necessarily positive definite).

(a) Describe a strategy of *symmetric pivoting* to preserve the hermitian structure while still leading to a unit lower-triangular matrix with entries $|\ell_{ij}| \leq 1$.

(b) What is the form of the matrix factorization computed by your algorithm?

(c) What is its asymptotic operation count?

21.6. Suppose $A \in \mathbb{C}^{m \times m}$ is *strictly column diagonally dominant*, which means that for each k,

$$|a_{kk}| > \sum_{j \neq k} |a_{jk}|. \qquad (21.11)$$

Show that if Gaussian elimination with partial pivoting is applied to A, no row interchanges take place.

21.7. In Lecture 20 the "two strokes of luck" were explained by the use of the vectors e_k and ℓ_k. Give an explanation based on these vectors for the "third stroke of luck" in the present lecture.

Lecture 22. Stability of Gaussian Elimination

Gaussian elimination with partial pivoting is explosively unstable for certain matrices, yet stable in practice. This apparent paradox has a statistical explanation.

Stability and the Size of L and U

The stability analysis of most algorithms of numerical linear algebra, including virtually all of those based on unitary operations, is straightforward. The stability analysis of Gaussian elimination with partial pivoting, however, is complicated, and has been a point of difficulty in numerical analysis since the 1950s. This is one of the reasons why we saved Gaussian elimination until the second half of this book.

In (20.9) we gave an example of a 2×2 matrix for which Gaussian elimination without pivoting was unstable. In that example, the factor L had an entry of size 10^{20}. An attempt to solve a system of equations based on L introduced rounding errors of relative order $\epsilon_{\text{machine}}$, hence absolute order $\epsilon_{\text{machine}} \times 10^{20}$. Not surprisingly, this destroyed the accuracy of the result.

It turns out that this example is, in a sense, entirely general. Instability in Gaussian elimination—with or without pivoting—can arise only if one or both of the factors L and U is large relative to the size of A. Thus the purpose of pivoting, from the point of view of stability, is to ensure that L and U are not too large. As long as all the intermediate quantities that arise during the

163

elimination are of manageable size, the rounding errors they emit are very small, and the algorithm is backward stable.

The following theorem makes this idea precise. It is stated for Gaussian elimination without pivoting, but it applies to elimination with pivoting too if A is taken to represent the original matrix with appropriately permuted rows and/or columns.

Theorem 22.1. *Let the factorization $A = LU$ of a nonsingular matrix $A \in \mathbb{C}^{m \times m}$ be computed by Gaussian elimination without pivoting (Algorithm 20.1) on a computer satisfying the axioms (13.5) and (13.7). If A has an LU factorization, then for all sufficiently small $\epsilon_{\mathrm{machine}}$, the factorization completes successfully in floating point arithmetic (no zero pivots are encountered), and the computed matrices $\tilde{L}$ and $\tilde{U}$ satisfy*

$$\tilde{L}\tilde{U} = A + \delta A, \qquad \frac{\|\delta A\|}{\|L\|\|U\|} = O(\epsilon_{\mathrm{machine}}) \qquad (22.1)$$

for some $\delta A \in \mathbb{C}^{m \times n}$.

As usual in numerical linear algebra, we make no claims about $\tilde{L} - L$ or $\tilde{U} - U$, only about $\tilde{L}\tilde{U} - LU$.

At first glance this estimate may look like half a dozen others in this book, such as (16.3) or (17.3). What makes it different is that the quantity in the denominator is $\|L\|\|U\|$, not $\|A\|$. If $\|L\|\|U\| = O(\|A\|)$, then (22.1) asserts that Gaussian elimination is backward stable. If $\|L\|\|U\| \neq O(\|A\|)$, however, we must expect backward instability.

For Gaussian elimination without pivoting, both L and U can be unboundedly large. That algorithm is unstable by any standard, and we shall not discuss it further. Instead, from now on, we shall confine our attention to Gaussian elimination with partial pivoting.

Growth Factors

Consider Gaussian elimination with partial pivoting. Because each pivot selection involves maximization over a column, this algorithm produces a matrix L with entries of absolute value ≤ 1 everywhere below the diagonal. This implies $\|L\| = O(1)$ in any norm. Therefore, for Gaussian elimination with partial pivoting, (22.1) reduces to the condition $\|\delta A\|/\|U\| = O(\epsilon_{\mathrm{machine}})$. We conclude that the algorithm is backward stable provided $\|U\| = O(\|A\|)$.

There is a standard reformulation of this conclusion that is perhaps more vivid. Gaussian elimination reduces a full matrix A to an upper-triangular matrix U. We have just seen that the key question for stability is whether amplification of the entries takes place during this reduction. In particular, let the *growth factor* for A be defined as the ratio

$$\rho = \frac{\max_{i,j} |u_{ij}|}{\max_{i,j} |a_{ij}|}. \qquad (22.2)$$

If ρ is of order 1, not much growth has taken place, and the elimination process is stable. If ρ is bigger than this, we must expect instability. Specifically, since $\|L\| = O(1)$, and since (22.2) implies $\|U\| = O(\rho\|A\|)$, the following result is a corollary of Theorem 22.1.

Theorem 22.2. *Let the factorization $PA = LU$ of a matrix $A \in \mathbb{C}^{m \times m}$ be computed by Gaussian elimination with partial pivoting (Algorithm 21.1) on a computer satisfying the axioms (13.5) and (13.7). Then the computed matrices $\tilde{P}, \tilde{L},$ and $\tilde{U}$ satisfy*

$$\tilde{L}\tilde{U} = \tilde{P}A + \delta A, \qquad \frac{\|\delta A\|}{\|A\|} = O(\rho\, \epsilon_{\text{machine}}) \qquad (22.3)$$

for some $\delta A \in \mathbb{C}^{m \times n}$, where ρ is the growth factor for A. If $|\ell_{ij}| < 1$ for each $i > j$, implying that there are no ties in the selection of pivots in exact arithmetic, then $\tilde{P} = P$ for all sufficiently small $\epsilon_{\text{machine}}$.

Is Gaussian elimination backward stable? According to Theorem 22.2 and our definition (14.5) of backward stability, the answer is yes if $\rho = O(1)$ uniformly for all matrices of a given dimension m, and otherwise no.

And now, the complications begin.

Worst-Case Instability

For certain matrices A, despite the beneficial effects of pivoting, ρ turns out to be huge. For example, suppose A is the matrix

$$A = \begin{bmatrix} 1 & & & & 1 \\ -1 & 1 & & & 1 \\ -1 & -1 & 1 & & 1 \\ -1 & -1 & -1 & 1 & 1 \\ -1 & -1 & -1 & -1 & 1 \end{bmatrix}. \qquad (22.4)$$

At the first step, no pivoting takes place, but entries $2, 3, \dots, m$ in the final column are doubled from 1 to 2. Another doubling occurs at each subsequent elimination step. At the end we have

$$U = \begin{bmatrix} 1 & & & & 1 \\ & 1 & & & 2 \\ & & 1 & & 4 \\ & & & 1 & 8 \\ & & & & 16 \end{bmatrix}. \qquad (22.5)$$

The final $PA = LU$ factorization looks like this:

$$\begin{bmatrix} 1 & & & & 1 \\ -1 & 1 & & & 1 \\ -1 & -1 & 1 & & 1 \\ -1 & -1 & -1 & 1 & 1 \\ -1 & -1 & -1 & -1 & 1 \end{bmatrix} = \begin{bmatrix} 1 & & & & \\ -1 & 1 & & & \\ -1 & -1 & 1 & & \\ -1 & -1 & -1 & 1 & \\ -1 & -1 & -1 & -1 & 1 \end{bmatrix} \begin{bmatrix} 1 & & & & 1 \\ & 1 & & & 2 \\ & & 1 & & 4 \\ & & & 1 & 8 \\ & & & & 16 \end{bmatrix}.$$

For this 5×5 matrix, the growth factor is $\rho = 16$. For an $m \times m$ matrix of the same form, it is $\rho = 2^{m-1}$. (This is as large as ρ can get; see Exercise 22.1.)

A growth factor of order 2^m corresponds to a loss of on the order of m bits of precision, which is catastrophic for a practical computation. Since a typical computer represents floating point numbers with just sixty-four bits, whereas matrix problems of dimensions in the hundreds or thousands are solved all the time, a loss of m bits of precision is intolerable for real computations.

This brings us to an awkward point. Here, in the discussion of Gaussian elimination with pivoting—for the only time in this book—the definitions of stability presented in Lecture 14 fail us. According to the definitions, all that matters in determining stability or backward stability is the existence of a certain bound applicable uniformly to all matrices *for each fixed dimension m*. Uniformity with respect to m is not required. Here, for each m, we have a uniform bound involving the constant 2^{m-1}. Thus, according to our definitions, Gaussian elimination is backward stable.

稳定的 **Theorem 22.3.** *According to the definitions of Lecture 14, Gaussian elimination with partial pivoting is backward stable.*

This conclusion is absurd, however, in view of the vastness of 2^{m-1} for practical values of m.

For the remainder of this lecture, we ask the reader to put aside our formal definitions of stability and accept a more informal (and more standard) use of words. Gaussian elimination for certain matrices is explosively unstable, as can be confirmed by numerical experiments with MATLAB, LINPACK, LAPACK, or other software packages of impeccable reputation (Exercise 22.2).

Stability in Practice

If Gaussian elimination is unstable, why is it so famous and so popular? This brings us to a point that is not just an artifact of definitions but a fundamental fact about the behavior of this algorithm. *Despite examples like (22.4), Gaussian elimination with partial pivoting is utterly stable in practice. Large factors U like (22.5) never seem to appear in real applications. In fifty years of computing, no matrix problems that excite an explosive instability are known to have arisen under natural circumstances.*

This is a curious situation indeed. How can an algorithm that fails for certain matrices be entirely trustworthy in practice? The answer seems to be that although some matrices cause instability, these represent such an extraordinarily small proportion of the set of all matrices that they "never" arise in practice simply for statistical reasons.

One can learn more about this phenomenon by considering random matrices. Of course, the matrices that arise in applications are not random in any ordinary sense. They have all kinds of special properties, and if one tried to describe them as random samples from some distribution, it would have to be a curious distribution indeed. It would certainly be unreasonable to expect that any particular distribution of random matrices should match the behavior of the matrices arising in practice in a close quantitative way.

However, the phenomenon to be explained is not a matter of precise quantities. Matrices with large growth factors are vanishingly rare in applications. If we can show that they are vanishingly rare among random matrices in some well-defined class, the mechanisms involved must surely be the same. The argument does not depend on one measure of "vanishingly" agreeing with the other to any particular factor such as 2 or 10 or 100.

Figures 22.1 and 22.2 present experiments with random matrices as defined in Exercise 12.3: each entry is an independent sample from the real normal distribution of mean 0 and standard deviation $m^{-1/2}$. In Figure 22.1, a collection of random matrices of various dimensions have been factored and the growth factors presented as a scatter plot. Only two of the matrices gave a growth factor as large as $m^{1/2}$. In Figure 22.2, the results of factoring one million matrices each of dimensions $m = 8$, 16, and 32 are shown. Here, the growth factors have been collected in bins of width 0.2 and the resulting data plotted as a probability density distribution. The probability density of growth factors appears to decrease exponentially with size. Among these three million matrices, though the maximum growth factor in principle might have been 2,147,483,648, the maximum actually encountered was 11.99.

Similar results are obtained with random matrices defined by other probability distributions, such as uniformly distributed entries in $[-1, 1]$ (Exercise 22.3). If you pick a billion matrices at random, you will almost certainly not find one for which Gaussian elimination is unstable.

Explanation

We shall not attempt to give a full explanation of why the matrices for which Gaussian elimination is unstable are so rare. This would not be possible, as the matter is not yet fully understood. But we shall present an outline of an explanation.

If $PA = LU$, then $U = L^{-1}PA$. It follows that if Gaussian elimination is unstable when applied to the matrix A, implying that ρ is large, then L^{-1} must be large too. Now, as it happens, random triangular matrices tend

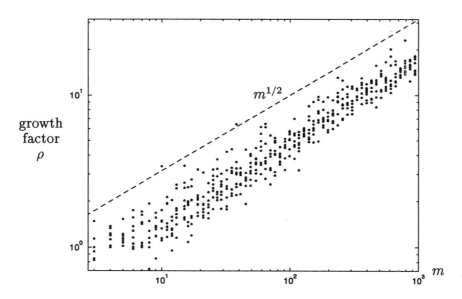

Figure 22.1. *Growth factors for Gaussian elimination with partial pivoting applied to 496 random matrices (independent, normally distributed entries) of various dimensions. The typical size of ρ is of order $m^{1/2}$, much less than the maximal possible value 2^{m-1}.*

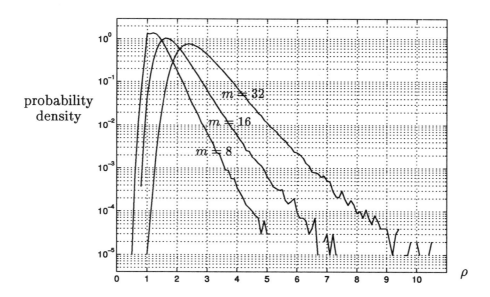

Figure 22.2. *Probability density distributions for growth factors of random matrices of dimensions $m = 8$, 16, 32, based on sample sizes of one million for each dimension. The density appears to decrease exponentially with ρ. The chatter near the end of each curve is an artifact of the finite sample sizes.*

to have huge inverses, exponentially large as a function of the dimension m (Exercise 12.3(d)). In particular, this is true for random triangular matrices of the form delivered by Gaussian elimination with partial pivoting, with 1 on the diagonal and entries ≤ 1 in absolute value below.

When Gaussian elimination is applied to random matrices A, however, the resulting factors L are anything but random. Correlations appear among the signs of the entries of L that render these matrices extraordinarily well-conditioned. A typical entry of L^{-1}, far from being exponentially large, is usually less than 1 in absolute value. Figure 22.3 presents evidence of this phenomenon based on a single (but typical) matrix of dimension $m = 128$.

We thus arrive at the question: why do the matrices L delivered by Gaussian elimination almost never have large inverses?

The answer lies in the consideration of column spaces. Since U is upper-triangular and $PA = LU$, the column spaces of PA and L are the same. By this we mean that the first column of PA spans the same space as the first column of L, the first two columns of PA span the same space as the first two columns of L, and so on. If A is random, its column spaces are randomly oriented, and it follows that the same must be true of the column spaces of $P^{-1}L$. However, this condition is incompatible with L^{-1} being large. It can be shown that if L^{-1} is large, then the column spaces of L, or of any permutation $P^{-1}L$, must be skewed in a fashion that is very far from random.

Figure 22.4 gives evidence of this. The figure shows "where the energy is" in the successive column spaces of the same two matrices as in Figure 22.3. The device for doing this is a Q portrait, defined by the MATLAB commands

$$[\texttt{Q,R}] = \texttt{qr(A)}, \qquad \texttt{spy(abs(Q) > 1/sqrt(m))}. \qquad (22.6)$$

These commands first compute a QR factorization of the matrix A, then plot a dot at each position of Q corresponding to an entry larger than the standard deviation, $m^{-1/2}$. The figure illustrates that for a random A, even after row interchanges to the form PA, the column spaces are oriented nearly randomly, whereas for a matrix A that gives a large growth factor, the orientations are very far from random. It is likely that by quantifying this argument, it can be proved that growth factors larger than order $m^{1/2}$ are exponentially rare among random matrices in the sense that for any $\alpha > 1/2$ and $M > 0$, the probability of the event $\rho > m^{\alpha}$ is smaller than m^{-M} for all sufficiently large m. As of this writing, however, such a theorem has not yet been proved.

Let us summarize the stability of Gaussian elimination with partial pivoting. This algorithm is highly unstable for certain matrices A. For instability to occur, however, the column spaces of A must be skewed in a very special fashion, one that is exponentially rare in at least one class of random matrices. Decades of computational experience suggest that matrices whose column spaces are skewed in this fashion arise very rarely in applications.

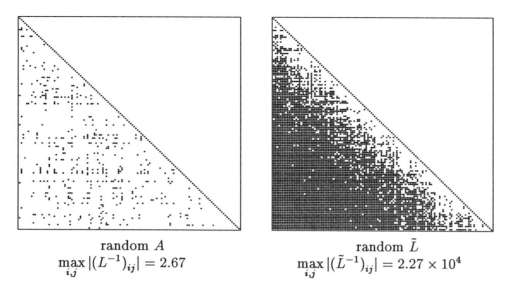

random A
$$\max_{i,j} |(L^{-1})_{ij}| = 2.67$$

random $\tilde{L}$
$$\max_{i,j} |(\tilde{L}^{-1})_{ij}| = 2.27 \times 10^4$$

Figure 22.3. *Let A be a random 128×128 matrix with factorization $PA = LU$. On the left, L^{-1} is shown: the dots represent entries with magnitude ≥ 1. On the right, a similar picture for $\tilde{L}^{-1}$, where $\tilde{L}$ is the same as L except that the signs of its subdiagonal entries have been randomized. Gaussian elimination tends to produce matrices L that are extraordinarily well-conditioned.*

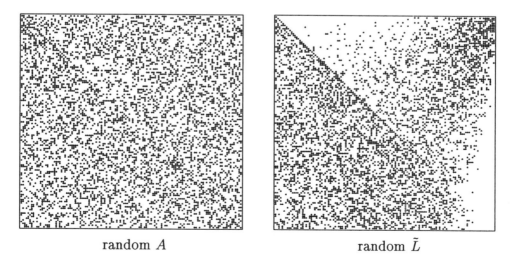

random A random $\tilde{L}$

Figure 22.4. *Q portraits (22.6) of the same two matrices. On the left, the random matrix A after permutation to the form PA, or equivalently, the factor L. On the right, the matrix $\tilde{L}$ with randomized signs. The column spaces of $\tilde{L}$ are skewed in a manner exponentially unlikely to arise in typical classes of random matrices.*

Exercises

22.1. Show that for Gaussian elimination with partial pivoting applied to any matrix $A \in \mathbb{C}^{m \times m}$, the growth factor (22.2) satisfies $\rho \leq 2^{m-1}$.

22.2. Experiment with solving 60×60 systems of equations $Ax = b$ by Gaussian elimination with partial pivoting, with A having the form (22.4). Do you observe that the results are useless because of the growth factor of order 2^{60}? At your first attempt you may not observe this, because the integer entries of A may prevent any rounding errors from occurring. If so, find a way to modify your problem slightly so that the growth factor is the same or nearly so and catastrophic rounding errors really do take place.

22.3. Reproduce the figures of this lecture, approximately if not in full detail, but based on random matrices with entries uniformly distributed in $[-1, 1]$ rather than normally distributed. Do you see any significant differences?

22.4. (a) Suppose $PA = LU$ (LU factorization with partial pivoting) and $A = QR$ (QR factorization). Describe a relationship between the last row of L^{-1} and the last column of Q.

(b) Show that if A is random in the sense of having independent, normally distributed entries, then its column spaces are randomly oriented, so that in particular, the last column of Q is a random unit vector.

(c) Combine the results of (a) and (b) to make a statement about the final row of L^{-1} in Gaussian elimination applied to a random matrix A.

Lecture 23. Cholesky Factorization

Hermitian positive definite matrices can be decomposed into triangular factors twice as quickly as general matrices. The standard algorithm for this, Cholesky factorization, is a variant of Gaussian elimination that operates on both the left and the right of the matrix at once, preserving and exploiting symmetry.

Hermitian Positive Definite Matrices

symmetric A real matrix $A \in \mathbb{R}^{m \times m}$ is *symmetric* if it has the same entries below the diagonal as above: $a_{ij} = a_{ji}$ for all i, j, hence $A = A^T$. Such a matrix satisfies $x^T A y = y^T A x$ for all vectors $x, y \in \mathbb{R}^m$.

hermitian For a complex matrix $A \in \mathbb{C}^{m \times m}$, the analogous property is that A is *hermitian*. A hermitian matrix has entries below the diagonal that are complex conjugates of those above the diagonal: $a_{ij} = \overline{a_{ji}}$, hence $A = A^*$. (These definitions appeared already in Lecture 2.) Note that this means that the diagonal entries of a hermitian matrix must be real.

HPD A hermitian matrix A satisfies $x^* A y = \overline{y^* A x}$ for all $x, y \in \mathbb{C}^m$. This means in particular that for any $x \in \mathbb{C}^m$, $x^* A x$ is real. If in addition $x^* A x > 0$ for all $x \neq 0$, then A is said to be *hermitian positive definite* (or sometimes just *positive definite*). Many matrices that arise in physical systems are hermitian positive definite because of fundamental physical laws.

If A is an $m \times m$ hermitian positive definite matrix and X is an $m \times n$ matrix of full rank with $m \geq n$, then the matrix $X^* A X$ is also hermitian positive definite. It is hermitian because $(X^* A X)^* = X^* A^* X = X^* A X$, and

172

it is positive definite because, for any vector $x \neq 0$, we have $Xx \neq 0$ and thus $x^*(X^*AX)x = (Xx)^*A(Xx) > 0$. By choosing X to be an $m \times n$ matrix with a 1 in each column and zeros elsewhere, we can write any $n \times n$ principal submatrix of A in the form X^*AX. Therefore, any principal submatrix of A must be positive definite. In particular, every diagonal entry of A is a positive real number.

The eigenvalues of a hermitian positive definite matrix are also positive real numbers. If $Ax = \lambda x$ for $x \neq 0$, we have $x^*Ax = \lambda x^*x > 0$ and therefore $\lambda > 0$. Conversely, it can be shown that if a hermitian matrix has all positive eigenvalues, then it is positive definite.

Eigenvectors that correspond to distinct eigenvalues of a hermitian matrix are orthogonal. (As discussed in the next lecture, hermitian matrices are normal.) Suppose $Ax_1 = \lambda_1 x_1$ and $Ax_2 = \lambda_2 x_2$ with $\lambda_1 \neq \lambda_2$. Then

$$\lambda_2 x_1^* x_2 = x_1^* A x_2 = \overline{x_2^* A x_1} = \overline{\lambda_1 x_2^* x_1} = \lambda_1 x_1^* x_2,$$

so $(\lambda_1 - \lambda_2)x_1^* x_2 = 0$. Since $\lambda_1 \neq \lambda_2$, we have $x_1^* x_2 = 0$.

Symmetric Gaussian Elimination

We turn now to the problem of decomposing a hermitian positive definite matrix into triangular factors. To begin, consider what happens if a single step of Gaussian elimination is applied to a hermitian matrix A with a 1 in the upper-left position:

$$A = \begin{bmatrix} 1 & w^* \\ w & K \end{bmatrix} = \begin{bmatrix} 1 & 0 \\ w & I \end{bmatrix} \begin{bmatrix} 1 & w^* \\ 0 & K - ww^* \end{bmatrix}.$$

As described in Lecture 20, zeros have been introduced into the first column of the matrix by an elementary lower-triangular operation on the left that subtracts multiples of the first row from subsequent rows.

Gaussian elimination would now continue the reduction to triangular form by introducing zeros in the second column. However, in order to maintain symmetry, Cholesky factorization first introduces zeros in the first row to match the zeros just introduced in the first column. We can do this by a right upper-triangular operation that subtracts multiples of the first column from the subsequent ones:

$$\begin{bmatrix} 1 & w^* \\ 0 & K - ww^* \end{bmatrix} = \begin{bmatrix} 1 & 0 \\ 0 & K - ww^* \end{bmatrix} \begin{bmatrix} 1 & w^* \\ 0 & I \end{bmatrix}.$$

Note that this upper-triangular operation is exactly the adjoint of the lower-triangular operation that we used to introduce zeros in the first column.

Combining the operations above, we find that the matrix A has been factored into three terms:

$$A = \begin{bmatrix} 1 & w^* \\ w & K \end{bmatrix} = \begin{bmatrix} 1 & 0 \\ w & I \end{bmatrix} \begin{bmatrix} 1 & 0 \\ 0 & K - ww^* \end{bmatrix} \begin{bmatrix} 1 & w^* \\ 0 & I \end{bmatrix}. \tag{23.1}$$

The idea of Cholesky factorization is to continue this process, zeroing one column and one row of A symmetrically until it is reduced to the identity.

Cholesky Factorization

In order for the symmetric triangular reduction to work in general, we need a factorization that works for any $a_{11} > 0$, not just $a_{11} = 1$. The generalization of (23.1) is accomplished by adjusting some of the elements of R_1 by a factor of $\sqrt{a_{11}}$. Let $\alpha = \sqrt{a_{11}}$ and observe:

$$
A = \begin{bmatrix} a_{11} & w^* \\ w & K \end{bmatrix} \quad \text{关键}
$$

$$
= \begin{bmatrix} \alpha & 0 \\ w/\alpha & I \end{bmatrix} \begin{bmatrix} 1 & 0 \\ 0 & K - ww^*/a_{11} \end{bmatrix} \begin{bmatrix} \alpha & w^*/\alpha \\ 0 & I \end{bmatrix} = R_1^* A_1 R_1.
$$

This is the basic step that is applied repeatedly in Cholesky factorization. If the upper-left entry of the submatrix $K - ww^*/a_{11}$ is positive, the same formula can be used to factor it; we then have $A_1 = R_2^* A_2 R_2$ and thus $A = R_1^* R_2^* A_2 R_2 R_1$. The process is continued down to the bottom-right corner, giving us eventually a factorization

$$
A = \underbrace{R_1^* R_2^* \cdots R_m^*}_{R^*} \underbrace{R_m \cdots R_2 R_1}_{R}. \tag{23.2}
$$

This equation has the form

$$
A = R^* R, \qquad r_{jj} > 0, \tag{23.3}
$$

where R is upper-triangular. A reduction of this kind of a hermitian positive definite matrix is known as a *Cholesky factorization*.

    The description above left one item dangling. How do we know that the upper-left entry of the submatrix $K - ww^*/a_{11}$ is positive? The answer is that it must be positive because $K - ww^*/a_{11}$ is positive definite, since it is the $(m-1) \times (m-1)$ lower-right principal submatrix of the positive definite matrix $R_1^{-*} A R_1^{-1}$. By induction, the same argument shows that all the submatrices A_j that appear in the course of the factorization are positive definite, and thus the process cannot break down. We can formalize this conclusion as follows.

Theorem 23.1. *Every hermitian positive definite matrix $A \in \mathbb{C}^{m \times m}$ has a unique Cholesky factorization* (23.3).

Proof. Existence is what we just discussed; a factorization exists since the algorithm cannot break down. In fact, the algorithm also establishes uniqueness. At each step (23.2), the value $\alpha = \sqrt{a_{11}}$ is determined by the form of the R^*R factorization, and once α is determined, the first row of R_1^* is determined too. Since the analogous quantities are determined at each step of the reduction, the entire factorization is unique.   □

The Algorithm

When Cholesky factorization is implemented, only half of the matrix being operated on needs to be represented explicitly. This simplification allows half of the arithmetic to be avoided. A formal statement of the algorithm (only one of many possibilities) is given below. The input matrix A represents the superdiagonal half of the $m \times m$ hermitian positive definite matrix to be factored. (In practical software, a compressed storage scheme may be used to avoid wasting half the entries of a square array.) The output matrix R represents the upper-triangular factor for which $A = R^*R$. Each outer iteration corresponds to a single elementary factorization: the upper-triangular part of the submatrix $R_{k:m,k:m}^*$ represents the superdiagonal part of the hermitian matrix being factored at step k.

Algorithm 23.1. Cholesky Factorization

$R = A$
for $k = 1$ **to** m
$\quad$ **for** $j = k + 1$ **to** m
$\qquad R_{j,j:m} = R_{j,j:m} - R_{k,j:m}R_{kj}/R_{kk}$
$\quad R_{k,k:m} = R_{k,k:m}/\sqrt{R_{kk}}$

Operation Count

The arithmetic done in Cholesky factorization is dominated by the inner loop. A single execution of the line

$$R_{j,j:m} = R_{j,j:m} - R_{k,j:m}R_{kj}/R_{kk}$$

requires one division, $m - j + 1$ multiplications, and $m - j + 1$ subtractions, for a total of $\sim 2(m - j)$ flops. This calculation is repeated once for each j from $k + 1$ to m, and that loop is repeated for each k from 1 to m. The sum is straightforward to evaluate:

$$\sum_{k=1}^{m} \sum_{j=k+1}^{m} 2(m - j) \; \sim \; 2 \sum_{k=1}^{m} \sum_{j=1}^{k} j \; \sim \; \sum_{k=1}^{m} k^2 \; \sim \; \frac{1}{3}m^3 \text{ flops.}$$

Thus, Cholesky factorization involves only half as many operations as Gaussian elimination, which would require $\sim \frac{2}{3}m^3$ flops to factor the same matrix.

As usual, the operation count can also be determined graphically. For each k, two floating point operations are carried out (one multiplication and one subtraction) at each position of a triangular layer. The entire algorithm corresponds to stacking m layers:

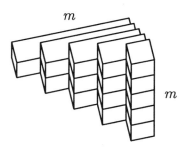

As $m \to \infty$, the solid converges to a tetrahedron with volume $\frac{1}{6}m^3$. Since each unit cube corresponds to two floating point operations, we obtain again

$$\text{Work for Cholesky factorization:} \quad \sim \frac{1}{3}m^3 \text{ flops.} \qquad (23.4)$$

Stability

All of the subtleties of the stability analysis of Gaussian elimination vanish for Cholesky factorization. This algorithm is always stable. Intuitively, the reason is that the factors R can never grow large. In the 2-norm, for example, we have $\|R\| = \|R^*\| = \|A\|^{1/2}$ (proof: SVD), and in other p-norms with $1 \le p \le \infty$, $\|R\|$ cannot differ from $\|A\|^{1/2}$ by more than a factor of $\sqrt{m}$. Thus, numbers much larger than the entries of A can never arise.

Note that the stability of Cholesky factorization is achieved without the need for any pivoting. Intuitively, one may observe that this is related to the fact that most of the weight of a hermitian positive definite matrix is on the diagonal. For example, it is not hard to show that the largest entry must appear on the diagonal, and this property carries over to the positive definite submatrices constructed in the inductive process (23.2).

An analysis of the stability of the Cholesky process leads to the following backward stability result.

Theorem 23.2. *Let $A \in \mathbb{C}^{m \times m}$ be hermitian positive definite, and let a Cholesky factorization of A be computed by Algorithm 23.1 on a computer satisfying (13.5) and (13.7). For all sufficiently small $\epsilon_{\text{machine}}$, this process is guaranteed to run to completion (i.e., no zero or negative corner entries r_{kk} will arise), generating a computed factor $\tilde{R}$ that satisfies*

$$\tilde{R}^* \tilde{R} = A + \delta A, \qquad \frac{\|\delta A\|}{\|A\|} = O(\epsilon_{\text{machine}}) \qquad (23.5)$$

for some $\delta A \in \mathbb{C}^{m \times m}$.

Like so many algorithms of numerical linear algebra, this one would look much worse if we tried to carry out a forward error analysis rather than a backward one. If A is ill-conditioned, $\tilde{R}$ will not generally be close to R; the best we can say is $\|\tilde{R} - R\|/\|R\| = O(\kappa(A)\epsilon_{\text{machine}})$. (In other words, Cholesky factorization is in general an ill-conditioned problem.) It is only the product $\tilde{R}^*R$ that satisfies the much better error bound (23.5). Thus the errors introduced in $\tilde{R}$ by rounding are large but "diabolically correlated," just as we saw in Lecture 16 for QR factorization.

Solution of $Ax = b$

If A is hermitian positive definite, the standard way to solve a system of equations $Ax = b$ is by Cholesky factorization. Algorithm 23.1 reduces the system to $R^*Rx = b$, and we then solve two triangular systems in succession: first $R^*y = b$ for the unknown y, then $Rx = y$ for the unknown x. Each triangular solution requires just $\sim m^2$ flops, so the total work is again $\sim \frac{1}{3}m^3$ flops.

By reasoning analogous to that of Lecture 16, it can be shown that this process is backward stable.

Theorem 23.3. *The solution of hermitian positive definite systems $Ax = b$ via Cholesky factorization (Algorithm 23.1) is backward stable, generating a computed solution $\tilde{x}$ that satisfies*

$$(A + \Delta A)\tilde{x} = b, \qquad \frac{\|\Delta A\|}{\|A\|} = O(\epsilon_{\text{machine}}) \qquad (23.6)$$

for some $\Delta A \in \mathbb{C}^{m \times m}$.

Exercises

23.1. Let A be a nonsingular square matrix and let $A = QR$ and $A^*A = U^*U$ be QR and Cholesky factorizations, respectively, with the usual normalizations $r_{jj}, u_{jj} > 0$. Is it true or false that $R = U$?

23.2. Using the proof of Theorem 16.2 as a guide, derive Theorem 23.3 from Theorems 23.2 and 17.1.

23.3. *Reverse Software Engineering of "\ ".* The following MATLAB session records a sequence of tests of the elapsed times for various computations on a workstation manufactured in 1991. For each part, try to explain: (i) Why was this experiment carried out? (ii) Why did the result came out as it did? Your

answers should refer to formulas from the text for flop counts. The MATLAB
queries help chol and help slash may help in your detective work.

(a) ```
m = 200; Z = randn(m,m);
A = Z'*Z; b = randn(m,1);
tic; x = A\b; toc;
        elapsed_time = 1.0368
```

(b) ```
tic; x = A\b; toc;
        elapsed_time = 1.0303
```

(c) ```
A2 = A; A2(m,1) = A2(m,1)/2;
tic; x = A2\b; toc;
        elapsed_time = 2.0361
```

(d) ```
I = eye(m,m); emin = min(eig(A));
A3 = A - .9*emin*I;
tic; x = A3\b; toc;
        elapsed_time = 1.0362
```

(e) ```
A4 = A - 1.1*emin*I;
tic; x = A4\b; toc;
        elapsed_time = 2.9624
```

(f) ```
A5 = triu(A);
tic; x = A5\b; toc;
        elapsed_time = 0.1261
```

(g) ```
A6 = A5; A6(m,1) = A5(1,m);
tic; x = A6\b; toc;
        elapsed_time = 2.0012
```
```

Part V

Eigenvalues

Lecture 24. Eigenvalue Problems

Eigenvalue problems are particularly interesting in scientific computing, because the best algorithms for finding eigenvalues are powerful, yet particularly far from obvious. Here, we review the mathematics of eigenvalues and eigenvectors. Algorithms are discussed in later lectures.

Eigenvalues and Eigenvectors

Let $A \in \mathbb{C}^{m \times m}$ be a square matrix. A nonzero vector $x \in \mathbb{C}^m$ is an *eigenvector* of A, and $\lambda \in \mathbb{C}$ is its corresponding *eigenvalue*, if

$$Ax = \lambda x. \tag{24.1}$$

The idea here is that the action of a matrix A on a subspace S of $\mathbb{C}^m$ may sometimes mimic scalar multiplication. When this happens, the special subspace S is called an *eigenspace*, and any nonzero $x \in S$ is an eigenvector.

The set of all the eigenvalues of a matrix A is the *spectrum* of A, a subset of $\mathbb{C}$ denoted by $\Lambda(A)$.

Eigenvalue problems have a very different character from the problems involving square or rectangular linear systems of equations discussed in the previous lectures. For a system of equations, the domain of A could be one space and the range could be a different one. In Example 1.1, for example, A mapped n-vectors of polynomial coefficients to m-vectors of sampled polynomial values. To ask about the eigenvalues of such an A would be meaningless. Eigenvalue problems make sense only when the range and the domain spaces are the same. This reflects the fact that in applications, eigenvalues are generally used where a matrix is to be compounded iteratively, either explicitly as a power A^k or implicitly in a functional form such as e^{tA}.

Broadly speaking, eigenvalues and eigenvectors are useful for two reasons, one algorithmic, the other physical. Algorithmically, eigenvalue analysis can simplify solutions of certain problems by reducing a coupled system to a collection of scalar problems. Physically, eigenvalue analysis can give insight into the behavior of evolving systems governed by linear equations. The most familiar examples in this latter class are the study of *resonance* (e.g., of musical instruments when struck or plucked or bowed) and of *stability* (e.g., of fluid flows subjected to small perturbations). In such cases eigenvalues tend to be particularly useful for analyzing behavior for large times t. See Exercise 24.3.

Eigenvalue Decomposition

An *eigenvalue decomposition* of a square matrix A, already mentioned in (5.1), is a factorization

$$A = X\Lambda X^{-1}. \tag{24.2}$$

(As we discuss below, such a factorization does not always exist.) Here X is nonsingular and Λ is diagonal.

This definition can be rewritten

$$AX = X\Lambda, \tag{24.3}$$

that is,

$$\left[\begin{array}{c} \\ A \\ \\ \end{array}\right]\left[\begin{array}{c|c|c|c} \\ x_1 & x_2 & \cdots & x_m \\ \\ \end{array}\right] = \left[\begin{array}{c|c|c|c} \\ x_1 & x_2 & \cdots & x_m \\ \\ \end{array}\right]\left[\begin{array}{cccc} \lambda_1 & & & \\ & \lambda_2 & & \\ & & \ddots & \\ & & & \lambda_m \end{array}\right].$$

This makes it clear that if x_j is the jth column of X and λ_j is the jth diagonal entry of Λ, then $Ax_j = \lambda_j x_j$. Thus the jth column of X is an eigenvector of A and the jth entry of Λ is the corresponding eigenvalue.

The eigenvalue decomposition expresses a change of basis to "eigenvector coordinates." If $Ax = b$ and $A = X\Lambda X^{-1}$, we have

$$(X^{-1}b) = \Lambda(X^{-1}x). \tag{24.4}$$

Thus, to compute Ax, we can expand x in the basis of columns of X, apply Λ, and interpret the result as a vector of coefficients of a linear combination of the columns of X.

Geometric Multiplicity

As stated above, the set of eigenvectors corresponding to a single eigenvalue, together with the zero vector, forms a subspace of $\mathbb{C}^m$ known as an *eigenspace*. If λ is an eigenvalue of A, let us denote the corresponding eigenspace by E_λ. An eigenspace E_λ is an example of an *invariant subspace* of A; that is, $AE_\lambda \subseteq E_\lambda$.

The dimension of E_λ can be interpreted as the maximum number of linearly independent eigenvectors that can be found, all with the same eigenvalue λ. This number is known as the *geometric multiplicity* of λ. The geometric multiplicity can also be described as the dimension of the nullspace of $A - \lambda I$, since that nullspace is again E_λ.

Characteristic Polynomial

The *characteristic polynomial* of $A \in \mathbb{C}^{m \times m}$, denoted by p_A or simply p, is the degree m polynomial defined by

$$p_A(z) = \det(zI - A). \tag{24.5}$$

Thanks to the placement of the minus sign, p is *monic:* the coefficient of its degree m term is 1.

Theorem 24.1. *λ is an eigenvalue of A if and only if $p_A(\lambda) = 0$.*

Proof. This follows from the definition of an eigenvalue:

λ is an eigenvalue $\iff$ there is a nonzero vector x such that $\lambda x - Ax = 0$
$\iff \lambda I - A$ is singular
$\iff \det(\lambda I - A) = 0.$ $\qquad\qquad\square$

Theorem 24.1 has an important consequence. *Even if a matrix is real, some of its eigenvalues may be complex.* Physically, this is related to the phenomenon that real dynamical systems can have motions that oscillate as well as grow or decay. Algorithmically, it means that even if the input to a matrix eigenvalue problem is real, the output may have to be complex.

Algebraic Multiplicity

By the fundamental theorem of algebra, we can write p_A in the form

$$p_A(z) = (z - \lambda_1)(z - \lambda_2) \cdots (z - \lambda_m) \tag{24.6}$$

for some numbers $\lambda_j \in \mathbb{C}$. By Theorem 24.1, each λ_j is an eigenvalue of A, and all eigenvalues of A appear somewhere in this list. In general, an eigenvalue might appear more than once. We define the *algebraic multiplicity* of an eigenvalue λ of A to be its multiplicity as a root of p_A. An eigenvalue is *simple* if its algebraic multiplicity is 1.

The characteristic polynomial gives us an easy way to count the number of eigenvalues of a matrix.

Theorem 24.2. *If $A \in \mathbb{C}^{m \times m}$, then A has m eigenvalues, counted with algebraic multiplicity. In particular, if the roots of p_A are simple, then A has m distinct eigenvalues.*

Note that in particular, every matrix has at least one eigenvalue.

The algebraic multiplicity of an eigenvalue is always at least as great as its geometric multiplicity. To prove this, we need to know something about similarity transformations.

Similarity Transformations

If $X \in \mathbb{C}^{m \times m}$ is nonsingular, then the map $A \mapsto X^{-1}AX$ is called a *similarity transformation* of A. We say that two matrices A and B are *similar* if there is a similarity transformation relating one to the other, i.e., if there exists a nonsingular $X \in \mathbb{C}^{m \times m}$ such that $B = X^{-1}AX$. As described above in the special case of the diagonalization (24.2), any similarity transformation is a change of basis operation.

Many properties are shared by similar matrices A and $X^{-1}AX$.

Theorem 24.3. *If X is nonsingular, then A and $X^{-1}AX$ have the same characteristic polynomial, eigenvalues, and algebraic and geometric multiplicities.*

Proof. The proof that the characteristic polynomials match is a straightforward computation:

$$
\begin{aligned}
p_{X^{-1}AX}(z) &= \det(zI - X^{-1}AX) = \det(X^{-1}(zI - A)X) \\
&= \det(X^{-1}) \det(zI - A) \det(X) = \det(zI - A) = p_A(z).
\end{aligned}
$$

From the agreement of the characteristic polynomials, the agreement of the eigenvalues and algebraic multiplicities follows. Finally, to prove that the geometric multiplicities agree, we can verify that if E_λ is an eigenspace for A, then $X^{-1}E_\lambda$ is an eigenspace for $X^{-1}AX$, and conversely. □

We can now relate geometric multiplicity to algebraic multiplicity.

Theorem 24.4. *The algebraic multiplicity of an eigenvalue λ is at least as great as its geometric multiplicity.*

Proof. Let n be the geometric multiplicity of λ for the matrix A. Form an $m \times n$ matrix $\hat{V}$ whose n columns constitute an orthonormal basis of the eigenspace $\{x : Ax = \lambda x\}$. Then, extending $\hat{V}$ to a square unitary matrix V, we obtain V^*AV in the form

$$B = V^*AV = \begin{bmatrix} \lambda I & C \\ 0 & D \end{bmatrix}, \qquad (24.7)$$

where I is the $n \times n$ identity, C is $n \times (m-n)$, and D is $(m-n) \times (m-n)$. By the definition of the determinant, $\det(zI - B) = \det(zI - \lambda I)\det(zI - D) = (z-\lambda)^n \det(zI - D)$. Therefore the algebraic multiplicity of λ as an eigenvalue of B is at least n. Since similarity transformations preserve multiplicities, the same is true for A. $\qquad \square$

Defective Eigenvalues and Matrices

Although a generic matrix has algebraic and geometric multiplicities that are equal (namely, all 1), this is by no means true of every matrix.

Example 24.1. Consider the matrices

$$A = \begin{bmatrix} 2 & & \\ & 2 & \\ & & 2 \end{bmatrix}, \qquad B = \begin{bmatrix} 2 & 1 & \\ & 2 & 1 \\ & & 2 \end{bmatrix}.$$

Both A and B have characteristic polynomial $(z-2)^3$, so there is a single eigenvalue $\lambda = 2$ of algebraic multiplicity 3. In the case of A, we can choose three independent eigenvectors, for example e_1, e_2, and e_3, so the geometric multiplicity is also 3. For B, on the other hand, we can find only a single independent eigenvector (a scalar multiple of e_1), so the geometric multiplicity of the eigenvalue is only 1. $\qquad \square$

An eigenvalue whose algebraic multiplicity exceeds its geometric multiplicity is a *defective eigenvalue*. A matrix that has one or more defective eigenvalues is a *defective matrix*.

Any diagonal matrix is nondefective. For such a matrix, both the algebraic and the geometric multiplicities of an eigenvalue λ are equal to the number of its occurrences along the diagonal.

Diagonalizability

The class of nondefective matrices is precisely the class of matrices that have an eigenvalue decomposition (24.2).

Theorem 24.5. *An $m \times m$ matrix A is nondefective if and only if it has an eigenvalue decomposition $A = X\Lambda X^{-1}$.*

Proof. ($\Longleftarrow$) Given an eigenvalue decomposition $A = X\Lambda X^{-1}$, we know by Theorem 24.3 that Λ is similar to A, with the same eigenvalues and the same multiplicities. Since Λ is a diagonal matrix, it is nondefective, and thus the same holds for A.

($\Longrightarrow$) A nondefective matrix must have m linearly independent eigenvectors, because eigenvectors with different eigenvalues must be linearly independent, and each eigenvalue can contribute as many linearly independent eigenvectors as its multiplicity. If these m independent eigenvectors are formed into the columns of a matrix X, then X is nonsingular and we have $A = X\Lambda X^{-1}$. $\square$

In view of this result, another term for nondefective is *diagonalizable*.

Does a diagonalizable matrix A in some sense "behave like" its diagonal equivalent Λ? The answer depends on what aspect of behavior one measures and on the condition number of X, the matrix of eigenvectors. If X is highly ill-conditioned, then a great deal of information may be discarded in passing from A to Λ. See "A Note of Caution: Nonnormality" in Lecture 34.

Determinant and Trace

The *trace* of $A \in \mathbb{C}^{m \times m}$ is the sum of its diagonal elements: $\text{tr}(A) = \sum_{j=1}^{m} a_{jj}$. Both the trace and the determinant are related simply to the eigenvalues.

Theorem 24.6. *The determinant* $\det(A)$ *and trace* $\text{tr}(A)$ *are equal to the product and the sum of the eigenvalues of A, respectively, counted with algebraic multiplicity:*

$$\det(A) = \prod_{j=1}^{m} \lambda_j, \qquad \text{tr}(A) = \sum_{j=1}^{m} \lambda_j. \tag{24.8}$$

Proof. From (24.5) and (24.6), we compute

$$\det(A) = (-1)^m \det(-A) = (-1)^m p_A(0) = \prod_{j=1}^{m} \lambda_j.$$

This establishes the first formula. As for the second, from (24.5), it follows that the coefficient of the z^{m-1} term of p_A is the negative of the sum of the diagonal elements of A, or $-\text{tr}(A)$. On the other hand, from (24.6), this coefficient is also equal to $-\sum_{j=1}^{m} \lambda_j$. Thus $\text{tr}(A) = \sum_{j=1}^{m} \lambda_j$. $\square$

Unitary Diagonalization

It sometimes happens that not only does an $m \times m$ matrix A have m linearly independent eigenvectors, but these can be chosen to be orthogonal. In such a case, A is *unitarily diagonalizable*, that is, there exists a unitary matrix Q such that

$$A = Q\Lambda Q^*. \tag{24.9}$$

This factorization is both an eigenvalue decomposition and a singular value decomposition, aside from the matter of the signs (possibly complex) of the entries of Λ.

We have already seen a class of matrices that are unitarily diagonalizable: the hermitian matrices. The following result follows from Theorem 24.9, below.

Theorem 24.7. *A hermitian matrix is unitarily diagonalizable, and its eigenvalues are real.*

The hermitian matrices are not the only ones that are unitarily diagonalizable. Other examples include skew-hermitian matrices, unitary matrices, circulant matrices, and any of these plus a multiple of the identity. In general, the class of matrices that are unitarily diagonalizable have an elegant characterization. By definition, we say that a matrix A is *normal* if $A^*A = AA^*$. The following result is well known.

Theorem 24.8. *A matrix is unitarily diagonalizable if and only if it is normal.*

Schur Factorization

One final matrix factorization is actually the one that is most useful in numerical analysis, because all matrices, including defective ones, can be factored in this way. A *Schur factorization* of a matrix A is a factorization

$$A = QTQ^*, \tag{24.10}$$

where Q is unitary and T is upper-triangular. Note that since A and T are similar, the eigenvalues of A necessarily appear on the diagonal of T.

Theorem 24.9. *Every square matrix A has a Schur factorization.*

Proof. We proceed by induction on the dimension m of A. The case $m = 1$ is trivial, so suppose $m \geq 2$. Let x be any eigenvector of A, with corresponding eigenvalue λ. Take x to be normalized and let it be the first column of a unitary matrix U. Then, just as in (24.7), it is easily checked that the product U^*AU has the form

$$U^*AU = \begin{bmatrix} \lambda & B \\ 0 & C \end{bmatrix}.$$

By the inductive hypothesis, there exists a Schur factorization VTV^* of C. Now write

$$Q = U \begin{bmatrix} 1 & 0 \\ 0 & V \end{bmatrix}.$$

This is a unitary matrix, and we have

$$Q^*AQ = \begin{bmatrix} \lambda & BV \\ 0 & T \end{bmatrix}.$$

This is the Schur factorization we seek.   □

Eigenvalue-Revealing Factorizations

In the preceding pages we have described three examples of *eigenvalue-revealing factorizations*, factorizations of a matrix that reduce it to a form in which the eigenvalues are explicitly displayed. We can summarize these as follows.

A diagonalization $A = X\Lambda X^{-1}$ exists if and only if A is nondefective.

A unitary diagonalization $A = Q\Lambda Q^*$ exists if and only if A is normal.

A unitary triangularization (Schur factorization) $A = QTQ^*$ always exists.

To compute eigenvalues, we shall construct one of these factorizations. In general, this will be the Schur factorization, since this applies without restriction to all matrices. Moreover, since unitary transformations are involved, the algorithms that result tend to be numerically stable. If A is normal, then the Schur form comes out diagonal, and in particular, if A is hermitian, then we can take advantage of this symmetry throughout the computation and reduce A to diagonal form with half as much work or less than is required for general A.

Exercises

24.1. For each of the following statements, prove that it is true or give an example to show it is false. Throughout, $A \in \mathbb{C}^{m \times m}$ unless otherwise indicated, and "ew" stands for eigenvalue. (This comes from the German "Eigenwert." The corresponding abbreviation for eigenvector is "ev," from "Eigenvektor.")

(a) If λ is an ew of A and $\mu \in \mathbb{C}$, then $\lambda - \mu$ is an ew of $A - \mu I$.

(b) If A is real and λ is an ew of A, then so is $-\lambda$.

(c) If A is real and λ is an ew of A, then so is $\overline{\lambda}$.

(d) If λ is an ew of A and A is nonsingular, then λ^{-1} is an ew of A^{-1}.

(e) If all the ew's of A are zero, then $A = 0$.

(f) If A is hermitian and λ is an ew of A, then $|\lambda|$ is a singular value of A.

(g) If A is diagonalizable and all its ew's are equal, then A is diagonal.

24.2. Here is *Gerschgorin's theorem*, which holds for any $m \times m$ matrix A, symmetric or nonsymmetric. *Every eigenvalue of A lies in at least one of the m circular disks in the complex plane with centers a_{ii} and radii $\sum_{j \neq i} |a_{ij}|$. Moreover, if n of these disks form a connected domain that is disjoint from the other $m - n$ disks, then there are precisely n eigenvalues of A within this domain.*

(a) Prove the first part of Gerschgorin's theorem. (Hint: Let λ be any eigenvalue of A, and x a corresponding eigenvector with largest entry 1.)

(b) Prove the second part. (Hint: Deform A to a diagonal matrix and use the fact that the eigenvalues of a matrix are continuous functions of its entries.)

(c) Give estimates based on Gerschgorin's theorem for the eigenvalues of

$$
A = \begin{pmatrix} 8 & 1 & 0 \\ 1 & 4 & \epsilon \\ 0 & \epsilon & 1 \end{pmatrix}, \qquad |\epsilon| < 1.
$$

(d) Find a way to establish the tighter bound $|\lambda_3 - 1| \leq \epsilon^2$ on the smallest eigenvalue of A. (Hint: Consider diagonal similarity transformations.)

24.3. Let A be a 10×10 random matrix with entries from the standard normal distribution, minus twice the identity. Write a program to plot $\|e^{tA}\|_2$ against t for $0 \leq t \leq 20$ on a log scale, comparing the result to the straight line $e^{t\alpha(A)}$, where $\alpha(A) = \max_j \operatorname{Re}(\lambda_j)$ is the *spectral abscissa* of A. Run the program for ten random matrices A and comment on the results. What property of a matrix leads to a $\|e^{tA}\|_2$ curve that remains oscillatory as $t \to \infty$?

24.4. For an arbitrary $A \in \mathbb{C}^{m \times m}$ and norm $\| \cdot \|$, prove using Theorem 24.9:

(a) $\lim_{n \to \infty} \|A^n\| = 0 \iff \rho(A) < 1$, where ρ is the spectral radius (Exercise 3.2).

(b) $\lim_{t \to \infty} \|e^{tA}\| = 0 \iff \alpha(A) < 0$, where α is the spectral abscissa.

Lecture 25. Overview of Eigenvalue Algorithms

This and the next five lectures describe some of the classical "direct" algorithms for computing eigenvalues and eigenvectors, as well as a few modern variants. Most of these algorithms proceed in two phases: first, a preliminary reduction from full to structured form; then, an iterative process for the final convergence. This lecture outlines the two-phase approach and explains why it is advantageous.

Shortcomings of Obvious Algorithms

Although eigenvalues and eigenvectors have simple definitions and elegant characterizations, the best ways to compute them are not obvious.

Perhaps the first method one might think of would be to compute the coefficients of the characteristic polynomial and use a rootfinder to extract its roots. Unfortunately, as mentioned in Lecture 15, this strategy is a bad one, because polynomial rootfinding is an ill-conditioned problem in general, even when the underlying eigenvalue problem is well-conditioned. (In fact, polynomial rootfinding is by no means a mainstream topic in scientific computing—precisely because it is so rarely the best way to solve applied problems.)

Another idea would be to take advantage of the fact that the sequence

$$\frac{x}{\|x\|}, \quad \frac{Ax}{\|Ax\|}, \quad \frac{A^2x}{\|A^2x\|}, \quad \frac{A^3x}{\|A^3x\|}, \quad \cdots$$

converges, under certain assumptions, to an eigenvector corresponding to the largest eigenvalue of A in absolute value. This method for finding an eigenvector is called *power iteration*. Unfortunately, although power iteration is famous, it is by no means an effective tool for general use. Except for special matrices, it is very slow.

Instead of ideas like these, the best general purpose eigenvalue algorithms are based on a different principle: the computation of an eigenvalue-revealing factorization of A, where the eigenvalues appear as entries of one of the factors. We saw three eigenvalue-revealing factorizations in the last lecture: diagonalization, unitary diagonalization, and unitary triangularization (Schur factorization). In practice, eigenvalues are usually computed by constructing one of these factorizations. Conceptually, what must be done to achieve this is to apply a sequence of transformations to A to introduce zeros in the necessary places, just as in the algorithms we have considered in the preceding lectures of this book. Thus we see that finding eigenvalues ends up rather similar in flavor to solving systems of equations or least squares problems. The algorithms of numerical linear algebra are mainly built upon one technique used over and over again: putting zeros into matrices.

A Fundamental Difficulty

Though the flavors are related, however, a new spice appears in the dish when it comes to computing eigenvalues. What is new is that it would appear that algebraic considerations must preclude the success of any algorithm of this kind.

To see the difficulty, note that just as eigenvalue problems can be reduced to polynomial rootfinding problems, conversely, any polynomial rootfinding problem can be stated as an eigenvalue problem. Suppose we have the monic polynomial

$$p(z) = z^m + a_{m-1}z^{m-1} + \cdots + a_1 z + a_0. \qquad (25.1)$$

By expanding in minors, it is not hard to verify that $p(z)$ is equal to $(-1)^m$ times the determinant of the $m \times m$ matrix

$$
\begin{bmatrix}
-z & & & & & -a_0 \\
1 & -z & & & & -a_1 \\
 & 1 & -z & & & -a_2 \\
 & & 1 & \ddots & & \vdots \\
 & & & \ddots & -z & -a_{m-2} \\
 & & & & 1 & (-z - a_{m-1})
\end{bmatrix}. \qquad (25.2)
$$

This means that the roots of p are equal to the eigenvalues of the matrix

$$
A = \begin{bmatrix}
0 & & & & & & -a_0 \\
1 & 0 & & & & & -a_1 \\
& 1 & 0 & & & & -a_2 \\
& & 1 & \ddots & & & \vdots \\
& & & \ddots & 0 & & -a_{m-2} \\
& & & & 1 & & -a_{m-1}
\end{bmatrix}. \tag{25.3}
$$

(We can also get to (25.3) directly, without passing through (25.2), by noting that if z is a root of p, then it follows from (25.1) that $(1, z, z^2, \ldots, z^{m-1})$ is a left eigenvector of A with eigenvalue z.) A is called a *companion matrix* corresponding to p.

Now the difficulty is apparent. It is well known that no formula exists for expressing the roots of an arbitrary polynomial, given its coefficients. This impossibility result is one of the crowning achievements of a body of mathematical work carried out by Abel, Galois, and others in the nineteenth century. Abel proved in 1824 that no analogue of the quadratic formula can exist for polynomials of degree 5 or more.

Theorem 25.1. *For any $m \geq 5$, there is a polynomial $p(z)$ of degree m with rational coefficients that has a real root $p(r) = 0$ with the property that r cannot be written using any expression involving rational numbers, addition, subtraction, multiplication, division, and kth roots.*

This theorem implies that even if we could work in exact arithmetic, there could be no computer program that would produce the exact roots of an arbitrary polynomial in a finite number of steps. It follows that the same conclusion applies to the more general problem of computing eigenvalues of matrices.

This does not mean that we cannot write a good eigenvalue solver. It does mean, however, that such a solver cannot be based on the same kind of techniques that we have used so far for solving linear systems. Methods like Householder reflections and Gaussian elimination would solve linear systems of equations exactly in a finite number of steps if they could be implemented in exact arithmetic. By contrast,

Any eigenvalue solver must be iterative.

The goal of an eigenvalue solver is to produce *sequences of numbers that converge rapidly towards eigenvalues*. In this respect eigenvalue computations are more representative of scientific computing than solutions of linear systems of equations; see the Appendix.

The need to iterate may seem discouraging at first, but the algorithms available in this field converge extraordinarily quickly. In most cases it is possible to compute sequences of numbers that double or triple the numbers of digits of accuracy at every step. Thus, although computing eigenvalues is an "unsolvable" problem in principle, in practice it differs from the solution of linear systems by only a small constant factor, typically closer to 1 than 10. Theoretically speaking, the dependence of the operation count on $\epsilon_{\text{machine}}$ involves terms as weak as $\log(|\log(\epsilon_{\text{machine}})|)$; see Exercise 25.2.

Schur Factorization and Diagonalization

Most of the general purpose eigenvalue algorithms in use today proceed by computing the Schur factorization. We compute a Schur factorization $A = QTQ^*$ by transforming A by a sequence of elementary unitary similarity transformations $X \mapsto Q_j^* X Q_j$, so that the product

$$\underbrace{Q_j^* \cdots Q_2^* Q_1^*}_{Q^*} A \underbrace{Q_1 Q_2 \cdots Q_j}_{Q} \tag{25.4}$$

converges to an upper-triangular matrix T as $j \to \infty$.

If A is real but not symmetric, then in general it may have complex eigenvalues in conjugate pairs, in which case its Schur form will be complex. Thus an algorithm that computes the Schur factorization will have to be capable of generating complex outputs from real inputs. This can certainly be done; after all, zerofinders for polynomials with real coefficients have the same property. Alternatively, it is possible to carry out the entire computation in real arithmetic if one computes what is known as a *real Schur factorization*. Here, T is permitted to have 2×2 blocks along the diagonal, one for each complex conjugate pair of eigenvalues. This option is important in practice, and is included in all the software libraries, but we shall not give details here.

On the other hand, suppose A is hermitian. Then $Q_j^* \cdots Q_2^* Q_1^* A Q_1 Q_2 \cdots Q_j$ is also hermitian, and thus the limit of the converging sequence is both triangular and hermitian, hence diagonal. This implies that the same algorithms that compute a unitary triangularization of a general matrix also compute a unitary diagonalization of a hermitian matrix. In practice, this is essentially how the hermitian case is typically handled, although various modifications are introduced to take special advantage of the hermitian structure at each step.

Two Phases of Eigenvalue Computations

Whether or not A is hermitian, the sequence (25.4) is usually split into two phases. In the first phase, a direct method is applied to produce an *upper-Hessenberg* matrix H, that is, a matrix with zeros below the first subdiagonal. In the second phase, an iteration is applied to generate a formally infinite sequence of Hessenberg matrices that converge to a triangular form. Schematically, the process looks like this:

$$
\begin{bmatrix}
\times & \times & \times & \times & \times \\
\times & \times & \times & \times & \times \\
\times & \times & \times & \times & \times \\
\times & \times & \times & \times & \times \\
\times & \times & \times & \times & \times
\end{bmatrix}
\xrightarrow{\text{Phase 1}}
\begin{bmatrix}
\times & \times & \times & \times & \times \\
\times & \times & \times & \times & \times \\
 & \times & \times & \times & \times \\
 & & \times & \times & \times \\
 & & & \times & \times
\end{bmatrix}
\xrightarrow{\text{Phase 2}}
\begin{bmatrix}
\times & \times & \times & \times & \times \\
 & \times & \times & \times & \times \\
 & & \times & \times & \times \\
 & & & \times & \times \\
 & & & & \times
\end{bmatrix}.
$$
$$A \neq A^* \qquad\qquad\qquad H \qquad\qquad\qquad T$$

The first phase, a direct reduction, requires $O(m^3)$ flops. The second, iterative phase never terminates in principle, and if left to run forever would require an infinite number of flops. However, in practice, convergence to machine precision is achieved in $O(m)$ iterations. Each iteration requires $O(m^2)$ flops, and thus the total work requirement is $O(m^3)$ flops. These figures explain the importance of Phase 1. Without that preliminary step, each iteration of Phase 2 would involve a full matrix, requiring $O(m^3)$ work, and this would bring the total to $O(m^4)$—or higher, since convergence might also sometimes require more than $O(m)$ iterations.

If A is hermitian, the two-phase approach becomes even faster. The intermediate matrix is now a hermitian Hessenberg matrix, that is, *tridiagonal*. The final result is a hermitian triangular matrix, that is, diagonal, as mentioned above. Schematically:

$$
\begin{bmatrix}
\times & \times & \times & \times & \times \\
\times & \times & \times & \times & \times \\
\times & \times & \times & \times & \times \\
\times & \times & \times & \times & \times \\
\times & \times & \times & \times & \times
\end{bmatrix}
\xrightarrow{\text{Phase 1}}
\begin{bmatrix}
\times & \times & & & \\
\times & \times & \times & & \\
 & \times & \times & \times & \\
 & & \times & \times & \times \\
 & & & \times & \times
\end{bmatrix}
\xrightarrow{\text{Phase 2}}
\begin{bmatrix}
\times & & & & \\
 & \times & & & \\
 & & \times & & \\
 & & & \times & \\
 & & & & \times
\end{bmatrix}.
$$
$$A = A^* \qquad\qquad\qquad T \qquad\qquad\qquad D$$

In this hermitian case we shall see that if only eigenvalues are required (not eigenvectors), then each step of Phase 2 can be carried out with only $O(m)$ flops, bringing the total work estimate for Phase 2 to $O(m^2)$ flops. Thus, for hermitian eigenvalue problems, we are in the paradoxical situation that the "infinite" part of the algorithm is in practice not merely as fast as the "finite" part, but an order of magnitude faster.

Exercises

25.1. (a) Let $A \in \mathbb{C}^{m \times m}$ be tridiagonal and hermitian, with all its sub- and superdiagonal entries nonzero. Prove that the eigenvalues of A are distinct. (Hint: Show that for any $\lambda \in \mathbb{C}$, $A - \lambda I$ has rank at least $m - 1$.)

(b) On the other hand, let A be upper-Hessenberg, with all its subdiagonal entries nonzero. Give an example that shows that the eigenvalues of A are not necessarily distinct.

25.2. Let $e_1, e_2, e_3, \ldots$ be a sequence of nonnegative numbers representing errors in some iterative process that converge to zero, and suppose there are a constant C and an exponent α such that for all sufficiently large k, $e_{k+1} \leq C(e_k)^\alpha$. Various algorithms for "Phase 2" of an eigenvalue calculation exhibit *cubic convergence* ($\alpha = 3$), *quadratic convergence* ($\alpha = 2$), or *linear convergence* ($\alpha = 1$ with $C < 1$), which is also, perhaps confusingly, known as *geometric convergence*.

(a) Suppose we want an answer of accuracy $O(\epsilon_{\text{machine}})$. Assuming the amount of work for each step is $O(1)$, show that the total work requirement in the case of linear convergence is $O(\log(\epsilon_{\text{machine}}))$. How does the constant C enter into your work estimate?

(b) Show that in the case of superlinear convergence, i.e., $\alpha > 1$, the work requirement becomes $O(\log(|\log(\epsilon_{\text{machine}})|))$. (Hint: The problem may be simplified by defining a new error measure $f_k = C^{1/(\alpha-1)} e_k$.) How does the exponent α enter into your work estimate?

25.3. Suppose we have a 3×3 matrix and wish to introduce zeros by left- and/or right-multiplications by unitary matrices Q_j such as Householder reflectors or Givens rotations. Consider the following three matrix structures:

$$\text{(a)} \begin{bmatrix} \times & \times & 0 \\ 0 & \times & \times \\ 0 & 0 & \times \end{bmatrix}, \qquad \text{(b)} \begin{bmatrix} \times & \times & 0 \\ \times & 0 & \times \\ 0 & \times & \times \end{bmatrix}, \qquad \text{(c)} \begin{bmatrix} \times & \times & 0 \\ 0 & 0 & \times \\ 0 & 0 & \times \end{bmatrix}.$$

For each one, decide which of the following situations holds, and justify your claim.

(i) Can be obtained by a sequence of left-multiplications by matrices Q_j;

(ii) Not (i), but can be obtained by a sequence of left- and right-multiplications by matrices Q_j;

(iii) Cannot be obtained by any sequence of left- and right-multiplications by matrices Q_j.

Lecture 26. Reduction to Hessenberg or Tridiagonal Form

We now describe the first of the two computational phases outlined in the previous lecture: reduction of a full matrix to Hessenberg form by a sequence of unitary similarity transformations. If the original matrix is hermitian, the result is tridiagonal.

A Bad Idea

To compute the Schur factorization $A = QTQ^*$, we would like to apply unitary similarity transformations to A in such a way as to introduce zeros below the diagonal. A natural first idea might be to attempt direct triangularization by using Householder reflectors to introduce these zeros, one after another.

The first Householder reflector Q_1^*, multiplied on the left of A, would introduce zeros below the diagonal in the first column of A. In the process it will change all of the rows of A. In this and the following diagrams, as usual, entries that are changed at each step are written in boldface:

$$
\begin{bmatrix}
\times & \times & \times & \times & \times \\
\times & \times & \times & \times & \times \\
\times & \times & \times & \times & \times \\
\times & \times & \times & \times & \times \\
\times & \times & \times & \times & \times
\end{bmatrix}
\quad \xrightarrow{\;Q_1^*\cdot\;} \quad
\begin{bmatrix}
\mathbf{\times} & \mathbf{\times} & \mathbf{\times} & \mathbf{\times} & \mathbf{\times} \\
\mathbf{0} & \mathbf{\times} & \mathbf{\times} & \mathbf{\times} & \mathbf{\times} \\
\mathbf{0} & \mathbf{\times} & \mathbf{\times} & \mathbf{\times} & \mathbf{\times} \\
\mathbf{0} & \mathbf{\times} & \mathbf{\times} & \mathbf{\times} & \mathbf{\times} \\
\mathbf{0} & \mathbf{\times} & \mathbf{\times} & \mathbf{\times} & \mathbf{\times}
\end{bmatrix} .
$$

$$A \qquad\qquad\qquad\qquad Q_1^* A$$

Unfortunately, to complete the similarity transformation, we must also multiply by Q_1 on the right of A:

$$
\begin{bmatrix}
\times & \times & \times & \times & \times \\
 & \times & \times & \times & \times \\
 & \times & \times & \times & \times \\
 & \times & \times & \times & \times \\
 & \times & \times & \times & \times
\end{bmatrix}
\quad \overset{\cdot Q_1}{\longrightarrow} \quad
\begin{bmatrix}
\times & \times & \times & \times & \times \\
\times & \times & \times & \times & \times \\
\times & \times & \times & \times & \times \\
\times & \times & \times & \times & \times \\
\times & \times & \times & \times & \times
\end{bmatrix} .
$$

$$
\qquad\quad Q_1^*A \qquad\qquad\qquad\qquad Q_1^*AQ_1
$$

This has the effect of replacing each column of the matrix by a linear combination of all the columns. The result is that the zeros that were previously introduced are destroyed; we are no better off than when we started.

Of course, with hindsight we know that this idea had to fail, because of the "fundamental difficulty" described in the previous lecture. No finite process can reveal the eigenvalues of A exactly.

Curiously, this too-simple strategy, which appears futile as we have discussed it, does have the effect, typically, of reducing the size of the entries below the diagonal, even if it does not make them zero. We shall return to this "bad idea" when we discuss the QR algorithm.

A Good Idea

The right strategy for introducing zeros in Phase 1 is to be less ambitious and operate on fewer entries of the matrix. We shall only conquer territory we are sure we can defend.

At the first step, we select a Householder reflector Q_1^* that leaves the first row unchanged. When it is multiplied on the left of A, it forms linear combinations of only rows $2,\ldots,m$ to introduce zeros into rows $3,\ldots,m$ of the first column. Then, when Q_1 is multiplied on the right of Q_1^*A, it leaves the first column unchanged. It forms linear combinations of columns $2,\ldots,m$ and does not alter the zeros that have been introduced:

$$
\begin{bmatrix}
\times & \times & \times & \times & \times \\
\times & \times & \times & \times & \times \\
\times & \times & \times & \times & \times \\
\times & \times & \times & \times & \times \\
\times & \times & \times & \times & \times
\end{bmatrix}
\overset{Q_1^* \cdot}{\longrightarrow}
\begin{bmatrix}
\times & \times & \times & \times & \times \\
\times & \times & \times & \times & \times \\
0 & \times & \times & \times & \times \\
0 & \times & \times & \times & \times \\
0 & \times & \times & \times & \times
\end{bmatrix}
\overset{\cdot Q_1}{\longrightarrow}
\begin{bmatrix}
\times & \times & \times & \times & \times \\
\times & \times & \times & \times & \times \\
 & \times & \times & \times & \times \\
 & \times & \times & \times & \times \\
 & \times & \times & \times & \times
\end{bmatrix} .
$$

$$
\qquad A \qquad\qquad\qquad\quad Q_1^*A \qquad\qquad\qquad Q_1^*AQ_1
$$

This idea is repeated to introduce zeros into subsequent columns. For example, the second Householder reflector, Q_2, leaves the first and second rows and columns unchanged:

$$
\begin{bmatrix}
\times & \times & \times & \times & \times \\
\times & \times & \times & \times & \times \\
  & \times & \times & \times & \times \\
  & \times & \times & \times & \times \\
  & \times & \times & \times & \times
\end{bmatrix}
\xrightarrow{\;Q_2^*\cdot\;}
\begin{bmatrix}
\times & \times & \times & \times & \times \\
\times & \times & \times & \times & \times \\
  & \mathbf{\times} & \mathbf{\times} & \mathbf{\times} & \mathbf{\times} \\
  & \mathbf{0} & \mathbf{\times} & \mathbf{\times} & \mathbf{\times} \\
  & \mathbf{0} & \mathbf{\times} & \mathbf{\times} & \mathbf{\times}
\end{bmatrix}
\xrightarrow{\;\cdot Q_2\;}
\begin{bmatrix}
\times & \times & \mathbf{\times} & \mathbf{\times} & \mathbf{\times} \\
\times & \times & \mathbf{\times} & \mathbf{\times} & \mathbf{\times} \\
  & \times & \mathbf{\times} & \mathbf{\times} & \mathbf{\times} \\
  & & \mathbf{\times} & \mathbf{\times} & \mathbf{\times} \\
  & & \mathbf{\times} & \mathbf{\times} & \mathbf{\times}
\end{bmatrix}.
$$

$$
Q_1^*AQ_1 \qquad\qquad\qquad Q_2^*Q_1^*AQ_1 \qquad\qquad\qquad Q_2^*Q_1^*AQ_1Q_2
$$

After repeating this process $m-2$ times, we have a product in Hessenberg form, as desired:

$$
\begin{bmatrix}
\times & \times & \times & \times & \times \\
\times & \times & \times & \times & \times \\
  & \times & \times & \times & \times \\
  & & \times & \times & \times \\
  & & & \times & \times
\end{bmatrix}
$$

$$
\underbrace{Q_{m-2}^* \cdots Q_2^* Q_1^*}_{Q^*} \, A \, \underbrace{Q_1 Q_2 \cdots Q_{m-2}}_{Q} = H.
$$

The algorithm is formulated below; compare Algorithm 10.1.

Algorithm 26.1. Householder Reduction to Hessenberg Form

for $k = 1$ **to** $m-2$

$\qquad x = A_{k+1:m,k}$

$\qquad v_k = \mathrm{sign}(x_1)\|x\|_2 e_1 + x$

$\qquad v_k = v_k/\|v_k\|_2$

$\qquad A_{k+1:m,k:m} = A_{k+1:m,k:m} - 2v_k(v_k^* A_{k+1:m,k:m})$

$\qquad A_{1:m,k+1:m} = A_{1:m,k+1:m} - 2(A_{1:m,k+1:m} v_k)v_k^*$

Just as in Algorithm 10.1, here the matrix $Q = \prod_{k=1}^{m-2} Q_k$ is never formed explicitly. The reflection vectors v_k are saved instead, and can be used to multiply by Q or reconstruct Q later if necessary. For details, see Lecture 10.

Operation Count

The number of operations required by Algorithm 26.1 can be counted with the same geometric reasoning we have used before. The rule of thumb is that unitary operations require four flops for each element operated upon.

The work is dominated by the two updates of submatrices of A. The first loop applies a Householder reflector on the left of the matrix. The kth such reflector operates on the last $m-k$ rows. Since at the time the reflector is applied, these rows have zeros in the first $k-1$ columns, arithmetic has to be performed only on the last $m - k + 1$ entries of each row. The picture is as follows:

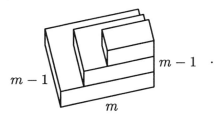

As $m \to \infty$, the volume converges to $\frac{1}{3}m^3$. At four flops per element, the amount of work in this loop is $\sim \frac{4}{3}m^3$ flops.

The second inner loop applies a Householder reflector on the right of the matrix. At the kth step, the reflector operates by forming linear combinations of the last $m - k$ columns. This loop involves more work than the first one because there are no zeros that can be ignored. Arithmetic must be performed on all of the m entries of each of the columns operated upon, a total of $m(m-k)$ entries for a single value of k. The picture looks like this:

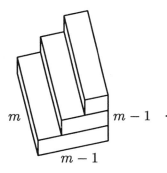

The volume converges as $m \to \infty$ to $\frac{1}{2}m^3$, so, at four flops per element, this second loop requires $\sim 2m^3$ flops.

All together, the total amount of work for unitary reduction of an $m \times m$ matrix to Hessenberg form is:

$$\text{Work for Hessenberg reduction:} \quad \sim \frac{10}{3}m^3 \text{ flops.} \quad (26.1)$$

The Hermitian Case: Reduction to Tridiagonal Form

If A is hermitian, the algorithm just described will reduce A to tridiagonal form (at least, in the absence of rounding errors). This is easy to see: since A is hermitian, Q^*AQ is also hermitian, and any hermitian Hessenberg matrix is tridiagonal.

Since zeros are now introduced in rows as well as columns, additional arithmetic can be avoided by ignoring these additional zeros. With this optimization, applying a Householder reflector on the right is as cheap as applying the reflector on the left, and the total cost of applying the right reflectors is reduced from $2m^3$ to $\frac{4}{3}m^3$ flops. We have two pyramids to add up instead of a pyramid and a prism, and the total amount of arithmetic is reduced to $\frac{8}{3}m^3$ flops.

This saving, however, is based only on sparsity, not symmetry. In fact, at every stage of the computation, the matrix being operated upon is hermitian. This gives another factor of two that can be taken advantage of, bringing the total work estimate to

$$\text{Work for tridiagonal reduction:} \quad \sim \frac{4}{3}m^3 \text{ flops.} \qquad (26.2)$$

We shall not give the details of the implementation.

Stability

Like the Householder algorithm for QR factorization, the algorithm just described is backward stable. Recall from Theorem 16.1 that, for any $A \in \mathbb{C}^{m \times n}$, the Householder algorithm for QR factorization computes reflection vectors equivalent to an implicit, exactly unitary factor $\tilde{Q}$ (16.2), as well as an explicit upper-triangular factor $\tilde{R}$, such that

$$\tilde{Q}\tilde{R} = A + \delta A, \qquad \frac{\|\delta A\|}{\|A\|} = O(\epsilon_{\text{machine}}).$$

The same kind of error estimate can be established for Algorithm 26.1. Let $\tilde{H}$ be the actual Hessenberg matrix computed in floating point arithmetic, and let $\tilde{Q}$, as before, be the exactly unitary matrix (16.2) corresponding to the reflection vectors $\tilde{v}_k$ computed in floating point arithmetic. The following result can be proved.

Theorem 26.1. *Let the Hessenberg reduction $A = QHQ^*$ of a matrix $A \in \mathbb{C}^{m \times m}$ be computed by Algorithm 26.1 on a computer satisfying the axioms (13.5) and (13.7), and let the computed factors $\tilde{Q}$ and $\tilde{H}$ be defined as indicated above. Then we have*

$$\tilde{Q}\tilde{H}\tilde{Q}^* = A + \delta A, \qquad \frac{\|\delta A\|}{\|A\|} = O(\epsilon_{\text{machine}}) \qquad (26.3)$$

for some $\delta A \in \mathbb{C}^{m \times m}$.

Exercises

26.1. Theorem 26.1 and its successors in later lectures show that we can compute eigenvalues $\{\tilde{\lambda}_k\}$ of A numerically that are the exact eigenvalues of a matrix $A + \delta A$ with $\|\delta A\|/\|A\| = O(\epsilon_{\text{machine}})$. Does this mean they are close to the exact eigenvalues $\{\lambda_k\}$ of A? This is a question of eigenvalue perturbation theory.

One can approach such problems geometrically as follows. Given $A \in \mathbb{C}^{m \times m}$ with spectrum $\Lambda(A) \subseteq \mathbb{C}$ and $\epsilon > 0$, define the 2-norm ϵ-*pseudospectrum* of A, $\Lambda_\epsilon(A)$, to be the set of numbers $z \in \mathbb{C}$ satisfying any of the following conditions:

   (i) z is an eigenvalue of $A + \delta A$ for some δA with $\|\delta A\|_2 \leq \epsilon$;

   (ii) There exists a vector $u \in \mathbb{C}^m$ with $\|(A - zI)u\|_2 \leq \epsilon$ and $\|u\|_2 = 1$;

   (iii) $\sigma_m(zI - A) \leq \epsilon$;

   (iv) $\|(zI - A)^{-1}\|_2 \geq \epsilon^{-1}$.

The matrix $(zI - A)^{-1}$ in (iv) is known as the *resolvent* of A at z; if z is an eigenvalue of A, we use the convention $\|(zI - A)^{-1}\|_2 = \infty$. In (iii), σ_m denotes the smallest singular value.

Prove that conditions (i)–(iv) are equivalent.

26.2. Let A be the 32×32 matrix with -1 on the main diagonal, 1 on the first and second superdiagonals, and 0 elsewhere.

(a) Using an SVD algorithm built into MATLAB or another software system, together with contour plotting software, generate a plot of the boundaries of the 2-norm ϵ-pseudospectra of A for $\epsilon = 10^{-1}, 10^{-2}, \ldots, 10^{-8}$.

(b) Produce a semilogy plot of $\|e^{tA}\|_2$ against t for $0 \leq t \leq 50$. What is the initial growth rate of the curve before the eventual decay sets in? Can you relate this to your plot of pseudospectra? (Compare Exercise 24.3.)

26.3. One of the best known results of eigenvalue perturbation theory is the *Bauer–Fike theorem*. Suppose $A \in \mathbb{C}^{m \times m}$ is diagonalizable with $A = V \Lambda V^{-1}$, and let $\delta A \in \mathbb{C}^{m \times m}$ be arbitrary. Then every eigenvalue of $A + \delta A$ lies in at least one of the m circular disks in the complex plane of radius $\kappa(V)\|\delta A\|_2$ centered at the eigenvalues of A, where κ is the 2-norm condition number. (Compare Exercise 24.2.)

(a) Prove the Bauer–Fike theorem by using the equivalence of conditions (i) and (iv) of Exercise 26.1.

(b) Suppose A is normal. Show that for each eigenvalue $\tilde{\lambda}_j$ of $A + \delta A$, there is an eigenvalue λ_j of A such that

$$|\tilde{\lambda}_j - \lambda_j| \leq \|\delta A\|_2. \tag{26.4}$$

Lecture 27. Rayleigh Quotient, Inverse Iteration

In this lecture we present some classical eigenvalue algorithms. Individually, these tools are useful in certain circumstances—especially inverse iteration, which is the standard method for determining an eigenvector when the corresponding eigenvalue is known. Combined, they are the ingredients of the celebrated QR algorithm, described in the next two lectures.

Restriction to Real Symmetric Matrices

Throughout numerical linear algebra, most algorithmic ideas are applicable either to general matrices or, with certain simplifications, to hermitian matrices. For the topics discussed in this and the next three lectures, this continues to be at least partly true, but some of the differences between the general and the hermitian cases are rather sizable. Therefore, in these four lectures, we simplify matters by considering only matrices that are real and symmetric. We also assume throughout that $\| \cdot \| = \| \cdot \|_2$.

Thus, for these four lectures: $A = A^T \in \mathbb{R}^{m \times m}$, $x \in \mathbb{R}^m$, $x^* = x^T$, $\|x\| = \sqrt{x^T x}$. In particular, this means that A has real eigenvalues and a complete set of orthogonal eigenvectors. We use the following notation:

$$\text{real eigenvalues: } \lambda_1, \dots, \lambda_m,$$
$$\text{orthonormal eigenvectors: } q_1, \dots, q_m.$$

The eigenvectors are presumed normalized by $\|q_j\| = 1$, and the ordering of the eigenvalues will be specified as necessary.

Most of the ideas to be described in the next few lectures pertain to Phase 2 of the two phases described in Lecture 25. This means that by the time we come to applying these ideas, A will be not just real and symmetric, but tridiagonal. This tridiagonal structure is occasionally of mathematical importance, for example in choosing shifts for the QR algorithm, and it is always of algorithmic importance, reducing many steps from $O(m^3)$ to $O(m)$ flops, as discussed at the end of the lecture.

Rayleigh Quotient

The *Rayleigh quotient* of a vector $x \in \mathbb{R}^m$ is the scalar

$$r(x) = \frac{x^T A x}{x^T x}. \tag{27.1}$$

Notice that if x is an eigenvector, then $r(x) = \lambda$ is the corresponding eigenvalue. One way to motivate this formula is to ask: given x, what scalar α "acts most like an eigenvalue" for x in the sense of minimizing $\|Ax - \alpha x\|_2$? This is an $m \times 1$ least squares problem of the form $x\alpha \approx Ax$ (x is the matrix, α is the unknown vector, Ax is the right-hand side). By writing the normal equations (11.9) for this system, we obtain the answer: $\alpha = r(x)$. Thus $r(x)$ is a natural eigenvalue estimate to consider if x is close to, but not necessarily equal to, an eigenvector.

To make these ideas quantitative, it is fruitful to view $x \in \mathbb{R}^m$ as a variable, so that r is a function $\mathbb{R}^m \to \mathbb{R}$. We are interested in the local behavior of $r(x)$ when x is near an eigenvector. One way to approach this question is to calculate the partial derivatives of $r(x)$ with respect to the coordinates x_j:

$$\frac{\partial r(x)}{\partial x_j} = \frac{\frac{\partial}{\partial x_j}(x^T A x)}{x^T x} - \frac{(x^T A x)\frac{\partial}{\partial x_j}(x^T x)}{(x^T x)^2}$$

$$= \frac{2(Ax)_j}{x^T x} - \frac{(x^T A x)2x_j}{(x^T x)^2} = \frac{2}{x^T x}(Ax - r(x)x)_j.$$

If we collect these partial derivatives into an m-vector, we find we have calculated the *gradient* of $r(x)$, denoted by $\nabla r(x)$. We have shown:

$$\nabla r(x) = \frac{2}{x^T x}(Ax - r(x)x). \tag{27.2}$$

From this formula we see that at an eigenvector x of A, the gradient of $r(x)$ is the zero vector. Conversely, if $\nabla r(x) = 0$ with $x \neq 0$, then x is an eigenvector and $r(x)$ is the corresponding eigenvalue.

Geometrically speaking, the eigenvectors of A are the *stationary points* of the function $r(x)$, and the eigenvalues of A are the values of $r(x)$ at these

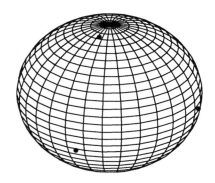

Figure 27.1. *The Rayleigh quotient $r(x)$ is a continuous function on the unit sphere $\|x\| = 1$ in $\mathbb{R}^m$, and the stationary points of $r(x)$ are the normalized eigenvectors of A. In this example with $m = 3$, there are three orthogonal stationary points (as well as their antipodes).*

stationary points. Actually, since $r(x)$ is independent of the scale of x, these stationary points lie along lines through the origin in $\mathbb{R}^m$. If we normalize by restricting attention to the unit sphere $\|x\| = 1$, they become isolated points (assuming that the eigenvalues of A are simple), as suggested in Figure 27.1.

Let q_J be one of the eigenvectors of A. From the fact that $\nabla r(q_J) = 0$, together with the smoothness of the function $r(x)$ (everywhere except at the origin $x = 0$), we derive an important consequence:

$$r(x) - r(q_J) = O(\|x - q_J\|^2) \quad \text{as } x \to q_J. \tag{27.3}$$

Thus the Rayleigh quotient is a *quadratically accurate* estimate of an eigenvalue. Herein lies its power.

A more explicit way to derive (27.3) is to expand x as a linear combination of the eigenvectors $q_1, \ldots, q_m$ of A. If $x = \sum_{j=1}^m a_j q_j$, then $r(x) = \sum_{j=1}^m a_j^2 \lambda_j / \sum_{j=1}^m a_j^2$. Thus $r(x)$ is a weighted mean of the eigenvalues of A, with the weights equal to the squares of the coordinates of x in the eigenvector basis. Because of this squaring of the coordinates, it is not hard to see that if $|a_j/a_J| \leq \epsilon$ for all $j \neq J$, then $r(x) - r(q_J) = O(\epsilon^2)$.

Power Iteration

Now we switch tacks. Suppose $v^{(0)}$ is a vector with $\|v^{(0)}\| = 1$. The following process, *power iteration*, was cited as a not especially good idea at the beginning of Lecture 25. It may be expected to produce a sequence $v^{(i)}$ that converges to an eigenvector corresponding to the largest eigenvalue of A.

Algorithm 27.1. Power Iteration

$v^{(0)}$ = some vector with $\|v^{(0)}\| = 1$
for $k = 1, 2, \ldots$
   $w = Av^{(k-1)}$ apply A
   $v^{(k)} = w/\|w\|$ normalize
   $\lambda^{(k)} = (v^{(k)})^T A v^{(k)}$ Rayleigh quotient

In this and the algorithms to follow, we give no attention to termination conditions, describing the loop only by the suggestive expression "for $k = 1, 2, \ldots$." Of course, in practice, termination conditions are very important, and this is one of the points where top-quality software such as can be found in LAPACK or MATLAB is likely to be superior to a program an individual might write.

We can analyze power iteration easily. Write $v^{(0)}$ as a linear combination of the orthonormal eigenvectors q_i:

$$v^{(0)} = a_1 q_1 + a_2 q_2 + \cdots + a_m q_m.$$

Since $v^{(k)}$ is a multiple of $A^k v^{(0)}$, we have for some constants c_k

$$
\begin{aligned}
v^{(k)} &= c_k A^k v^{(0)} \\
&= c_k(a_1 \lambda_1^k q_1 + a_2 \lambda_2^k q_2 + \cdots + a_m \lambda_m^k q_m) \\
&= c_k \lambda_1^k \left(a_1 q_1 + a_2 (\lambda_2/\lambda_1)^k q_2 + \cdots + a_m (\lambda_m/\lambda_1)^k q_m \right). \quad (27.4)
\end{aligned}
$$

From here we obtain the following conclusion.

Theorem 27.1. *Suppose $|\lambda_1| > |\lambda_2| \geq \cdots \geq |\lambda_m| \geq 0$ and $q_1^T v^{(0)} \neq 0$. Then the iterates of Algorithm 27.1 satisfy*

$$\|v^{(k)} - (\pm q_1)\| = O\left(\left| \frac{\lambda_2}{\lambda_1} \right|^k \right), \qquad |\lambda^{(k)} - \lambda_1| = O\left(\left| \frac{\lambda_2}{\lambda_1} \right|^{2k} \right) \quad (27.5)$$

as $k \to \infty$. The $\pm$ sign means that at each step k, one or the other choice of sign is to be taken, and then the indicated bound holds.

Proof. The first equation follows from (27.4), since $a_1 = q_1^T v^{(0)} \neq 0$ by assumption. The second follows from this and (27.3). If $\lambda_1 > 0$, then the $\pm$ signs are all $+$ or all $-$, whereas if $\lambda_1 < 0$, they alternate. $\square$

The $\pm$ signs in (27.5) and in similar equations below are not very appealing. There is an elegant way to avoid these complications, which is to speak of convergence of subpaces, not vectors—to say that $\langle v^{(k)} \rangle$ converges to $\langle q_1 \rangle$, for example. However, we shall not do this, in order to avoid getting into the details of how convergence of subspaces can be made precise.

On its own, power iteration is of limited use, for several reasons. First, it can find only the eigenvector corresponding to the largest eigenvalue. Second, the convergence is linear, reducing the error only by a constant factor $\approx |\lambda_2/\lambda_1|$ at each iteration. Finally, the quality of this factor depends on having a largest eigenvalue that is significantly larger than the others. If the largest two eigenvalues are close in magnitude, the convergence will be very slow.

Fortunately, there is a way to amplify the differences between eigenvalues.

Inverse Iteration

For any $\mu \in \mathbb{R}$ that is not an eigenvalue of A, the eigenvectors of $(A - \mu I)^{-1}$ are the same as the eigenvectors of A, and the corresponding eigenvalues are $\{(\lambda_j - \mu)^{-1}\}$, where $\{\lambda_j\}$ are the eigenvalues of A. This suggests an idea. Suppose μ is close to an eigenvalue λ_J of A. Then $(\lambda_J - \mu)^{-1}$ may be much larger than $(\lambda_j - \mu)^{-1}$ for all $j \neq J$. Thus, if we apply power iteration to $(A - \mu I)^{-1}$, the process will converge rapidly to q_J. This idea is called *inverse iteration*.

Algorithm 27.2. Inverse Iteration

$v^{(0)} =$ some vector with $\|v^{(0)}\| = 1$
for $k = 1, 2, \ldots$
    Solve $(A - \mu I)w = v^{(k-1)}$ for w apply $(A - \mu I)^{-1}$
    $v^{(k)} = w/\|w\|$ normalize
    $\lambda^{(k)} = (v^{(k)})^T A v^{(k)}$ Rayleigh quotient

---

What if μ is an eigenvalue of A, so that $A - \mu I$ is singular? What if it is nearly an eigenvalue, so that $A - \mu I$ is so ill-conditioned that an accurate solution of $(A - \mu I)w = v^{(k-1)}$ cannot be expected? These apparent pitfalls of inverse iteration cause no trouble at all; see Exercise 27.5.

Like power iteration, inverse iteration exhibits only linear convergence. Unlike power iteration, however, we can choose the eigenvector that will be found by supplying an estimate μ of the corresponding eigenvalue. Furthermore, the rate of linear convergence can be controlled, for it depends on the quality of μ. If μ is much closer to one eigenvalue of A than to the others, then the largest eigenvalue of $(A - \mu I)^{-1}$ will be much larger than the rest. Using the same reasoning as with power iteration, we obtain the following theorem.

Theorem 27.2. *Suppose λ_J is the closest eigenvalue to μ and λ_K is the second closest, that is, $|\mu - \lambda_J| < |\mu - \lambda_K| \leq |\mu - \lambda_j|$ for each $j \neq J$. Furthermore,* *suppose $q_J^T v^{(0)} \neq 0$. Then the iterates of Algorithm 27.2 satisfy*

$$\|v^{(k)} - (\pm q_J)\| = O\left(\left|\frac{\mu - \lambda_J}{\mu - \lambda_K}\right|^k\right), \qquad |\lambda^{(k)} - \lambda_J| = O\left(\left|\frac{\mu - \lambda_J}{\mu - \lambda_K}\right|^{2k}\right)$$

as $k \to \infty$, where the $\pm$ sign has the same meaning as in Theorem 27.1.

Inverse iteration is one of the most valuable tools of numerical linear algebra, for it is the standard method of calculating one or more eigenvectors of a matrix if the eigenvalues are already known. In this case Algorithm 27.2 is applied as written, except that the calculation of the Rayleigh quotient is dispensed with.

Rayleigh Quotient Iteration

So far in this lecture, we have presented one method for obtaining an eigenvalue estimate from an eigenvector estimate (the Rayleigh quotient), and another method for obtaining an eigenvector estimate from an eigenvalue estimate (inverse iteration). The possibility of combining these ideas is irresistible:

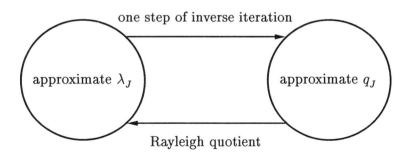

(The figure is oversimplified; to get from an approximate λ_J to an approximate q_J by a step of inverse iteration, one also needs a preliminary approximation to q_J.) The idea is to use continually improving eigenvalue estimates to increase the rate of convergence of inverse iteration at every step. This algorithm is called *Rayleigh quotient iteration*.

Algorithm 27.3. Rayleigh Quotient Iteration

$v^{(0)} =$ some vector with $\|v^{(0)}\| = 1$
$\lambda^{(0)} = (v^{(0)})^T A v^{(0)} =$ corresponding Rayleigh quotient
for $k = 1, 2, \ldots$
        Solve $(A - \lambda^{(k-1)}I)w = v^{(k-1)}$ for w apply $(A - \lambda^{(k-1)}I)^{-1}$
        $v^{(k)} = w/\|w\|$ normalize
        $\lambda^{(k)} = (v^{(k)})^T A v^{(k)}$ Rayleigh quotient

The convergence of this algorithm is spectacular: each iteration triples the number of digits of accuracy.

Theorem 27.3. *Rayleigh quotient iteration converges to an eigenvalue/eigenvector pair for all except a set of measure zero of starting vectors $v^{(0)}$. When it converges, the convergence is ultimately cubic in the sense that if λ_J is an eigenvalue of A and $v^{(0)}$ is sufficiently close to the eigenvector q_J, then*

$$\|v^{(k+1)} - (\pm q_J)\| = O(\|v^{(k)} - (\pm q_J)\|^3) \qquad (27.6)$$

and

$$|\lambda^{(k+1)} - \lambda_J| = O(|\lambda^{(k)} - \lambda_J|^3) \qquad (27.7)$$

as $k \to \infty$. The $\pm$ signs are not necessarily the same on the two sides of (27.6).

Proof. We shall not prove the assertion about convergence for almost all starting vectors. Here, however, is a proof that if convergence occurs, it is ultimately cubic. For simplicity, we assume that the eigenvalue λ_J is simple. By (27.3), if $\|v^{(k)} - q_J\| \le \epsilon$ for sufficiently small ϵ, then the Rayleigh quotient yields an eigenvalue estimate $\lambda^{(k)}$ with $|\lambda^{(k)} - \lambda_J| = O(\epsilon^2)$. By the argument used to prove Theorem 27.2, if we now take one step of inverse iteration to obtain a new $v^{(k+1)}$ from $v^{(k)}$ and $\lambda^{(k)}$, then

$$\|v^{(k+1)} - q_J\| = O(|\lambda^{(k)} - \lambda_J|\,\|v^{(k)} - q_J\|) = O(\epsilon^3).$$

Moreover, the constants implicit in the O symbols are uniform throughout sufficiently small neighborhoods of λ_J and q_J. Thus we have convergence in the following pattern:

$$
\begin{array}{ccc}
\|v^{(k)} - (\pm q_J)\| & & |\lambda^{(k)} - \lambda_J| \\
\epsilon & \to & O(\epsilon^2) \\
\downarrow & \nearrow & \\
O(\epsilon^3) & \to & O(\epsilon^6) \\
\downarrow & \nearrow & \\
O(\epsilon^9) & \to & O(\epsilon^{18}) \\
\vdots & & \vdots
\end{array}
$$

The estimates (27.6)–(27.7) follow from the uniformity just mentioned. $\square$

Example 27.1. Cubic convergence is so fast that we must give a numerical example. Consider the symmetric matrix

$$A = \begin{bmatrix} 2 & 1 & 1 \\ 1 & 3 & 1 \\ 1 & 1 & 4 \end{bmatrix},$$

and let $v^{(0)} = (1, 1, 1)^T/\sqrt{3}$ be the initial eigenvector estimate. When Rayleigh quotient iteration is applied to A, the following values $\lambda^{(k)}$ are computed by the first three iterations:

$$\lambda^{(0)} = 5, \quad \lambda^{(1)} = 5.2131\ldots, \quad \lambda^{(2)} = 5.214319743184\ldots.$$

The actual value of the eigenvalue corresponding to the eigenvector closest to $v^{(0)}$ is $\lambda = 5.214319743377$. After only three iterations, Rayleigh quotient iteration has produced a result accurate to ten digits. Three more iterations would increase this figure to about 270 digits, if our machine precision were high enough.   □

Operation Counts

We close this lecture with a note on the amount of work required to execute each step of the three iterations we have described.

First, suppose $A \in \mathbb{R}^{m \times m}$ is a full matrix. Then each step of power iteration involves a matrix-vector multiplication, requiring $O(m^2)$ flops. Each step of inverse iteration involves the solution of a linear system, which might seem to require $O(m^3)$ flops, but this figure reduces to $O(m^2)$ if the matrix is processed in advance by LU or QR factorization or another method. In the case of Rayleigh quotient iteration, the matrix to be inverted changes at each step, and beating $O(m^3)$ flops per step is not so straightforward.

These figures improve greatly if A is tridiagonal. Now, all three iterations require just $O(m)$ flops per step. For the analogous iterations involving non-symmetric matrices, incidentally, we must deal with Hessenberg instead of tridiagonal structure, and this figure increases to $O(m^2)$.

Exercises

27.1. Let $A \in \mathbb{C}^{m \times m}$ be given, not necessarily hermitian. Show that a number $z \in \mathbb{C}$ is a Rayleigh quotient of A if and only if it is a diagonal entry of Q^*AQ for some unitary matrix Q. Thus Rayleigh quotients are just diagonal entries of matrices, once you transform orthogonally to the right coordinate system.

27.2. Again let $A \in \mathbb{C}^{m \times m}$ be arbitrary. The set of all Rayleigh quotients of A, corresponding to all nonzero vectors $x \in \mathbb{C}^m$, is known as the *field of values* or *numerical range* of A, a subset of the complex plane denoted by $W(A)$.
(a) Show that $W(A)$ contains the convex hull of the eigenvalues of A.
(b) Show that if A is normal, then $W(A)$ is equal to the convex hull of the eigenvalues of A.

27.3. Show that for a nonhermitian matrix $A \in \mathbb{C}^{m \times m}$, the Rayleigh quotient $r(x)$ gives an eigenvalue estimate whose accuracy is generally linear, not quadratic. Explain what convergence rate this suggests for the Rayleigh quotient iteration applied to nonhermitian matrices.

27.4. Every real symmetric square matrix can be orthogonally diagonalized, and the developments of this lecture are invariant under orthogonal changes of coordinates. Thus it would have been sufficient to carry out each derivation of this lecture under the assumption that A is a diagonal matrix with entries ordered by decreasing absolute value. Making this assumption, describe the form taken by (27.4), (27.5), and Algorithm 27.3.

27.5. As mentioned in the text, inverse iteration depends on the solution of a system of equations that may be exceedingly ill-conditioned, with condition number on the order of $\epsilon_{\text{machine}}^{-1}$. We know that it is impossible in general to solve ill-conditioned systems accurately. Is this not a fatal flaw in the algorithm?

Show as follows that the answer is no—that ill-conditioning is not a problem in inverse iteration. Suppose A is a real symmetric matrix with one eigenvalue much smaller than the others in absolute value (without loss of generality, we are taking $\mu = 0$). Suppose v is a vector with components in the directions of all the eigenvalues $q_1, \ldots, q_m$ of A, and suppose $Aw = v$ is solved backward stably, yielding a computed vector $\tilde{w}$. Making use of the calculation on p. 95, show that although $\tilde{w}$ may be far from w, $\tilde{w}/\|\tilde{w}\|$ will not be far from $w/\|w\|$.

27.6. What happens to Figure 27.1 if two of the eigenvalues of A are equal?

Lecture 28. QR Algorithm without Shifts

The QR algorithm, dating to the early 1960s, is one of the jewels of numerical analysis. Here we show that in its simplest form, this algorithm can be viewed as a stable procedure for computing QR factorizations of the matrix powers A, A^2, A^3, ...

The QR Algorithm

The most basic version of the QR algorithm seems impossibly simple.

Algorithm 28.1. "Pure" QR Algorithm

$A^{(0)} = A$
for $k = 1, 2, \ldots$
   $Q^{(k)} R^{(k)} = A^{(k-1)}$ QR factorization of $A^{(k-1)}$
   $A^{(k)} = R^{(k)} Q^{(k)}$ Recombine factors in reverse order

---

All we do is take a QR factorization, multiply the computed factors Q and R together in the reverse order RQ, and repeat. Yet under suitable assumptions, this simple algorithm converges to a Schur form for the matrix A—upper-triangular if A is arbitrary, diagonal if A is hermitian. Here, to keep the discussion simple, we shall continue to assume as in the last lecture that A is real and symmetric, with real eigenvalues λ_j and orthonormal eigenvectors q_j. Thus our interest is in the convergence of the matrices $A^{(k)}$ to diagonal form.

211

For convergence to diagonal form to be useful for finding eigenvalues, of course, the operations involved must be similarity transformations. This is easily verified: the QR algorithm first triangularizes $A^{(k)}$ by forming $R^{(k)} = (Q^{(k)})^T A^{(k-1)}$, and the multiplication on the right by $Q^{(k)}$ then gives $A^{(k)} = (Q^{(k)})^T A^{(k-1)} Q^{(k)}$. In fact, we have seen this similarity transformation before: it is the "bad idea" mentioned in Lecture 26. Although this transformation is a bad idea when trying to reduce A to triangular form in a single step, it turns out to be quite powerful as the basis of an iteration.

Like the Rayleigh quotient iteration, the QR algorithm for real symmetric matrices converges cubically. To achieve this performance, however, the algorithm as presented above must be modified by the introduction of shifts at each step. The use of shifts is one of three modifications of Algorithm 28.1 that are required to bring it closer to a practical algorithm:

1. Before starting the iteration, A is reduced to tridiagonal form, as discussed in Lecture 26.
2. Instead of $A^{(k)}$, a shifted matrix $A^{(k)} - \mu^{(k)} I$ is factored at each step, where $\mu^{(k)}$ is some eigenvalue estimate.
3. Whenever possible, and in particular whenever an eigenvalue is found, the problem is "deflated" by breaking $A^{(k)}$ into submatrices.

A QR algorithm incorporating these modifications has the following outline.

Algorithm 28.2. "Practical" QR Algorithm

$(Q^{(0)})^T A^{(0)} Q^{(0)} = A$ $A^{(0)}$ is a tridiagonalization of A

for $k = 1, 2, \ldots$

    Pick a shift $\mu^{(k)}$ e.g., choose $\mu^{(k)} = A_{mm}^{(k-1)}$

    $Q^{(k)} R^{(k)} = A^{(k-1)} - \mu^{(k)} I$ QR factorization of $A^{(k-1)} - \mu^{(k)} I$

    $A^{(k)} = R^{(k)} Q^{(k)} + \mu^{(k)} I$ Recombine factors in reverse order

    If any off-diagonal element $A_{j,j+1}^{(k)}$ is sufficiently close to zero,

        set $A_{j,j+1} = A_{j+1,j} = 0$ to obtain

$$\begin{bmatrix} A_1 & 0 \\ 0 & A_2 \end{bmatrix} = A^{(k)}$$

        and now apply the QR algorithm to A_1 and A_2.

This algorithm, the QR algorithm with well-chosen shifts, has been the standard method for computing all the eigenvalues of a matrix since the early 1960s. Only in the 1990s has a competitor emerged, the divide-and-conquer algorithm described in Lecture 30.

Tridiagonalization was discussed in Lecture 26, shifts are discussed in the next lecture, and deflation is not discussed further in this book. For now, let us confine our attention to the "pure" QR algorithm and explain how it finds eigenvalues.

Unnormalized Simultaneous Iteration

Our approach will be to relate the QR algorithm to another method called *simultaneous iteration*, whose behavior is more obvious.

The idea of simultaneous iteration is to apply the power iteration to several vectors at once. (An equivalent term is *block power iteration*.) Suppose we start with a set of n linearly independent vectors $v_1^{(0)}, \ldots, v_n^{(0)}$. It seems plausible that just as $A^k v_1^{(0)}$ converges as $k \to \infty$ (under suitable assumptions) to the eigenvector corresponding to the largest eigenvalue of A in absolute value, the space $\langle A^k v_1^{(0)}, \ldots, A^k v_n^{(0)} \rangle$ should converge (again under suitable assumptions) to the space $\langle q_1, \ldots, q_n \rangle$ spanned by the eigenvectors $q_1, \ldots, q_n$ of A corresponding to the n largest eigenvalues in absolute value.

In matrix notation, we might proceed like this. Define $V^{(0)}$ to be the $m \times n$ initial matrix

$$V^{(0)} = \left[\begin{array}{c|c|c} \, & & \, \\ v_1^{(0)} & \cdots & v_n^{(0)} \\ \, & & \, \end{array} \right], \tag{28.1}$$

and define $V^{(k)}$ to be the result after k applications of A:

$$V^{(k)} = A^k V^{(0)} = \left[\begin{array}{c|c|c} \, & & \, \\ v_1^{(k)} & \cdots & v_n^{(k)} \\ \, & & \, \end{array} \right]. \tag{28.2}$$

Since our interest is in the column space of $V^{(k)}$, let us extract a well-behaved basis for this space by computing a reduced QR factorization of $V^{(k)}$:

$$\hat{Q}^{(k)} \hat{R}^{(k)} = V^{(k)}. \tag{28.3}$$

Here $\hat{Q}^{(k)}$ and $\hat{R}^{(k)}$ have dimensions $m \times n$ and $n \times n$, respectively. It seems plausible that as $k \to \infty$, under suitable assumptions, the successive columns of $\hat{Q}^{(k)}$ should converge to the eigenvectors $\pm q_1, \pm q_2, \ldots, \pm q_n$.

This expectation can be justified by an analysis analogous to that of the last lecture. If we expand $v_j^{(0)}$ and $v_j^{(k)}$ in the eigenvectors of A, we have

$$v_j^{(0)} = a_{1j} q_1 + \cdots + a_{mj} q_m,$$
$$v_j^{(k)} = \lambda_1^k a_{1j} q_1 + \cdots + \lambda_m^k a_{mj} q_m.$$

As in the last section, simple convergence results will hold provided that two conditions are satisfied. The first assumption we make is that the leading $n+1$ eigenvalues are distinct in absolute value:

$$|\lambda_1| > |\lambda_2| > \cdots > |\lambda_n| > |\lambda_{n+1}| \geq |\lambda_{n+2}| \geq \cdots \geq |\lambda_m|. \tag{28.4}$$

Our second assumption is that the collection of expansion coefficients a_{ij} is in an appropriate sense nonsingular. Define $\hat{Q}$ to be the $m \times n$ matrix whose columns are the eigenvectors $q_1, q_2, \ldots, q_n$. (Thus $\hat{Q}$, a matrix of eigenvectors, is entirely different from $\hat{Q}^{(k)}$, a factor in a reduced QR factorization.) We assume the following:

$$\text{All the leading principal minors of } \hat{Q}^T V^{(0)} \text{ are nonsingular.} \qquad (28.5)$$

By the leading principal minors of $\hat{Q}^T V^{(0)}$, we mean its upper-left square submatrices of dimensions $1 \times 1, 2 \times 2, \ldots, n \times n$. (The condition (28.5) happens to be equivalent to the condition that $\hat{Q}^T V^{(0)}$ has an LU factorization; see Exercise 20.1.)

Theorem 28.1. *Suppose that the iteration (28.1)–(28.3) is carried out and that assumptions (28.4) and (28.5) are satisfied. Then as $k \to \infty$, the columns of the matrices $\hat{Q}^{(k)}$ converge linearly to the eigenvectors of A:*

$$\|q_j^{(k)} - \pm q_j\| = O(C^k) \qquad (28.6)$$

for each j with $1 \le j \le n$, where $C < 1$ is the constant $\max_{1 \le k \le n} |\lambda_{k+1}|/|\lambda_k|$. As in the theorems of the last lecture, the $\pm$ sign means that at each step k, one or the other choice of sign is to be taken, and then the indicated bound holds.

Proof. Extend $\hat{Q}$ to a full $m \times m$ orthogonal matrix Q of eigenvectors of A, and let Λ be the corresponding diagonal matrix of eigenvalues; thus $A = Q\Lambda Q^T$. Just as $\hat{Q}$ is the leading $m \times n$ section of Q, define $\hat{\Lambda}$ (still diagonal) to be the leading $n \times n$ section of Λ. Then we have

$$V^{(k)} = A^k V^{(0)} = Q\Lambda^k Q^T V^{(0)} = \hat{Q}\hat{\Lambda}^k \hat{Q}^T V^{(0)} + O(|\lambda_{n+1}|^k)$$

as $k \to \infty$. If (28.5) holds, then in particular, $\hat{Q}^T V^{(0)}$ is nonsingular, so we can multiply the term $O(|\lambda_{n+1}|^k)$ on the right by $(\hat{Q}^T V^{(0)})^{-1}\hat{Q}^T V^{(0)}$ to transform this equation to

$$V^{(k)} = \left(\hat{Q}\hat{\Lambda}^k + O(|\lambda_{n+1}|^k)\right) \hat{Q}^T V^{(0)}.$$

Since $\hat{Q}^T V^{(0)}$ is nonsingular, the column space of this matrix is the same as the column space of

$$\hat{Q}\hat{\Lambda}^k + O(|\lambda_{n+1}|^k).$$

From the form of $\hat{Q}\hat{\Lambda}^k$ and the assumption (28.4), it is clear that this column space converges linearly to that of $\hat{Q}$. This convergence can be quantified, for example, by defining angles between subspaces; we omit the details.

Now in fact, we have assumed that not only is $\hat{Q}^T V^{(0)}$ nonsingular, but so are all of its leading principal minors. It follows that the argument above also applies to leading subsets of the columns of $V^{(k)}$ and $\hat{Q}$: the first columns, the first and second columns, the first and second and third columns, and so on. In each case we conclude that the space spanned by the indicated columns of $V^{(k)}$ converges linearly to the space spanned by the corresponding columns of $\hat{Q}$. From this convergence of all the successive column spaces, together with the definition of the QR factorization (28.3), (28.6) follows.   □

Simultaneous Iteration

As $k \to \infty$, the vectors $v_1^{(k)}, \ldots, v_n^{(k)}$ in the algorithm (28.1)–(28.3) all converge to multiples of the same dominant eigenvector q_1 of A. Thus, although the space they span, $\langle v_1^{(k)}, \ldots, v_j^{(k)} \rangle$, converges to something useful, these vectors constitute a highly ill-conditioned basis of that space. If we actually carried out simultaneous iteration in floating point arithmetic as just described, the desired information would quickly be lost to rounding errors.

The remedy is simple: one must orthonormalize at each step rather than once and for all. Thus we shall not construct $V^{(k)}$ as defined above, but a different sequence of matrices $Z^{(k)}$ with the same column spaces.

Algorithm 28.3. Simultaneous Iteration

Pick $\hat{Q}^{(0)} \in \mathbb{R}^{m \times n}$ with orthonormal columns.
for $k = 1, 2, \ldots$
$\qquad Z = A\hat{Q}^{(k-1)}$
$\qquad \hat{Q}^{(k)} \hat{R}^{(k)} = Z$ $\qquad\qquad$ reduced QR factorization of Z

From the form of this algorithm, it is clear that the column spaces of $\hat{Q}^{(k)}$ and $Z^{(k)}$ are the same, both being equal to the column space of $A^k \hat{Q}^{(0)}$. Thus, mathematically speaking, this new formulation of simultaneous iteration converges under the same conditions as the old one.

Theorem 28.2. *Algorithm* 28.3 *generates the same matrices* $\hat{Q}^{(k)}$ *as the iteration* (28.1)–(28.3) *considered in Theorem* 28.1 *(assuming that the initial matrices* $\hat{Q}^{(0)}$ *are the same), and under the same assumptions* (28.4) *and* (28.5), *it converges as described in that theorem.*

Simultaneous Iteration $\Longleftrightarrow$ QR Algorithm

Now we can explain the QR algorithm. It is equivalent to simultaneous iteration applied to a full set of $n = m$ initial vectors, namely, the identity, $\hat{Q}^{(0)} = I$. Since the matrices $\hat{Q}^{(k)}$ are now square, we are dealing with full QR factorizations and can drop the hats on $\hat{Q}^{(k)}$ and $\hat{R}^{(k)}$. In fact, we shall replace $\hat{R}^{(k)}$ by $R^{(k)}$ but $\hat{Q}^{(k)}$ by $\underline{Q}^{(k)}$ in order to distinguish the Q matrices of simultaneous iteration from those of the QR algorithm.

Here are the three formulas that define simultaneous iteration with $\underline{Q}^{(0)} = I$, followed by a fourth formula that we shall take as a definition of an $m \times m$ matrix $A^{(k)}$:

Simultaneous Iteration:

$$\underline{Q}^{(0)} = I, \tag{28.7}$$

$$Z = A\underline{Q}^{(k-1)}, \tag{28.8}$$

$$Z = \underline{Q}^{(k)}R^{(k)}, \tag{28.9}$$

$$A^{(k)} = (\underline{Q}^{(k)})^T A \underline{Q}^{(k)}. \tag{28.10}$$

And here are the three formulas that define the pure QR algorithm, followed by a fourth formula that we shall take as a definition of an $m \times m$ matrix $\underline{Q}^{(k)}$:

Unshifted QR Algorithm:

$$A^{(0)} = A, \tag{28.11}$$

$$A^{(k-1)} = Q^{(k)}R^{(k)}, \tag{28.12}$$

$$A^{(k)} = R^{(k)}Q^{(k)}, \tag{28.13}$$

$$\underline{Q}^{(k)} = Q^{(1)}Q^{(2)} \cdots Q^{(k)}. \tag{28.14}$$

Additionally, for both algorithms, let us define one further $m \times m$ matrix $\underline{R}^{(k)}$,

$$\underline{R}^{(k)} = R^{(k)}R^{(k-1)} \cdots R^{(1)}. \tag{28.15}$$

We can now exhibit the equivalence of these two algorithms.

Theorem 28.3. *The processes (28.7)–(28.10) and (28.11)–(28.14) generate identical sequences of matrices $\underline{R}^{(k)}$, $\underline{Q}^{(k)}$, and $A^{(k)}$, namely, those defined by the QR factorization of the kth power of A,*

$$A^k = \underline{Q}^{(k)} \underline{R}^{(k)}, \tag{28.16}$$

together with the projection

$$A^{(k)} = (\underline{Q}^{(k)})^T A \underline{Q}^{(k)}. \tag{28.17}$$

Proof. We proceed by induction in k. The base case $k = 0$ is trivial. For both simultaneous iteration and the QR algorithm, equations (28.7)–(28.15) imply $A^0 = \underline{Q}^{(0)} = \underline{R}^{(0)} = I$ and $A^{(0)} = A$, from which (28.16) and (28.17) are immediate.

Consider now the case $k \geq 1$ for simultaneous iteration. Formula (28.17) is valid by virtue of the definition (28.10) (they are identical), so we need only verify (28.16), which can be done as follows:

$$A^k = A\underline{Q}^{(k-1)}\underline{R}^{(k-1)} = \underline{Q}^{(k)}R^{(k)}\underline{R}^{(k-1)} = \underline{Q}^{(k)}\underline{R}^{(k)}.$$

The first equality follows from the inductive hypothesis on (28.16), the second from (28.8) and (28.9), and the third from (28.15).

On the other hand, consider the case $k \geq 1$ for the QR algorithm. We can verify (28.16) by the sequence

$$A^k = A\underline{Q}^{(k-1)}\underline{R}^{(k-1)} = \underline{Q}^{(k-1)}A^{(k-1)}\underline{R}^{(k-1)} = \underline{Q}^{(k)}\underline{R}^{(k)}.$$

The first equality follows from the inductive hypothesis on (28.16), the second from the inductive hypothesis on (28.17), and the third from (28.12), together with (28.14) and (28.15). Finally, we can verify (28.17) by the sequence

$$A^{(k)} = (Q^{(k)})^T A^{(k-1)} Q^{(k)} = (\underline{Q}^{(k)})^T A \underline{Q}^{(k)}.$$

The first equality follows from (28.12) and (28.13), and the second from the inductive hypothesis on (28.17).   □

Convergence of the QR Algorithm

All the pieces are in place. We can now say a great deal about the convergence of the unshifted QR algorithm.

First, at the level of qualitative understanding: (28.16) and (28.17) are the key. The first of these explains why the QR algorithm can be expected to find eigenvectors: it constructs orthonormal bases for successive powers A^k. The second explains why the algorithm finds eigenvalues. From (28.17) it follows that the diagonal elements of $A^{(k)}$ are Rayleigh quotients of A corresponding to the columns of $\underline{Q}^{(k)}$ (see Exercise 27.1). As those columns converge to eigenvectors, the Rayleigh quotients converge (twice as fast, by (27.3)) to the corresponding eigenvalues. Meanwhile, (28.17) implies that the off-diagonal elements of $A^{(k)}$ correspond to generalized Rayleigh quotients involving approximations of distinct eigenvectors of A on the left and the right. Since these approximations must become orthogonal as they converge to distinct eigenvectors, the off-diagonal elements of $A^{(k)}$ must converge to zero.

We cannot emphasize too strongly how fundamental equations (28.16) and (28.17) are to an understanding of the unshifted QR algorithm. They are memorable; and from them, everything of importance follows.

As for a more quantitative understanding, we have the following consequence of Theorem 28.2.

Theorem 28.4. *Let the pure QR algorithm (Algorithm 28.1) be applied to a real symmetric matrix A whose eigenvalues satisfy $|\lambda_1| > |\lambda_2| > \cdots > |\lambda_m|$ and whose corresponding eigenvector matrix Q has all nonsingular leading principal minors.   Then as $k \to \infty$, $A^{(k)}$ converges linearly with constant $\max_k |\lambda_{k+1}|/|\lambda_k|$ to $\mathrm{diag}(\lambda_1, \ldots, \lambda_m)$, and $\underline{Q}^{(k)}$ (with the signs of its columns adjusted as necessary) converges at the same rate to Q.*

Exercises

28.1.   What happens if you apply the unshifted QR algorithm to an orthogonal matrix? Figure out the answer, and then explain how it relates to Theorem 28.4.

28.2.   The preliminary reduction to tridiagonal form would be of little use if the steps of the QR algorithm did not preserve this structure. Fortunately, they do.

(a) In the QR factorization $A = QR$ of a tridiagonal matrix A, which entries of R are in general nonzero? Which entries of Q? (In practice we do not form Q explicitly.)

(b) Show that the tridiagonal structure is recovered when the product RQ is formed.

(c) Explain how Givens rotations or 2×2 Householder reflections can be used in the computation of the QR factorization of a tridiagonal matrix, reducing the operation count far below what would be required for a full matrix.

28.3.   A real symmetric matrix A has an eigenvalue 1 of multiplicity 8, while all the rest of the eigenvalues are ≤ 0.1 in absolute value. Describe an algorithm for finding an orthonormal basis of the 8-dimensional eigenspace corresponding to the dominant eigenvalue.

28.4.   Consider one step of Algorithm 28.1 applied to a tridiagonal symmetric matrix $A \in \mathbb{R}^{m \times m}$.

(a) If only eigenvalues are desired, then only $A^{(k)}$ is needed at step k, not $Q^{(k)}$. Determine how many flops are required to get from $A^{(k-1)}$ to $A^{(k)}$ using standard methods described in this book.

(b) If all the eigenvectors are desired, then the matrix $\underline{Q}^{(k)} = Q^{(1)}Q^{(2)} \cdots Q^{(k)}$ will need to be accumulated too. Determine how many flops are now required to get from step $k-1$ to step k.

Lecture 29. QR Algorithm with Shifts

What makes the QR iteration fly is the introduction of shifts $A \to A - \mu I$ at each step. Here we explain how this idea leads to cubic convergence, thanks to an implicit connection with the Rayleigh quotient iteration.

Connection with Inverse Iteration

We continue to assume that $A \in \mathbb{R}^{m \times m}$ is real and symmetric, with real eigenvalues $\{\lambda_j\}$ and orthonormal eigenvectors $\{q_j\}$.

As we have seen, the "pure" QR algorithm (Algorithm 28.1) is equivalent to simultaneous iteration applied to the identity matrix, and in particular, the first column of the result evolves according to the power iteration applied to e_1. There is a dual to this observation. Algorithm 28.1 is also equivalent to *simultaneous inverse iteration* applied to a "flipped" identity matrix P, and in particular, the mth column of the result evolves according to inverse iteration applied to e_m.

We can establish this claim as follows. Let $Q^{(k)}$, as in the last lecture, be the orthogonal factor at the kth step of the QR algorithm. In the last lecture, we showed that the accumulated product (28.14) of these matrices,

$$
\underline{Q}^{(k)} = \prod_{j=1}^{k} Q^{(j)} = \left[\left. q_1^{(k)} \, \right| \, q_2^{(k)} \, \left| \, \cdots \, \right| \, q_m^{(k)} \right],
$$

is the same orthogonal matrix that appears at step k (28.9) of simultaneous iteration. Another way to put this was to say that $\underline{Q}^{(k)}$ is the orthogonal factor in a QR factorization (28.16),

$$A^k = \underline{Q}^{(k)} \underline{R}^{(k)}.$$

Now consider what happens if we invert this formula. We calculate

$$A^{-k} = (\underline{R}^{(k)})^{-1} \underline{Q}^{(k)T} = \underline{Q}^{(k)} (\underline{R}^{(k)})^{-T}; \qquad (29.1)$$

for the second equality we have used the fact that A^{-1} is symmetric. Let P denote the $m \times m$ permutation matrix that reverses row or column order:

$$P = \begin{bmatrix} & & & 1 \\ & & 1 & \\ & \cdots & & \\ 1 & & & \end{bmatrix}.$$

Since $P^2 = I$, (29.1) can be rewritten as

$$A^{-k}P = [\underline{Q}^{(k)}P][P(\underline{R}^{(k)})^{-T}P]. \qquad (29.2)$$

The first factor in this product, $\underline{Q}^{(k)}P$, is orthogonal. The second, $P(\underline{R}^{(k)})^{-T}P$, is upper-triangular (start with the lower-triangular matrix $(\underline{R}^{(k)})^{-T}$, flip it top-to-bottom, then flip it again left-to-right). Thus (29.2) can be interpreted as a QR factorization of $A^{-k}P$. In other words, we are effectively carrying out simultaneous iteration on A^{-1} applied to the initial matrix P, which is to say, simultaneous inverse iteration on A. In particular, the first column of $\underline{Q}^{(k)}P$—the last column of $\underline{Q}^{(k)}$—is the result of applying k steps of inverse iteration to the vector e_m.

Connection with Shifted Inverse Iteration

Thus the QR algorithm is both simultaneous iteration and simultaneous inverse iteration: the symmetry is perfect. But, as we saw in Lecture 27, there is a huge difference between power iteration and inverse iteration: the latter can be accelerated arbitrarily through the use of shifts. The better we can estimate an eigenvalue $\mu \approx \lambda_J$, the more we shall accomplish by a step of inverse iteration with the shifted matrix $A - \mu I$. Algorithm 28.2 showed how shifts are introduced into a step of the QR algorithm. Doing this corresponds exactly to shifts in the corresponding simultaneous iteration and inverse iteration processes, and their beneficial effect is therefore exactly the same.

Let $\mu^{(k)}$ denote the eigenvalue estimate chosen at the kth step of the QR algorithm. From Algorithm 28.2, the relationship between steps $k - 1$ and k of the shifted QR algorithm is

$$A^{(k-1)} - \mu^{(k)}I = Q^{(k)}R^{(k)},$$

$$A^{(k)} = R^{(k)}Q^{(k)} + \mu^{(k)}I.$$

This implies

$$A^{(k)} = (Q^{(k)})^T A^{(k-1)} Q^{(k)}, \tag{29.3}$$

and by induction,

$$A^{(k)} = (\underline{Q}^{(k)})^T A \underline{Q}^{(k)},$$

which is unchanged from (28.17). However, (28.16) no longer holds. Instead, we have the factorization

$$(A - \mu^{(k)}I)(A - \mu^{(k-1)}I) \cdots (A - \mu^{(1)}I) \;=\; \underline{Q}^{(k)} \underline{R}^{(k)}, \tag{29.4}$$

a shifted variation on simultaneous iteration (we omit the proof). In words, $\underline{Q}^{(k)} = \prod_{j=1}^{k} Q^{(j)}$ is an orthogonalization of $\prod_{j=k}^{1}(A - \mu^{(j)}I)$. The first column of $\underline{Q}^{(k)}$ is the result of applying shifted power iteration to e_1 using the shifts $\mu^{(j)}$, and the last column is the result of applying k steps of shifted inverse iteration to e_m with the same shifts. If the shifts are good eigenvalue estimates, this last column of $\underline{Q}^{(k)}$ converges quickly to an eigenvector.

Connection with Rayleigh Quotient Iteration

We have discovered a powerful tool hidden in the shifted QR algorithm: shifted inverse iteration. To complete the idea, we now need a way of choosing shifts to achieve fast convergence in the last column of $\underline{Q}^{(k)}$.

The Rayleigh quotient is a good place to start. To estimate the eigenvalue corresponding to the eigenvector approximated by the last column of $\underline{Q}^k$, it is natural to apply the Rayleigh quotient to this last column. This gives us

$$\mu^{(k)} \;=\; \frac{(q_m^{(k)})^T A \, q_m^{(k)}}{(q_m^{(k)})^T q_m^{(k)}} \;=\; (q_m^{(k)})^T A \, q_m^{(k)}. \tag{29.5}$$

If this number is chosen as the shift at every step, the eigenvalue and eigenvector estimates $\mu^{(k)}$ and $q_m^{(k)}$ are identical to those that are computed by the Rayleigh quotient iteration starting with e_m. Therefore, the QR algorithm has cubic convergence in the sense that $q_m^{(k)}$ converges cubically to an eigenvector.

Notice that, in the QR algorithm, the Rayleigh quotient $r(q_m^{(k)})$ appears as the m, m entry of $A^{(k)}$—so it comes for free! We mentioned this at the end of the last lecture, but here is an explicit derivation for emphasis. Starting with (29.3), we have

$$A_{mm}^{(k)} = e_m^T A^{(k)} e_m = e_m^T \underline{Q}^{(k)T} A \underline{Q}^{(k)} e_m = q_m^{(k)T} A q_m^{(k)}. \tag{29.6}$$

Therefore, (29.5) is the same as simply setting $\mu^{(k)} = A_{mm}^{(k)}$. This is known as the *Rayleigh quotient shift*.

Wilkinson Shift

Although the Rayleigh quotient shift gives cubic convergence in the generic case, convergence is not guaranteed for all initial conditions. We can see this with a simple example. Consider the matrix

$$A = \begin{bmatrix} 0 & 1 \\ 1 & 0 \end{bmatrix}. \tag{29.7}$$

The unshifted QR algorithm does not converge at all:

$$A = Q^{(1)}R^{(1)} = \begin{bmatrix} 0 & 1 \\ 1 & 0 \end{bmatrix} \begin{bmatrix} 1 & 0 \\ 0 & 1 \end{bmatrix},$$

$$A^{(1)} = R^{(1)}Q^{(1)} = \begin{bmatrix} 1 & 0 \\ 0 & 1 \end{bmatrix} \begin{bmatrix} 0 & 1 \\ 1 & 0 \end{bmatrix} = A.$$

The Rayleigh quotient shift $\mu = A_{mm}$, however, has no effect either, since $A_{mm} = 0$. Thus it is clear that in the worst case, the QR algorithm with the Rayleigh quotient shift may fail.

The problem arises because of the symmetry of the eigenvalues. One eigenvalue is $+1$, and the other is -1, so when we attempt to improve the eigenvalue estimate 0, the tendency to favor each eigenvalue is equal, and the estimate is not improved. What is needed is an eigenvalue estimate that can break the symmetry. One such choice is defined as follows. Let B denote the lower-rightmost 2×2 submatrix of $A^{(k)}$:

$$B = \begin{bmatrix} a_{m-1} & b_{m-1} \\ b_{m-1} & a_m \end{bmatrix}.$$

The *Wilkinson shift* is defined as that eigenvalue of B that is closer to a_m, where in the case of a tie, one of the two eigenvalues of B is chosen arbitrarily. A numerically stable formula for the Wilkinson shift is

$$\mu = a_m - \text{sign}(\delta)b_{m-1}^2 \left/ \left(|\delta| + \sqrt{\delta^2 + b_{m-1}^2} \right) \right., \tag{29.8}$$

where $\delta = (a_{m-1} - a_m)/2$. If $\delta = 0$, $\text{sign}(\delta)$ can be arbitrarily set equal to 1 or -1.

Like the Rayleigh quotient shift, the Wilkinson shift achieves cubic convergence in the generic case. Moreover, it can be shown that it achieves at least quadratic convergence in the worst case. In particular, the QR algorithm with the Wilkinson shift always converges (in exact arithmetic).

In the example (29.7), the Wilkinson shift is either $+1$ or -1. Thus the symmetry is broken, and convergence takes place in one step.

Stability and Accuracy

This completes our discussion of the mechanics of the QR algorithm, though many practical details have been omitted, such as conditions for deflation and "implicit" strategies for shifting. It remains to say a word about stability and accuracy.

As one might expect from its use of orthogonal matrices, the QR algorithm is backward stable. As in previous lectures, the simplest way to formulate this result is to let $\tilde{\Lambda}$ denote the diagonalization of A as computed in floating point arithmetic, and $\tilde{Q}$ the exactly orthogonal matrix associated with the product of all the numerically computed Householder reflections (or Givens rotations) utilized along the way. Here is what can be proved.

Theorem 29.1. *Let a real, symmetric, tridiagonal matrix $A \in \mathbb{R}^{m \times m}$ be diagonalized by the QR algorithm (Algorithm 28.2) on a computer satisfying (13.5) and (13.7), and let $\tilde{\Lambda}$ and $\tilde{Q}$ be defined as indicated above. Then we have*

$$\tilde{Q}\tilde{\Lambda}\tilde{Q}^* = A + \delta A, \qquad \frac{\|\delta A\|}{\|A\|} = O(\epsilon_{\text{machine}}) \qquad (29.9)$$

for some $\delta A \in \mathbb{C}^{m \times m}$.

Like most of the algorithms in this book, then, the QR algorithm produces an exact solution of a slightly perturbed problem. Combining Theorems 26.1 and 29.1, we see that tridiagonal reduction followed by the QR algorithm is a backward stable algorithm for computing eigenvalues of matrices. To see what this implies about accuracy of the computed eigenvalues, we may combine this conclusion with the result (26.4) concerning perturbation of eigenvalues of real symmetric matrices (a special case of normal matrices). The conclusion is that the computed eigenvalues $\tilde{\lambda}_j$ satisfy

$$\frac{|\tilde{\lambda}_j - \lambda_j|}{\|A\|} = O(\epsilon_{\text{machine}}). \qquad (29.10)$$

This is not a bad result at all for an algorithm that requires just $\sim \frac{4}{3}m^3$ flops, two-thirds the cost of computing the product of a pair of $m \times m$ matrices!

Exercise

29.1. This five-part problem asks you to put together a MATLAB program that finds all the eigenvalues of a real symmetric matrix, using only elementary building blocks. It is not necessary to achieve optimal constant factors by exploiting symmetry or zero structure optimally. It is possible to solve the whole problem by a program about fifty lines long.

(a) Write a function T = tridiag(A) that reduces a real symmetric $m \times m$ matrix to tridiagonal form by orthogonal similarity transformations. Your program should use only elementary MATLAB operations—not the built-in function hess, for example. Your output matrix T should be symmetric and tridiagonal up to rounding errors. If you like, add a line that forces T at the end to be exactly symmetric and tridiagonal. For an example, apply your program to A = hilb(4).

(b) Write a function Tnew = qralg(T) that runs the unshifted QR algorithm on a real tridiagonal matrix T. For the QR factorization at each step, use programs [W,R] = house(A) and Q = formQ(W) of Exercise 10.2 if available, or MATLAB's command qr, or, for greater efficiency, a new code based on Givens rotations or 2×2 Householder reflections rather than $m \times m$ operations. Again, you may wish to enforce symmetry and tridiagonality at each step. Your program should stop and return the current tridiagonal matrix T as Tnew when the $m, m-1$ element satisfies $|t_{m,m-1}| < 10^{-12}$ (hardly an industrial strength convergence criterion!). Again, apply your program to A = hilb(4).

(c) Write a driver program which (i) calls tridiag, (ii) calls qralg to get one eigenvalue, (iii) calls qralg with a smaller matrix to get another eigenvalue, and so on until all of the eigenvalues of A are determined. Set things up so that the values of $|t_{m,m-1}|$ at every QR iteration are stored in a vector and so that at the end, your program generates a semilogy plot of these values as a function of the number of QR factorizations. (Here m will step from length(A) to length(A)-1 and so on down to 3 and finally 2 as the deflation proceeds, and the plot will be correspondingly sawtoothed.) Run your program for A = hilb(4). The output should be a set of eigenvalues and a "sawtooth plot."

(d) Modify qralg so that it uses the Wilkinson shift at each step. Turn in the new sawtooth plot for the same example.

(e) Rerun your program for the matrix A = diag(15:-1:1) + ones(15,15) and generate two sawtooth plots corresponding to shift and no shift. Discuss the rates of convergence observed here and for the earlier matrix. Is the convergence linear, superlinear, quadratic, cubic...? Is it meaningful to speak of a certain "number of QR iterations per eigenvalue?"

Lecture 30. Other Eigenvalue Algorithms

There is more to the computation of eigenvalues than the QR algorithm. In this lecture we briefly mention three famous alternatives for real symmetric eigenvalue problems: the Jacobi algorithm, for full matrices, and the bisection and divide-and-conquer algorithms, for tridiagonal matrices.

Jacobi

One of the oldest ideas for computing eigenvalues of matrices is the *Jacobi algorithm*, introduced by Jacobi in 1845. This method has attracted attention throughout the computer era, especially since the advent of parallel computing, though it has never quite managed to displace the competition.

The idea is as follows. For matrices of dimension 5 or larger, we know that eigenvalues can only be obtained by iteration (Lecture 25). However, smaller matrices than this can be handled in one step. Why not diagonalize a small submatrix of A, then another, and so on, hoping eventually to converge to a diagonalization of the full matrix?

The idea has been tried with 4×4 submatrices, but the standard approach is based on 2×2 submatrices. A 2×2 real symmetric matrix can be diagonalized in the form

$$J^T \begin{bmatrix} a & d \\ d & b \end{bmatrix} J = \begin{bmatrix} \neq 0 & 0 \\ 0 & \neq 0 \end{bmatrix}, \tag{30.1}$$

225

where J is orthogonal. Now there are several ways to choose J. One could take it to be a 2×2 Householder reflection of the form

$$F = \begin{bmatrix} -c & s \\ s & c \end{bmatrix}, \tag{30.2}$$

where $s = \sin\theta$ and $c = \cos\theta$ for some θ. Note that $\det F = -1$, the hallmark of a reflection. Alternatively, one can use not a reflection but a rotation,

$$J = \begin{bmatrix} c & s \\ -s & c \end{bmatrix}, \tag{30.3}$$

with $\det J = 1$. This is the standard approach for the Jacobi algorithm. It can be shown that the diagonalization (30.1) is accomplished if θ satisfies

$$\tan(2\theta) = \frac{2d}{b-a}, \tag{30.4}$$

and the matrix J based on this choice is called a *Jacobi rotation*. (It has the same form as a Givens rotation (Exercise 10.4); the only difference is that θ is chosen to make $J^T A J$ diagonal rather than $J^T A$ triangular.)

Now let $A \in \mathbb{R}^{m \times m}$ be symmetric. The Jacobi algorithm consists of the iterative application of transformations (30.1) based on matrices defined by (30.3) and (30.4). The matrix J is now enlarged to an $m \times m$ matrix that is the identity in all but four entries, where it has the form (30.3). Applying J^T on the left modifies two rows of A, and applying J on the right modifies two columns. At each step a symmetric pair of zeros is introduced into the matrix, but previous zeros are destroyed. Just as with the QR algorithm, however, the usual effect is that the magnitudes of these nonzeros shrink steadily.

Which off-diagonal entries a_{ij} should be zeroed at each step? The approach naturally fitted to hand computation is to pick the largest off-diagonal entry at each step. Analysis of convergence then becomes a triviality, for one can show that the sum of the squares of the off-diagonal entries decreases by at least the factor $1 - 2/(m^2 - m)$ at each step (Exercise 30.3). After $O(m^2)$ steps, each requiring $O(m)$ operations, the sum of squares must drop by a constant factor, and convergence to accuracy $\epsilon_{\text{machine}}$ is assured after $O(m^3 \log(\epsilon_{\text{machine}}))$ operations. In fact, it is known that the convergence is better than this, ultimately quadratic rather than linear, so the actual operation count is $O(m^3 \log(|\log(\epsilon_{\text{machine}})|))$ (Exercise 25.2).

On a computer, the off-diagonal entries are generally eliminated in a cyclic manner that avoids the $O(m^2)$ search for the largest. For example, if the $m(m-1)/2$ superdiagonal entries are eliminated in the simplest row-wise order, beginning with $a_{12}, a_{13}, \ldots$, then rapid asymptotic convergence is again guaranteed. After one *sweep* of 2×2 operations involving all of the $m(m-1)/2$ pairs of off-diagonal entries, the accuracy has generally improved by better than a constant factor, and again, the convergence is ultimately quadratic.

The Jacobi method is attractive because it deals only with pairs of rows and columns at a time, making it easily parallelizable (Exercise 30.4). The matrix is not tridiagonalized in advance; the Jacobi rotations would destroy that structure. Convergence for matrices of dimension $m \leq 1000$ is typically achieved in fewer than ten sweeps, and the final componentwise accuracy is generally even better than can be achieved by the QR algorithm. Unfortunately, even on parallel machines, the Jacobi algorithm is not usually as fast as tridiagonalization followed by the QR or divide-and-conquer algorithm (discussed below), though it usually comes within a factor of 10 (Exercise 30.2).

Bisection

Our next eigenvalue algorithm, the method of *bisection*, is of great practical importance. After a symmetric matrix has been tridiagonalized, this is the standard next step if one does not want all of the eigenvalues but just a subset of them. For example, bisection can find the largest 10% of the eigenvalues, or the smallest thirty eigenvalues, or all the eigenvalues in the interval $[1, 2]$. Once the desired eigenvalues are found, the corresponding eigenvectors can be obtained by one step of inverse iteration (Algorithm 27.2).

The starting point is elementary. Since the eigenvalues of a real symmetric matrix are real, we can find them by searching the real line for roots of the polynomial $p(x) = \det(A - xI)$. This sounds like a bad idea, for did we not mention in Lectures 15 and 25 that polynomial rootfinding is a highly unstable procedure for finding eigenvalues? The difference is that those remarks pertained to the idea of finding roots from the polynomial *coefficients*. Now, the idea is to find the roots by evaluating $p(x)$ at various points x, without ever looking at its coefficients, and applying the usual bisection process for nonlinear functions. This could be done, for example, by Gaussian elimination with pivoting (Exercise 21.1), and the resulting algorithm would be highly stable.

This much sounds useful enough, but not very exciting. What gives the bisection method its power and its appeal are some additional properties of eigenvalues and determinants that are not immediately obvious.

Given a symmetric matrix $A \in \mathbb{R}^{m \times m}$, let $A^{(1)}, \ldots, A^{(m)}$ denote its principal (i.e., upper-left) square submatrices of dimensions $1, \ldots, m$. It can be shown that the eigenvalues of these matrices *interlace*. Before defining this property, let us first sharpen it by assuming that A is tridiagonal and *irreducible* in the sense that all of its off-diagonal entries are nonzero:

$$A = \begin{bmatrix} a_1 & b_1 & & & & \\ b_1 & a_2 & b_2 & & & \\ & b_2 & a_3 & \ddots & & \\ & & \ddots & \ddots & b_{m-1} \\ & & & b_{m-1} & a_m \end{bmatrix}, \qquad b_j \neq 0. \qquad (30.5)$$

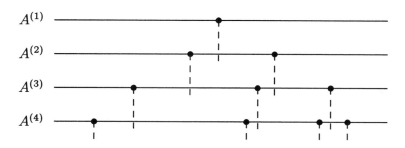

Figure 30.1. *Illustration of the strict eigenvalue interlace property* (30.6) *for the principal submatrices* $\{A^{(j)}\}$ *of an irreducible tridiagonal real symmetric matrix A. The eigenvalues of* $A^{(k)}$ *interlace those of* $A^{(k+1)}$. *The bisection algorithm takes advantage of this property.*

(If there are zeros on the off-diagonal, then the eigenvalue problem can be deflated, as in Algorithm 28.2.) By Exercise 25.1, the eigenvalues of $A^{(k)}$ are distinct; let them be denoted by $\lambda_1^{(k)} < \lambda_2^{(k)} < \cdots < \lambda_k^{(k)}$. The crucial property that makes bisection powerful is that these eigenvalues *strictly interlace*, satisfying the inequalities

$$\lambda_j^{(k+1)} < \lambda_j^{(k)} < \lambda_{j+1}^{(k+1)} \tag{30.6}$$

for $k = 1, 2, \ldots, m - 1$ and $j = 1, 2, \ldots, k - 1$. This behavior is sketched in Figure 30.1.

It is the interlacing property that makes it possible to count the exact number of eigenvalues of a matrix in a specified interval. For example, consider the 4×4 tridiagonal matrix

$$A = \begin{bmatrix} 1 & 1 & & \\ 1 & 0 & 1 & \\ & 1 & 2 & 1 \\ & & 1 & -1 \end{bmatrix}.$$

From the numbers

$$\det(A^{(1)}) = 1, \quad \det(A^{(2)}) = -1, \quad \det(A^{(3)}) = -3, \quad \det(A^{(4)}) = 4,$$

we know that $A^{(1)}$ has no negative eigenvalues, $A^{(2)}$ has one negative eigenvalue, $A^{(3)}$ has one negative eigenvalue, and $A^{(4)}$ has two negative eigenvalues. In general, for any symmetric tridiagonal $A \in \mathbb{R}^{m \times m}$, *the number of negative eigenvalues is equal to the number of sign changes in the sequence*

$$1, \ \det(A^{(1)}), \ \det(A^{(2)}), \ldots, \ \det(A^{(m)}), \tag{30.7}$$

which is known as a *Sturm sequence*. (This prescription works even if zero determinants are encountered along the way, if we define a "sign change" to

mean a transition from + or 0 to − or from − or 0 to + but not from + or
− to 0.) By shifting A by a multiple of the identity, we can determine the
number of eigenvalues in any interval $[a,b)$: it is the number of eigenvalues in
$(-\infty, b)$ minus the number in $(-\infty, a)$.

One more observation completes the description of the bisection algorithm:
for a tridiagonal matrix, the determinants of the matrices $\{A^{(k)}\}$ are related by
a three-term recurrence relation. Expanding $\det(A^{(k)})$ by minors with respect
to its entries b_{k-1} and a_k in row k gives, from (30.5),

$$\det(A^{(k)}) = a_k\det(A^{(k-1)}) - b_{k-1}^2\det(A^{(k-2)}). \tag{30.8}$$

Introducing the shift by xI and writing $p^{(k)}(x) = \det(A^{(k)} - xI)$, we get

$$p^{(k)}(x) = (a_k - x)p^{(k-1)}(x) - b_{k-1}^2 p^{(k-2)}(x). \tag{30.9}$$

If we define $p^{(-1)}(x) = 0$ and $p^{(0)}(x) = 1$, then this recurrence is valid for all
$k = 1, 2, \ldots, m$.

By applying (30.9) for a succession of values of x and counting sign changes
along the way, the bisection algorithm locates eigenvalues in arbitrarily small
intervals. The cost is $O(m)$ flops for each evaluation of the sequence, hence
$O(m\log(\epsilon_{\text{machine}}))$ flops in total to find an eigenvalue to relative accuracy
$\epsilon_{\text{machine}}$. If a small number of eigenvalues are needed, this is a distinct improvement over the $O(m^2)$ operation count for the QR algorithm. On a multiprocessor computer, multiple eigenvalues can be found independently on
separate processors.

Divide-and-Conquer

The divide-and-conquer algorithm, based on a recursive subdivision of a symmetric tridiagonal eigenvalue problem into problems of smaller dimension,
represents the most important advance in matrix eigenvalue algorithms since
the 1960s. First introduced by Cuppen in 1981, this method is more than
twice as fast as the QR algorithm if eigenvectors as well as eigenvalues are
required.

We shall give just the essential idea, omitting all details. But the reader
is warned that in this area, the details are particularly important, for the
algorithm is not fully stable unless they are gotten right—a matter that was
not well understood for a decade after Cuppen's original paper.

Let $T \in \mathbb{R}^{m \times m}$ with $m \geq 2$ be symmetric, tridiagonal, and irreducible in
the sense of having only nonzeros on the off-diagonal. (Otherwise, the problem
can be deflated.) Then for any n in the range $1 \leq n < m$, T can be split into

submatrices as follows:

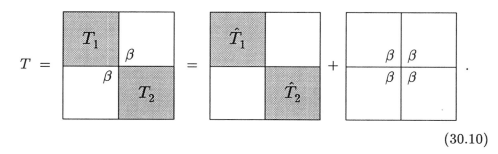

$$(30.10)$$

Here T_1 is the upper-left $n \times n$ principal submatrix of T, T_2 is the lower-right $(m - n) \times (m - n)$ principal submatrix, and $\beta = t_{n+1,n} = t_{n,n+1} \neq 0$. The only difference between T_1 and $\hat{T}_1$ is that the lower-right entry t_{nn} has been replaced by $t_{nn} - \beta$, and the only difference between T_2 and $\hat{T}_2$ is that the upper-left entry $t_{n+1,n+1}$ has been replaced by $t_{n+1,n+1} - \beta$. These modifications of two entries are introduced to make the rightmost matrix of (30.10) have rank one.

Here is how (30.10) might be expressed in words. *A tridiagonal matrix can be written as the sum of a 2×2 block-diagonal matrix with tridiagonal blocks and a rank-one correction.*

The divide-and-conquer algorithm proceeds as follows. Split the matrix T as in (30.10) with $n \approx m/2$. Suppose the eigenvalues of $\hat{T}_1$ and $\hat{T}_2$ are known. Since the correction matrix is of rank one, a nonlinear but rapid calculation can be used to get from the eigenvalues of $\hat{T}_1$ and $\hat{T}_2$ to those of T itself. Now recurse on this idea, finding the eigenvalues of $\hat{T}_1$ and $\hat{T}_2$ by further subdivisions with rank-one corrections, and so on. In this manner an $m \times m$ eigenvalue problem is reduced to a set of 1×1 eigenvalue problems together with a collection of rank-one corrections. (In practice, for maximal efficiency, it is customary to switch to the QR algorithm when the submatrices are of sufficiently small dimension rather than to carry the recursion all the way.)

In this process there is one key mathematical point. If the eigenvalues of $\hat{T}_1$ and $\hat{T}_2$ are known, how can those of T be found? To answer this, suppose that diagonalizations

$$\hat{T}_1 = Q_1 D_1 Q_1^T, \qquad \hat{T}_2 = Q_2 D_2 Q_2^T$$

have been computed. Then from (30.10) it follows that we have

$$T = \begin{bmatrix} Q_1 & \\ & Q_2 \end{bmatrix} \left(\begin{bmatrix} D_1 & \\ & D_2 \end{bmatrix} + \beta z z^T \right) \begin{bmatrix} Q_1^T & \\ & Q_2^T \end{bmatrix} \qquad (30.11)$$

with $z^T = (q_1^T, q_2^T)$, where q_1^T is the last row of Q_1 and q_2^T is the first row of Q_2. Since this equation is a similarity transformation, we have reduced the mathematical problem to the problem of finding the eigenvalues of a diagonal matrix plus a rank-one correction.

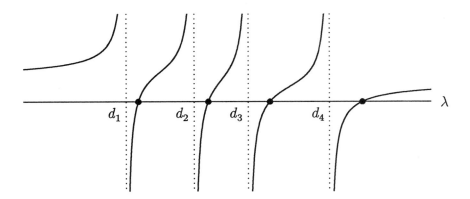

Figure 30.2. *Plot of the function $f(\lambda)$ of (30.12) for a problem of dimension 4. The poles of $f(\lambda)$ are the eigenvalues $\{d_j\}$ of D, and the roots of $f(\lambda)$ (solid dots) are the eigenvalues of $D + ww^T$. The rapid determination of these roots is the basis of each recursive step of the divide-and-conquer algorithm.*

To show how this is done, we simplify notation as follows. Suppose we wish to find the eigenvalues of $D + ww^T$, where $D \in \mathbb{R}^{m \times m}$ is a diagonal matrix with distinct diagonal entries $\{d_j\}$ and $w \in \mathbb{R}^m$ is a vector. (The choice of a plus sign corresponds to $\beta > 0$ above; for $\beta < 0$ we would consider $D - ww^T$.) We can assume $w_j \neq 0$ for all j, for otherwise, the problem is reducible. Then the eigenvalues of $D + ww^T$ are the roots of the rational function

$$f(\lambda) = 1 + \sum_{j=1}^{m} \frac{w_j^2}{d_j - \lambda}, \qquad (30.12)$$

as illustrated in Figure 30.2. This assertion can be justified by noting that if $(D + ww^T)q = \lambda q$ for some $q \neq 0$, then $(D - \lambda I)q + w(w^T q) = 0$, implying $q + (D - \lambda I)^{-1}w(w^T q) = 0$, that is, $w^T q + w^T (D - \lambda I)^{-1}w(w^T q) = 0$. This amounts to the equation $f(\lambda)(w^T q) = 0$, in which $w^T q$ must be nonzero, for otherwise q would be an eigenvector of D, hence nonzero in only one position, implying $w^T q \neq 0$ after all. We conclude that if q is an eigenvector of $D + ww^T$ with eigenvalue λ, then $f(\lambda)$ must be 0, and the converse follows because the form of $f(\lambda)$ guarantees that it has exactly m zeros. The equation $f(\lambda) = 0$ is known as the *secular equation*.

At each recursive step of the divide-and-conquer algorithm, the roots of (30.12) are found by a rapid iterative process related to Newton's method. Only $O(1)$ iterations are required for each root (or $O(\log(|\log(\epsilon_{\text{machine}})|))$ iterations if $\epsilon_{\text{machine}}$ is viewed as a variable), making the operation count $O(m)$ flops per root for an $m \times m$ matrix, or $O(m^2)$ flops all together. If we imagine a recursion in which a matrix of dimension m is split exactly in half at each step, the total operation count for finding eigenvalues of a tridiagonal matrix by the divide-and-conquer algorithm becomes

$$O\left(m^2 + 2\left(\frac{m}{2}\right)^2 + 4\left(\frac{m}{4}\right)^2 + 8\left(\frac{m}{8}\right)^2 + \cdots + m\left(\frac{m}{m}\right)^2\right), \qquad (30.13)$$

a series which converges to $O(m^2)$ (not $O(m^2 \log m)$) thanks to the squares in the denominators. Thus the operation count would appear to be of the same order $O(m^2)$ as for the QR algorithm.

So far, it is not clear why the divide-and-conquer algorithm is advantageous. Since the reduction of a full matrix to tridiagonal form ("Phase 1" in the terminology of Lecture 25) requires $4m^3/3$ flops (26.2), it would seem that any improvement in the $O(m^2)$ operation count for diagonalization of that tridiagonal matrix ("Phase 2") is hardly important. However, the economics change if one is computing eigenvectors as well as eigenvalues. Now, Phase 1 requires $8m^3/3$ flops but Phase 2 also requires $O(m^3)$ flops—for the QR algorithm, $\approx 6m^3$. The divide-and-conquer algorithm reduces this figure, ultimately because its nonlinear iterations involve just the scalar function (30.12), not the orthogonal matrices Q_j, whereas the QR algorithm must manipulate matrices Q_j at every iterative step.

An operation count reveals the following. The $O(m^3)$ part of the divide-and-conquer computation is the multiplication by Q_j and Q_j^T in (30.11). The total operation count, summed over all steps of the recursion, is $4m^3/3$ flops, a great improvement over $\approx 6m^3$ flops. Adding in the $8m^3/3$ flops for Phase 1 gives an improvement from $\approx 9m^3$ to $4m^3$.

Actually, the divide-and-conquer algorithm usually does even better than this, for a reason that is not elementary. For most matrices A, many of the vectors z and matrices Q_j that arise in (30.11) turn out to be numerically sparse in the sense that many of their entries have relative magnitudes less than machine precision. This sparsity allows a process of *numerical deflation*, whereby successive tridiagonal eigenvalue problems are reduced to uncoupled problems of smaller dimensions. In typical cases this reduces the Phase 2 operation count to an order less than m^3 flops, reducing the operation count for Phases 1 and 2 combined to $8m^3/3$. For eigenvalues alone, (30.13) becomes an overestimate and the Phase 2 operation count is reduced to an order lower than m^2 flops. The root of this fascinating phenomenon of deflation, which we shall not discuss further, is the fact that most of the eigenvectors of most tridiagonal matrices are "exponentially localized" (Exercise 30.7)—a fact that has been related by physicists to the phenomenon that glass is transparent.

We have spoken as if there is a single divide-and-conquer algorithm, but in fact, there are many variants. More complicated rank-one updates are often used for stability reasons, and rank-two updates are also sometimes used. Various methods are employed for finding the roots of $f(\lambda)$, and for large m, the fastest way to carry out the multiplications by Q_j is via multipole expansions rather than the obvious algorithm. A high-quality implementation of a divide-and-conquer algorithm can be found in the LAPACK library.

Exercises

30.1. Derive the formula (30.4), and give a precise geometric interpretation of the transformation (30.1) based on this choice of θ.

30.2. How many flops are required for one step (30.1) of the Jacobi algorithm? How many flops for $m(m-1)/2$ such steps, i.e., one sweep? How does the operation count for one sweep compare with the total operation count for tridiagonalizing a real symmetric matrix and finding its eigenvalues by the QR algorithm?

30.3. Show that if the largest off-diagonal entry is annihilated at each step of the Jacobi algorithm, then the sum of the squares of the off-diagonal entries decreases by at least the factor $1 - 2/(m^2 - m)$ at each step.

30.4. Suppose m is even and your computer has $m/2$ processors. Explain how $m/2$ transformations (30.1) can be carried out in parallel if they involve the disjoint row/column pairs $(1,2), (3,4), (5,6), \ldots, (m-1, m)$.

30.5. Write a program to find the eigenvalues of an $m \times m$ real symmetric matrix by the Jacobi algorithm with the standard row-wise ordering, plotting the sum of the squares of the off-diagonal entries on a log scale as a function of the number of sweeps. Apply your program to random matrices of dimensions 20, 40, and 80.

30.6. How many eigenvalues does

$$A = \begin{bmatrix} 1 & 1 & 0 & 0 \\ 1 & 1 & 1 & 0 \\ 0 & 1 & 2 & 1 \\ 0 & 0 & 1 & 3 \end{bmatrix}$$

have in the interval $[1, 2]$? Work out the answer on paper by bisection, making use of the recurrence (30.9).

30.7. Construct a random real symmetric tridiagonal matrix T of dimension 100 and compute its eigenvalue decomposition, $T = QDQ^T$. Plot a few of the eigenvectors on a log scale (the absolute values of a few columns of Q) and observe the phenomenon of localization. What proportion of the 10,000 entries of Q are greater than 10^{-10} in magnitude? What is the answer if instead of a random matrix, T is the discrete Laplacian with entries $1, -2, 1$?

Lecture 31. Computing the SVD

The computation of the SVD of an arbitrary matrix can be reduced to the computation of the eigenvalue decomposition of a hermitian square matrix, but the most obvious way of doing this is not stable. Instead, the standard methods for computing the SVD are based implicitly on another kind of reduction to hermitian form. For speed, the matrix is first unitarily bidiagonalized.

SVD of A and Eigenvalues of A^*A

As stated in Theorem 5.4, the SVD of the $m \times n$ matrix A $(m \geq n)$, $A = U\Sigma V^*$, is related to the eigenvalue decomposition of the matrix A^*A,

$$A^*A = V\Sigma^*\Sigma V^*.$$

Thus, mathematically speaking, we might calculate the SVD of A as follows:

1. Form A^*A;

2. Compute the eigenvalue decomposition $A^*A = V\Lambda V^*$;

3. Let Σ be the $m \times n$ nonnegative diagonal square root of Λ;

4. Solve the system $U\Sigma = AV$ for unitary U (e.g., via QR factorization).

This algorithm is frequently used, often by people who have rediscovered the SVD for themselves. The matrix A^*A is known as the *covariance matrix* of A, and it has familiar interpretations in statistics and other fields. The algorithm is unstable, however, because it reduces the SVD problem to an eigenvalue problem that may be much more sensitive to perturbations.

The difficulty can be explained as follows. We have seen that when a hermitian matrix A^*A is perturbed by δB, the absolute changes in each eigenvalue are bounded by the 2-norm of the perturbation. By Exercise 26.3(b), $|\lambda_k(A^*A + \delta B) - \lambda_k(A^*A)| \leq \|\delta B\|_2$. As is implied by equation (31.2) below, a similar bound holds for the singular values of A itself, $|\sigma_k(A + \delta A) - \sigma_k(A)| \leq \|\delta A\|_2$. Thus a backward stable algorithm for computing singular values would obtain $\tilde{\sigma}_k$ satisfying

$$\tilde{\sigma}_k = \sigma_k(A + \delta A), \qquad \frac{\|\delta A\|}{\|A\|} = O(\epsilon_{\mathrm{machine}}), \qquad (31.1)$$

which would imply

$$|\tilde{\sigma}_k - \sigma_k| = O(\epsilon_{\mathrm{machine}} \|A\|).$$

Now observe what happens if we proceed by computing $\lambda_k(A^*A)$. If $\lambda_k(A^*A)$ is computed stably, we must expect errors of the order

$$|\tilde{\lambda}_k - \lambda_k| = O(\epsilon_{\mathrm{machine}} \|A^*A\|) = O(\epsilon_{\mathrm{machine}} \|A\|^2).$$

Square-rooting to get σ_k, we find

$$|\tilde{\sigma}_k - \sigma_k| = O(|\tilde{\lambda}_k - \lambda_k| / \sqrt{\lambda_k}) = O(\epsilon_{\mathrm{machine}} \|A\|^2 / \sigma_k).$$

This is worse than the previous result by a factor $O(\|A\|/\sigma_k)$. This is no problem for the dominant singular values of A, with $\sigma_k \approx \|A\|$, but it is a big problem for any singular values with $\sigma_k \ll \|A\|$. For the smallest singular value σ_n, we must expect a loss of accuracy of order $\kappa(A)$—a "squaring of the condition number," just as in the use of the normal equations for certain least squares problems (Lecture 19).

A Different Reduction to an Eigenvalue Problem

There is an alternative, stable way to reduce the SVD to an eigenvalue problem. Assume that A is square, with $m = n$; this is no essential restriction, since we shall see that rectangular singular value problems can be reduced to square ones. Consider the $2m \times 2m$ hermitian matrix

$$H = \begin{bmatrix} 0 & A^* \\ A & 0 \end{bmatrix} \qquad (31.2)$$

mentioned earlier in Exercise 5.4. Since $A = U\Sigma V^*$ implies $AV = U\Sigma$ and $A^*U = V\Sigma^* = V\Sigma$, we have the block 2×2 equation

$$\begin{bmatrix} 0 & A^* \\ A & 0 \end{bmatrix} \begin{bmatrix} V & V \\ U & -U \end{bmatrix} = \begin{bmatrix} V & V \\ U & -U \end{bmatrix} \begin{bmatrix} \Sigma & 0 \\ 0 & -\Sigma \end{bmatrix}, \qquad (31.3)$$

which amounts to an eigenvalue decomposition of H. Thus we see that the singular values of A are the absolute values of the eigenvalues of H, and the singular vectors of A can be extracted from the eigenvectors of H.

Thus one could obtain the SVD of a square matrix A by forming the matrix H and computing its eigenvalue decomposition. In contrast to the use of AA^* or A^*A, this approach is stable. The standard algorithms for the SVD are based on this idea, though in a disguised manner in which no matrices of dimension as large as $m + n$ are formed explicitly. And the key step to make the process fast is an initial unitary reduction to bidiagonal form.

Two Phases

We have seen that hermitian eigenvalue problems are usually solved by a two-phase computation: first reduce the matrix to tridiagonal form, then diagonalize the tridiagonal matrix. Since the work of Golub, Kahan, and others in the 1960s, an analogous two-phase approach has been standard for the SVD. The matrix is brought into bidiagonal form, and then the bidiagonal matrix is diagonalized:

$$
\begin{bmatrix}
\times & \times & \times & \times \\
\times & \times & \times & \times \\
\times & \times & \times & \times \\
\times & \times & \times & \times \\
\times & \times & \times & \times \\
\times & \times & \times & \times
\end{bmatrix}
\quad
\begin{array}{c} \text{Phase 1} \\ \longrightarrow \end{array}
\quad
\begin{bmatrix}
\times & \times & & \\
& \times & \times & \\
& & \times & \times \\
& & & \times
\end{bmatrix}
\quad
\begin{array}{c} \text{Phase 2} \\ \longrightarrow \end{array}
\quad
\begin{bmatrix}
\times & & & \\
& \times & & \\
& & \times & \\
& & & \times
\end{bmatrix} .
$$
$$
\qquad A \qquad\qquad\qquad\qquad B \qquad\qquad\qquad\qquad \Sigma
$$

Phase 1 involves a finite number of operations, $O(mn^2)$ flops. Phase 2 in principle requires an infinite number of operations, but the standard algorithms converge superlinearly, and thus only $O(n \log(|\log(\epsilon_{\text{machine}})|))$ iterations are needed for convergence to order $\epsilon_{\text{machine}}$ (Exercise 25.2). In practice, we think of $\epsilon_{\text{machine}}$ as a constant and say that convergence is achieved in $O(n)$ iterations. Because the matrix operated on is bidiagonal, each of these iterations requires only $O(n)$ flops. Phase 2 therefore requires $O(n^2)$ flops all together (assuming singular values but not vectors are required). Thus, although Phase 1 is finite and Phase 2 is in principle infinite, in practice the latter is much the less expensive, just as we found for the symmetric eigenvalue problem.

Golub–Kahan Bidiagonalization

In Phase 1 of the SVD computation, we bring A into bidiagonal form by applying distinct unitary operations on the left and right. Note how this differs from the computation of eigenvalues, where the same unitary operations must be applied on both sides so that each step is a similarity transformation. In that case, it was only possible to introduce zeros below the first subdiagonal. Here, we are able to completely triangularize and also introduce zeros above the first superdiagonal.

The simplest method for accomplishing this, *Golub–Kahan bidiagonalization*, proceeds as follows. Householder reflectors are applied alternately on the left and the right. Each left reflection introduces a column of zeros below the diagonal. The right reflection that follows introduces a row of zeros to the right of the first superdiagonal, leaving intact the zeros just introduced in the column. For example, for a 6 × 4 matrix, the first two pairs of reflections look like this:

$$
\begin{bmatrix} \times & \times & \times & \times \\ \times & \times & \times & \times \\ \times & \times & \times & \times \\ \times & \times & \times & \times \\ \times & \times & \times & \times \\ \times & \times & \times & \times \end{bmatrix} \quad \overset{U_1^* \cdot}{\longrightarrow} \quad \begin{bmatrix} \mathbf{\times} & \mathbf{\times} & \mathbf{\times} & \mathbf{\times} \\ \mathbf{0} & \mathbf{\times} & \mathbf{\times} & \mathbf{\times} \\ \mathbf{0} & \mathbf{\times} & \mathbf{\times} & \mathbf{\times} \\ \mathbf{0} & \mathbf{\times} & \mathbf{\times} & \mathbf{\times} \\ \mathbf{0} & \mathbf{\times} & \mathbf{\times} & \mathbf{\times} \\ \mathbf{0} & \mathbf{\times} & \mathbf{\times} & \mathbf{\times} \end{bmatrix} \quad \overset{\cdot V_1}{\longrightarrow} \quad \begin{bmatrix} \times & \mathbf{\times} & \mathbf{0} & \mathbf{0} \\ \times & \mathbf{\times} & \mathbf{\times} & \mathbf{\times} \\ \times & \mathbf{\times} & \mathbf{\times} & \mathbf{\times} \\ \times & \mathbf{\times} & \mathbf{\times} & \mathbf{\times} \\ \times & \mathbf{\times} & \mathbf{\times} & \mathbf{\times} \\ \times & \mathbf{\times} & \mathbf{\times} & \mathbf{\times} \end{bmatrix}
$$

$$
A \qquad\qquad\qquad U_1^*A \qquad\qquad\qquad U_1^*AV_1
$$

$$
\overset{U_2^* \cdot}{\longrightarrow} \quad \begin{bmatrix} \times & \times & & & \\ & \mathbf{\times} & \mathbf{\times} & \mathbf{\times} & \\ & \mathbf{0} & \mathbf{\times} & \mathbf{\times} & \\ & \mathbf{0} & \mathbf{\times} & \mathbf{\times} & \\ & \mathbf{0} & \mathbf{\times} & \mathbf{\times} & \\ & \mathbf{0} & \mathbf{\times} & \mathbf{\times} & \end{bmatrix} \quad \overset{\cdot V_2}{\longrightarrow} \quad \begin{bmatrix} \times & \times & & \\ & \times & \mathbf{\times} & \mathbf{0} \\ & & \mathbf{\times} & \mathbf{\times} \\ & & \mathbf{\times} & \mathbf{\times} \\ & & \mathbf{\times} & \mathbf{\times} \\ & & \mathbf{\times} & \mathbf{\times} \end{bmatrix} .
$$

$$
U_2^*U_1^*AV_1 \qquad\qquad\qquad U_2^*U_1^*AV_1V_2
$$

The left-multiplication by U_1^* modifies rows 1 to m, introducing zeros in column 1 below the diagonal. The right-multiplication by V_1 modifies columns 2 to n, introducing zeros in row 1 without destroying the zeros in column 1. The process continues with operations on rows 2 to m, then columns 3 to n, and so on.

At the end of this process, n reflectors have been applied on the left and $n - 2$ on the right. The pattern of floating point operations resembles two Householder QR factorizations interleaved with each other, one operating on the $m \times n$ matrix A, the other on the $n \times m$ matrix A^*. The total operation count is therefore twice that of a QR factorization (10.9), i.e.,

$$
\text{Work for Golub–Kahan bidiagonalization:} \quad \sim 4mn^2 - \frac{4}{3}n^3 \text{ flops.} \quad (31.4)
$$

Faster Methods for Phase 1

For $m \gg n$, this operation count is unnecessarily large. A single QR factorization would introduce zeros everywhere below the diagonal, and for $m \gg n$, these are the great majority of the zeros that are needed. Yet the operation count for the Golub–Kahan method is twice as high. This observation suggests an alternative method for bidiagonalization with $m \gg n$, first proposed by Lawson and Hanson and later developed by Chan. The idea, *LHC bidiagonalization*, is illustrated as follows:

Lawson–Hanson–Chan bidiagonalization

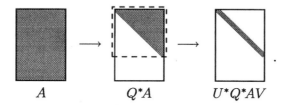

$$A \qquad\qquad Q^*A \qquad\qquad U^*Q^*AV$$

We begin by computing the QR factorization $A = QR$. Then we compute the Golub–Kahan bidiagonalization $B = U^*RV$ of R. The QR factorization requires $2mn^2 - \frac{2}{3}n^3$ flops (10.9), and the Golub–Kahan procedure, which now only has to operate on the upper $n \times n$ submatrix, requires $\frac{8}{3}n^3$ flops. The total operation count is

$$\text{Work for LHC bidiagonalization:} \quad \sim 2mn^2 + 2n^3 \text{ flops.} \tag{31.5}$$

This is cheaper than Golub–Kahan bidiagonalization for $m > \frac{5}{3}n$ (Exercise 31.1). Curiously, the LHC procedure creates zeros and then destroys them again (in the lower triangle of the upper $n \times n$ square of A), but there is a net gain.

The LHC procedure is advantageous only when $m > \frac{5}{3}n$, but the idea can be generalized so as to realize a saving for any $m > n$. The trick is to apply the QR factorization not at the beginning of the computation, but at a suitable point in the middle. This is advantageous because in the Golub–Kahan process, a matrix with $m > n$ becomes skinnier as the bidiagonalization proceeds. If the initial aspect ratio is, say, $m/n = 3/2$, it will steadily grow to $5/3$ and 2 and beyond. After step k, the aspect ratio of the remaining matrix is $(m - k)/(n - k)$, and when this figure gets sufficiently large, it makes sense to perform a QR factorization to reduce the problem to a square matrix.

Three-step bidiagonalization

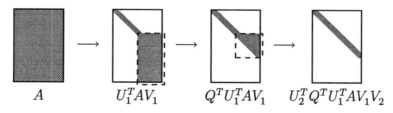

$$A \qquad U_1^T A V_1 \qquad Q^T U_1^T A V_1 \qquad U_2^T Q^T U_1^T A V_1 V_2$$

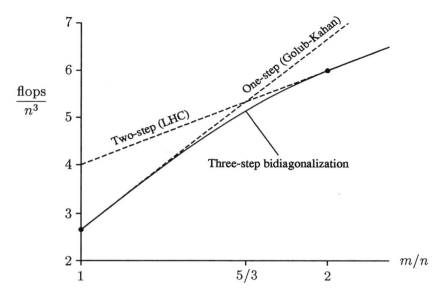

Figure 31.1. *Operation counts for three bidiagonalization algorithms applied to*
m × n matrices, from (31.4), (31.5), and (31.6). Three-step bidiagonalization
provides a pleasingly smooth interpolant between the other two methods, though
the improvement is hardly large.

When should the QR factorization be performed? If we aim solely to
minimize the operation count, the answer is simple: when the aspect ratio
reaches $(m - k)/(n - k) = 2$ (Exercise 31.2). This choice leads to the formula

$$\text{Work for three-step bidiagonalization:}\quad \sim 4mn^2 - \frac{4}{3}n^3 - \frac{2}{3}(m - n)^3 \text{ flops},$$

(31.6)

a modest improvement over the other two methods for $n < m < 2n$.

The operation counts for the three methods are plotted as a function of
m/n in Figure 31.1. It must be admitted that the improvement achieved by
the three-step method is small enough that in practice, other matters besides
the count may determine which method is best on a real machine (see p. 59).

Phase 2

In Phase 2 of the computation of the SVD, the SVD of the bidiagonal matrix
B is determined. From the 1960s to the 1990s, the standard algorithm for
this was a variant of the QR algorithm. More recently, divide-and-conquer
algorithms have also become competitive, and in the future, they are likely to
become the standard. We shall not give details.

Exercises

31.1. (a) Show that, as claimed in the text and illustrated in Figure 31.1, the crossover aspect ratio at which LHC bidiagonalization begins to beat Golub–Kahan bidiagonalization is $m/n = 5/3$.

(b) By what fraction is three-step bidiagonalization faster than the other two methods for $m/n = 5/3$?

31.2. Show that in three-step bidiagonalization, the optimal point at which to perform the QR factorization is when the matrix reaches an aspect ratio of 2.

31.3. Show that if the entries on both principal diagonals of a bidiagonal matrix are all nonzero, then the singular values of the matrix are distinct. (See Exercise 25.1.)

31.4. Let A be the $m \times m$ upper-triangular matrix with 0.1 on the main diagonal and 1 everywhere above the diagonal. Write a program to compute the smallest singular value of A in two ways: by calling a standard SVD software, and by forming A^*A and computing the square root of its smallest eigenvalue. Run your program for $1 \le m \le 30$ and plot the results as two curves on a log scale. Do the results conform to our general discussion of these algorithms?

31.5. Let A be an $m \times n$ matrix whose entries are independent samples from $N(0, 1)$, the normal distribution of mean 0, variance 1 (compare Exercise 12.3). Let B be a bidiagonal matrix

$$
B = \begin{bmatrix}
x_m & y_{n-1} & & & & \\
& x_{m-1} & y_{n-2} & & & \\
& & \ddots & \ddots & & \\
& & & x_{m-(n-2)} & y_1 & \\
& & & & x_{m-(n-1)}
\end{bmatrix},
$$

where each x or y is the positive square root of an independent sample from the χ^2 distribution with degree equal to the attached subscript. (The χ^2 distribution of degree k is equal to the distribution of the sum of squares of k independent variables from $N(0, 1)$.)

(a) Show that the distributions of the singular values of A and B are the same.

(b) Verify this result by an experiment. Specifically, take $m = 100$ and $n = 50$, construct random matrices A and B as indicated, and plot the singular values of A against those of B.

Part VI

Iterative Methods

Lecture 32. Overview of Iterative Methods

With this lecture the flavor of the book changes. We move from direct methods, a classical topic that is rather thoroughly understood, to the relatively untamed territory of iterative methods. These are the methods that seem likely to dominate the large-scale computations of the future.

Why Iterate?

The importance of iterative algorithms in linear algebra stems from a simple fact: noniterative or "direct" algorithms require $O(m^3)$ work. This is too much! It is too much both in the absolute sense that m^3 is huge when m is large, and in the relative sense that since the input to most matrix problems involves only $O(m^2)$ numbers, it seems unreasonable that $O(m^3)$ work must be expended in solving them.

The following table gives a thumbnail history of matrix computations over the years:

1950: $m = 20$	(Wilkinson)	
1965: $m = 200$	(Forsythe and Moler)	
1980: $m = 2000$	(LINPACK)	
1995: $m = 20000$	(LAPACK)	

These numbers represent a rough approximation to what dimensions might have been considered "very large" for a dense, direct matrix computation at the indicated dates. In the "Forsythe and Moler era" of the mid-1960s (named here after an influential textbook published in 1967), for example, a matrix of

243

dimension in the hundreds was large, stretching the limits of what could be calculated on available machines in a reasonable amount of time.

Evidently, in the course of forty-five years, the dimensions of tractable matrix problems have increased by a factor of 10^3. This progress is impressive, but it pales beside the progress achieved by computer hardware in the same period—a speedup by a factor of 10^9, from flops to gigaflops. In the fact that 10^9 is the cube of 10^3, we see played out in history the $O(m^3)$ bottleneck of direct matrix algorithms.

To put it another way, if matrix problems could be solved in $O(m^2)$ instead of $O(m^3)$ operations, some of the matrices being treated today might be 10 to 100 times larger. This is the aim, achieved for some matrices but not others, of matrix iterative methods.

Structure, Sparsity, and Black Boxes

Of course, it is not at all obvious that the $O(m^3)$ bottleneck can be beaten, and indeed, for "random" matrix problems, very likely it cannot. However, the large matrix problems that arise in practice are far from random, and there is a simple reason for this. Small matrices, say with dimension 3 or 30, may arise directly with more or less arbitrary entries in scientific problems—as representations of the relations between three forces in a structure, perhaps, or between thirty species in a chemical reaction. Large matrices, by contrast, usually arise indirectly in the discretization of differential or integral equations. One might say that if m is very large, it is probably an approximation to ∞. It follows that most large matrices of computational interest are simpler than their vast numbers of individual entries might suggest. They have some kind of structure, and as the years have gone by, ways have been found to exploit this structure in more and more contexts.

The most obvious structure that a large matrix may have is *sparsity*, i.e., preponderance of zero entries. (The opposite of *sparse* is *dense*.) For example, a finite difference discretization of a partial differential equation may lead to a matrix of dimension $m = 10^5$ with only $\nu = 10$ nonzero entries per row. This kind of structure is readily exploited by the iterative methods we shall discuss, for these algorithms use a matrix in the form of a *black box*:

$$x \longrightarrow \boxed{\begin{array}{c} \text{BLACK} \\ \text{BOX} \end{array}} \longrightarrow Ax.$$

The iterative algorithm requires nothing more than the ability to determine Ax for any x, which in a computer program will be effected by a procedure whose internal workings need be of no concern to the designer of the iterative algorithm. (Some iterative algorithms also require the computation of A^*x.) For the example of a sparse matrix A, it is easy to design a procedure to compute Ax in only $O(\nu m)$ rather than $O(m^2)$ operations. This is in marked contrast to the algorithms of direct linear algebra, such as Gaussian or House-holder triangularization, which explicitly manipulate matrix entries so as to introduce zeros, but in the process generally destroy sparsity.

Historically, sparsity has been the kind of structure exploited most often in iterative matrix computations. (Sparsity is also exploited by fast direct methods such as nested dissection or minimal degree reordering, not discussed in this book.) More recently, it has become apparent that other kinds of matrix structure may also be exploitable, even though the matrices involved are dense. For example, the solution of integral equations by numerical methods typically leads to dense matrix problems; in engineering these are called boundary element or panel methods. The coefficients of such matrices often have a great deal of regularity in them, and the quest for ways to take advantage of this regularity, for example by multipole methods or wavelet expansions, is an active area of research today. The black boxes involved in implementing such methods may have thousands of lines of code in them and be based on ideas that only specialists understand.

Projection into Krylov Subspaces

The iterative methods that occupy the remainder of this book are based on the idea of projecting an m-dimensional problem into a lower-dimensional Krylov subspace. Given a matrix A and a vector b, the associated *Krylov sequence* is the set of vectors $b, Ab, A^2 b, A^3 b, \ldots$, which can be computed by the black box in the form $b, Ab, A(Ab), A(A(Ab)), \ldots$. The corresponding *Krylov subspaces* are the spaces spanned by successively larger groups of these vectors.

Specifically, the algorithms that we shall discuss can be arranged in the following table:

	$Ax = b$	$Ax = \lambda x$
$A = A^*$	CG	Lanczos
$A \neq A^*$	GMRES CGN BCG et al.	Arnoldi

(This field is full of acronyms! CG, for example, stands for conjugate gradients, and, by the way, requires that A be positive definite as well as hermitian.) In each of these methods, the result of projection into the Krylov subspaces is that the original matrix problem is reduced to a sequence of matrix problems of dimensions $n = 1, 2, 3, \ldots$. When A is hermitian, the reduced matrices are tridiagonal, whereas in the nonhermitian case they have Hessenberg form.

Thus the Arnoldi iteration, for example, approximates eigenvalues of a large matrix by computing eigenvalues of certain Hessenberg matrices of successively larger dimensions.

Number of Steps, Work per Step, and Preconditioning

Gaussian elimination, QR factorization, and most other algorithms of dense linear algebra fit the following pattern: there are $O(m)$ steps, each requiring $O(m^2)$ work, for a total work estimate of $O(m^3)$. (Of course these figures, especially the second, may change on a parallel computer.) For iterative methods, the same figures still apply, but now they represent a typical worst-case behavior. When these methods succeed, they may do so by reducing one or both of these factors.

We shall see that the number of steps required for convergence to a satisfactory precision typically depends on spectral properties of the matrix A, if the word "spectral" is interpreted broadly. For example, the conjugate gradient iteration is guaranteed to solve a hermitian positive definite system $Ax = b$ quickly if the eigenvalues of A are clustered well away from the origin. Similarly, the Lanczos iteration is guaranteed to compute certain eigenvalues of a real hermitian matrix quickly if those eigenvalues are well separated from the rest of the spectrum (and if the initial vector that starts the iteration is suitably generic). The analysis of the convergence rates of these methods is a fascinating study that depends on the mathematical field of approximation theory. Specifically, the convergence of Krylov subspace iterative algorithms is closely related to problems of approximation of functions $f(z)$ by polynomials $p(z)$ on subsets of the real axis or the complex plane.

The work per step in a matrix iteration depends mainly on the structure of the matrix and on what advantage is taken of this structure in the $x \mapsto Ax$ black box.

The ideal iterative method in linear algebra reduces the number of steps from m to $O(1)$ and the work per step from $O(m^2)$ to $O(m)$, reducing the total work from $O(m^3)$ to $O(m)$. Such extraordinary speedups do occur in practical problems, but a more typical improvement is perhaps from $O(m^3)$ to $O(m^2)$. In a practical large-scale engineering computation of the mid-1990s, where iterative algorithms are successful, perhaps a typical result is that they beat direct algorithms by a factor on the order of 10. As machines get faster and m gets larger in the future, this factor will increase and iterative algorithms will become more important, illustrating the fundamental law of computer science: the faster the computer, the greater the importance of speed of algorithms.

Exact vs. Approximate Solutions

Matrix iterative methods are approximate in the sense that in principle they do not deliver exact answers, even in the absence of rounding errors, at least when carried to the number of iterative steps that is of practical interest. This property tends to make newcomers to these ideas uneasy; they may feel that iteration is an "engineering solution" of little elegance and doubtful reliability. This uneasiness tends to diminish as one gets to know these methods better. After all, even direct methods are inexact when carried out on a computer: one hopes for answers accurate to machine precision, no better. Since iterative methods too may be used to achieve the full accuracy of machine precision, the fact that they are in principle approximate need have little significance. As for elegance, the ideas that arise here are some of the most beautiful in numerical linear algebra.

These points are illustrated in Figure 32.1.

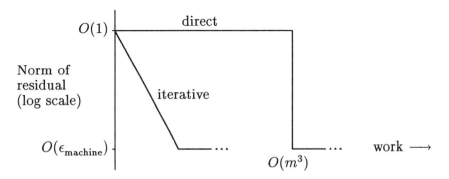

Figure 32.1. *Schematic illustration of convergence of direct and iterative methods. Under favorable circumstances, the iterative method converges geometrically until the residual is on the order of $\epsilon_{\text{machine}}$. The direct method makes no progress at all until $O(m^3)$ operations are completed, at which point the residual is again on the order of $\epsilon_{\text{machine}}$.*

Direct Methods That Beat $O(m^3)$

Finally, we must mention that there exist direct algorithms—finite, in principle exact—that solve $Ax = b$ and related problems in less than $O(m^3)$ operations. The first algorithm of this kind was discovered in 1969 by Volker Strassen, who reduced Gauss's exponent of 3 to $\log_2(7) \approx 2.81$, and subsequent improvements have reduced the best known exponent to its current value of ≈ 2.376 due to Coppersmith and Winograd. The history of these best known exponents is recorded in Figure 32.2.

So far, these fast algorithms have had negligible impact on practical computation, for two reasons. One is that in general, little is known about their stability properties. More fundamental is the fact that although the exponents in the fast algorithms are striking, the crossover values of m at which

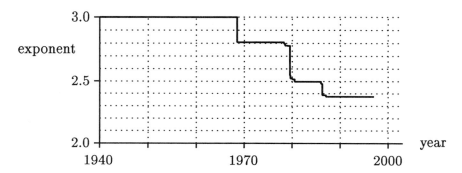

Figure 32.2. *Best known exponents for direct solution of $Ax = b$ (or equivalently, for computation of A^{-1}, AB, or $\det A$) for $m \times m$ matrices, as a function of time. Until* 1968, *the best known algorithms were of complexity $O(m^3)$. The currently best known algorithm solves $Ax = b$ in $O(m^{2.376})$ flops, but the constants are so large that this algorithm is impractical.*

they begin to beat standard methods are exceedingly high. Strassen's $m^{2.81}$ algorithm may be made to beat Gaussian elimination for values of m as low as 100, but because 2.81 is so close to 3, the victory never becomes dramatic for practical values of m. The existing methods with exponents much lower than this involve such large constant factors that they are slower than Gaussian elimination for the values of m attainable on current computers.

But what will happen in the future? The truth is that nobody knows. It is possible that tomorrow somebody will discover a "fast matrix inverse" that solves $Ax = b$ in $m^2 \log m$ floating point operations; or you, the reader, may do so this evening. Such a development would trigger the greatest upheaval in the history of numerical computation.

Exercises

32.1. An elliptic partial differential equation in three dimensions is discretized by a boundary element method. The result is a large dense linear system of equations in which each equation corresponds to a triangular surface element on a large sphere. To improve the accuracy, one must make the triangles smaller and thereby increase the number of equations, but the error shrinks only linearly in proportion to h, the diameter of the largest triangle.

A value of h is chosen, the system is solved by Gaussian elimination, and a solution accurate to two digits is obtained in one minute of computer time. It is decided that three digits of accuracy are needed. Assuming storage is not a constraint, approximately how much time will be required for the new computation on the same computer?

32.2. Consider the block matrix product

$$\begin{bmatrix} W & X \\ Y & Z \end{bmatrix} = \begin{bmatrix} A & B \\ C & D \end{bmatrix}\begin{bmatrix} E & F \\ G & H \end{bmatrix},$$

where, for simplicity, all the matrices $A, B, \ldots, Y, Z$ are assumed to be square and of the same dimension.

(a) Given $A, B, \ldots, G, H$, how many (i) matrix additions and (ii) matrix multiplications does it take to compute W, X, Y, Z by the obvious algorithm?

(b) Strassen showed that W, X, Y, Z can also be computed by the formulas

$$
\begin{aligned}
P_1 &= (A + D)(E + H), & P_5 &= (A + B)H, \\
P_2 &= (C + D)E, & P_6 &= (C - A)(E + F), \\
P_3 &= A(F - H), & P_7 &= (B - D)(G + H), \\
P_4 &= D(G - E), & & \\
W &= P_1 + P_4 - P_5 + P_7, & Y &= P_2 + P_4, \\
X &= P_3 + P_5, & Z &= P_1 + P_3 - P_2 + P_6.
\end{aligned}
$$

How many (i) matrix additions or subtractions and (ii) matrix multiplications are involved now?

(c) Show that by applying Strassen's formulas recursively, one can obtain an algorithm for multiplying matrices of dimension $m = 2^k$ with an operation count $O(m^{\log_2(7)})$ as $m \to \infty$.

(d) Write a recursive program that implements this idea, and give numerical evidence that your program works.

Lecture 33. The Arnoldi Iteration

Despite the many names and acronyms that have proliferated in the field of Krylov subspace matrix iterations, these algorithms are built upon a common foundation of a few fundamental ideas. One can take various approaches to describing this foundation. Ours will be to consider the Arnoldi process, a Gram–Schmidt-style iteration for transforming a matrix to Hessenberg form.

The Arnoldi/Gram–Schmidt Analogy

Suppose, to pass the time while marooned on a desert island, you challenged yourself to devise an algorithm to reduce a nonhermitian matrix to Hessenberg form by orthogonal similarity transformations, proceeding column by column from a prescribed first column q_1. To your surprise, you would probably find you could solve this problem in an hour and still have time to gather coconuts for dinner. The method you would come up with goes by the name of the Arnoldi iteration. If A is hermitian, the Hessenberg matrix becomes tridiagonal, an n-term recurrence relation becomes a three-term recurrence relation, and the name changes to the Lanczos iteration, to be discussed in Lecture 36.

Here is an analogy. For computing the QR factorization $A = QR$ of a matrix A, we have discussed two methods in this book: Householder reflections, which triangularize A by a succession of orthogonal operations, and Gram–Schmidt orthogonalization, which orthogonalizes A by a succession of triangular operations. Though Householder reflections lead to a more nearly orthogonal matrix Q in the presence of rounding errors, the Gram–Schmidt process has the advantage that it can be stopped part-way, leaving one with a reduced QR factorization of the first n columns of A. The problem of computing a Hessenberg reduction $A = QHQ^*$ of a matrix A is exactly analogous. There are two standard methods: Householder reflections (applied now on two sides of A rather than one) and the Arnoldi iteration. Thus Arnoldi is the analogue of Gram–Schmidt for similarity transformations to Hessenberg form rather than QR factorization. Like Gram–Schmidt, it has the advantage that it can be stopped part-way, leaving one with a partial reduction to Hessenberg form that is exploited in various manners to form iterative algorithms for eigenvalues or systems of equations.

Thus, this lecture is to Lecture 26 as Lecture 8 is to Lecture 10.

We can summarize the four algorithms just mentioned in a table:

	$A = QR$	$A = QHQ^*$
orthogonal structuring	Householder	Householder
structured orthogonalization	Gram–Schmidt	Arnoldi

For the remainder of this book, m and $n < m$ are positive integers, A is a real or complex $m \times m$ matrix, and $\| \cdot \| = \| \cdot \|_2$. In addition, one further character will now appear in the drama, an m-vector that we shall denote by b. The Arnoldi process needs this vector in order to get started. For applications to eigenvalue problems, we typically assume that b is random. For applications to systems of equations, as considered in later lectures, it will be the right-hand side, or more generally, the initial residual (see Exercise 35.5).

Mechanics of the Arnoldi Iteration

A complete reduction of A to Hessenberg form by an orthogonal similarity transformation might be written $A = QHQ^*$, or $AQ = QH$. However, in dealing with iterative methods we take the view that m is huge or infinite, so that computing the full reduction is out of the question. Instead we consider the first n columns of $AQ = QH$. Let Q_n be the $m \times n$ matrix whose columns are the first n columns of Q:

$$
Q_n = \begin{bmatrix} \big| & \big| & & \big| \\ q_1 & q_2 & \cdots & q_n \\ \big| & \big| & & \big| \end{bmatrix}.
$$

(33.1)

Here and in the lectures ahead, it would be consistent with our usage elsewhere in the book to put hats on the symbols Q_n, since these matrices are rectangular, but to keep the formulas uncluttered we do not do this.

Let $\tilde{H}_n$ be the $(n+1) \times n$ upper-left section of H, which is also a Hessenberg matrix:

$$\tilde{H}_n = \begin{bmatrix} h_{11} & \cdots & & h_{1n} \\ h_{21} & h_{22} & & \\ & \ddots & \ddots & \vdots \\ & & h_{n,n-1} & h_{nn} \\ & & & h_{n+1,n} \end{bmatrix}. \tag{33.2}$$

Then we have

$$AQ_n = Q_{n+1}\tilde{H}_n, \tag{33.3}$$

that is,

$$\begin{bmatrix} & & \\ & A & \\ & & \end{bmatrix} \begin{bmatrix} q_1 & \cdots & q_n \end{bmatrix} = \begin{bmatrix} q_1 & \cdots & q_{n+1} \end{bmatrix} \begin{bmatrix} h_{11} & \cdots & h_{1n} \\ h_{21} & & \vdots \\ & \ddots & \\ & & h_{n+1,n} \end{bmatrix}.$$

The nth column of this equation can be written as follows:

$$Aq_n = h_{1n}q_1 + \cdots + h_{nn}q_n + h_{n+1,n}q_{n+1}. \tag{33.4}$$

In words, q_{n+1} satisfies an $(n+1)$-term recurrence relation involving itself and the previous Krylov vectors.

The Arnoldi iteration is simply the modified Gram–Schmidt iteration that implements (33.4). The following algorithm should be compared with Algorithm 8.1.

Algorithm 33.1. Arnoldi Iteration

$b = $ arbitrary, $q_1 = b/\|b\|$
for $n = 1, 2, 3, \ldots$
$\quad v = Aq_n$
$\quad$ **for** $j = 1$ **to** n
$\quad\quad h_{jn} = q_j^* v$
$\quad\quad v = v - h_{jn}q_j$
$\quad h_{n+1,n} = \|v\|$ $\qquad$ [see Exercise 33.2 concerning $h_{n+1,n} = 0$]
$\quad q_{n+1} = v/h_{n+1,n}$

The reader can see at a glance how simple the Arnoldi process is. In a high-level language such as MATLAB, it can be implemented in less than a dozen lines. The matrix A appears only in the product Aq_n, which can be computed by a black box procedure as described in the last lecture.

QR Factorization of a Krylov Matrix

The power of the Arnoldi process lies in the various interpretations that can be made of it, and in the algorithms these suggest. For a first interpretation, consider the recurrence (33.4). It is evident from this formula that the vectors $\{q_j\}$ form bases of the successive *Krylov subspaces* generated by A and b, defined as follows:

$$\mathcal{K}_n = \langle b, Ab, \ldots, A^{n-1}b \rangle = \langle q_1, q_2, \ldots, q_n \rangle \subseteq \mathbb{C}^m. \qquad (33.5)$$

Moreover, since the vectors q_j are orthonormal, these are orthonormal bases. Thus the Arnoldi process can be described as the systematic construction of orthonormal bases for successive Krylov subspaces.

To express this observation in matrix form, let us define K_n to be the $m \times n$ *Krylov matrix*

$$K_n = \begin{bmatrix} & | & | & & | & \\ b & Ab & \cdots & A^{n-1}b \\ & | & | & & | & \end{bmatrix}. \qquad (33.6)$$

Then K_n must have a reduced QR factorization

$$K_n = Q_n R_n, \qquad (33.7)$$

where Q_n is the same matrix as above. In the Arnoldi process, neither K_n nor R_n is formed explicitly. Doing so would make for an unstable algorithm, since these are exceedingly ill-conditioned matrices in general, as the columns of K_n all tend to approximate the same dominant eigenvector of A. However, (33.6) and (33.7) give an intuitive explanation of why the Arnoldi process leads to effective methods for determining certain eigenvalues. Clearly K_n might be expected to contain good information about the eigenvalues of A with largest modulus, and the QR factorization might be expected to reveal this information by peeling off one approximate eigenvector after another, starting with the dominant one.

The explanation just given may remind the reader of a similar discussion that appeared earlier in this book. The relationship between (33.6)–(33.7) and the Arnoldi algorithm is analogous to that between simultaneous iteration and the QR algorithm for computing eigenvalues of matrices. One is easy to understand but unstable, the other is subtler but stabler. The difference is that, whereas the Arnoldi iteration is based upon the QR factorization (33.7) of the matrix whose columns are $b, Ab, \ldots, A^{n-1}b$, simultaneous iteration and the QR algorithm are based upon the QR factorization (28.16) of the matrix whose columns are $A^n e_1, \ldots, A^n e_m$. We can summarize this parallel in another table:

	quasi-direct	iterative
straightforward but unstable	simultaneous iteration	(33.6)–(33.7)
subtle but stable	QR algorithm	Arnoldi

Projection onto Krylov Subspaces

Another way to view the Arnoldi process is as a computation of projections onto successive Krylov subspaces. To see this, note that the product $Q_n^* Q_{n+1}$ is the $n \times (n+1)$ identity, i.e., the $n \times (n+1)$ matrix with 1 on the main diagonal and 0 elsewhere. Therefore $Q_n^* Q_{n+1} \tilde{H}_n$ is the $n \times n$ Hessenberg matrix obtained by removing the last row of $\tilde{H}_n$:

$$
H_n = \begin{bmatrix} h_{11} & \cdots & & h_{1n} \\ h_{21} & h_{22} & & \\ & \ddots & \ddots & \vdots \\ & & h_{n,n-1} & h_{nn} \end{bmatrix}. \tag{33.8}
$$

From (33.3) we accordingly have

$$
H_n = Q_n^* A Q_n. \tag{33.9}
$$

This matrix can be interpreted as the representation in the basis $\{q_1, \ldots, q_n\}$ of the orthogonal projection of A onto $\mathcal{K}_n$. Is it clear what this interpretation means? Here is a precise statement. Consider the linear operator $\mathcal{K}_n \rightarrow \mathcal{K}_n$ defined as follows: given $v \in \mathcal{K}_n$, apply A to it, then orthogonally project Av back into the space $\mathcal{K}_n$. Since the orthogonal projector of $\mathbb{C}^m$ onto $\mathcal{K}_n$ is $Q_n Q_n^*$, this operator can be written $Q_n Q_n^* A$ with respect to the standard basis of $\mathbb{C}^m$. With respect to the basis of columns of Q_n, it can therefore be written $Q_n^* A Q_n$.

The kind of projection just described comes up throughout applied and numerical mathematics. In another context it is known as the *Rayleigh–Ritz* procedure; not coincidentally, in the diagonal elements of H_n one recognizes the Rayleigh quotients of A with respect to the vectors q_j. This projection process is also one of the ideas underlying finite element methods for solution of partial differential equations, as well as their younger relatives known as spectral methods.

Since H_n is a projection of A, one might imagine that its eigenvalues would be related to those of A in a useful fashion. These n numbers,

$$\{\theta_j\} = \{\text{eigenvalues of } H_n\}, \tag{33.10}$$

are called the *Arnoldi eigenvalue estimates* (at step n) or *Ritz values* (with respect to $\mathcal{K}_n$) of A. In the next lecture we shall see that some of these numbers may be extraordinarily accurate approximations to some of the eigenvalues of A, even for $n \ll m$.

We summarize the developments of this lecture in a theorem, to be compared with Theorem 28.3.

Theorem 33.1. *The matrices Q_n generated by the Arnoldi iteration are reduced QR factors of the Krylov matrix (33.6):*

$$K_n = Q_n R_n. \tag{33.11}$$

The Hessenberg matrices H_n are the corresponding projections

$$H_n = Q_n^* A Q_n, \tag{33.12}$$

and the successive iterates are related by the formula

$$A Q_n = Q_{n+1} \tilde{H}_n. \tag{33.13}$$

Exercises

33.1. Let $A \in \mathbb{C}^{m \times m}$ and $b \in \mathbb{C}^m$ be arbitrary. Show that any $x \in K_n$ is equal to $p(A)b$ for some polynomial p of degree $\leq n - 1$.

33.2. Suppose Algorithm 33.1 is executed for a particular A and b until at some step n, an entry $h_{n+1,n} = 0$ is encountered.

(a) Show how (33.13) can be simplified in this case. What does this imply about the structure of a full $m \times m$ Hessenberg reduction $A = QHQ^*$ of A?

(b) Show that $\mathcal{K}_n$ is an invariant subspace of A, i.e., $A\mathcal{K}_n \subseteq \mathcal{K}_n$.

(c) Show that if the Krylov subspaces of A generated by b are defined by $\mathcal{K}_k = \langle b, Ab, \ldots, A^{k-1}b \rangle$ as in (33.5), then $\mathcal{K}_n = \mathcal{K}_{n+1} = \mathcal{K}_{n+2} = \cdots$.

(d) Show that each eigenvalue of H_n is an eigenvalue of A.

(e) Show that if A is nonsingular, then the solution x to the system of equations $Ax = b$ lies in $\mathcal{K}_n$.

The appearance of an entry $h_{n+1,n} = 0$ is called a *breakdown* of the Arnoldi iteration, but it is a breakdown of a benign sort. For applications in computing eigenvalues (Lecture 34) or solving systems of equations (Lecture 35), because of (d) and (e), a breakdown usually means that convergence has occurred and the iteration can be terminated. Alternatively, a new orthonormal vector q_{n+1} could be selected at random and the iteration then continued.

33.3. (a) Suppose Algorithm 33.1 is executed for a particular A and b and runs to completion ($n = m$), with no breakdown of the kind described in the last exercise. Show that this implies that the minimal polynomial of A is of degree m.

(b) Conversely, suppose that the minimal polynomial of A is of degree m. Show that this does not imply that for a particular choice of b, Algorithm 33.1 will necessarily run to completion.

(c) Explain why the result of (a) does not contradict Exercise 25.1(b).

Lecture 34. How Arnoldi Locates Eigenvalues

The Arnoldi iteration is two things: the basis of many of the iterative algorithms of numerical linear algebra and, more specifically, a technique for finding eigenvalues of nonhermitian matrices. Here we consider this second, specific role, and in the process, we describe a connection with polynomial approximation theory that is of broad importance.

Computing Eigenvalues by the Arnoldi Iteration

The use of the Arnoldi iteration for computing eigenvalues proceeds as follows. The iteration is carried out as described in the last lecture (Algorithm 33.1). At each step n, or at occasional steps, the eigenvalues of the Hessenberg matrix H_n are computed by standard methods such as the QR algorithm. (In practice this means a call to software such as provided by EISPACK, LAPACK, or MATLAB.) These are the "Arnoldi estimates" or "Ritz values" (33.10). Some of these numbers are typically observed to converge rapidly, often geometrically (i.e., linearly; see Exercise 25.2), and when they do, one may assume with reasonable confidence that the converged values are eigenvalues of A.

Since $n \ll m$ for a feasible computation, one cannot of course expect to compute all of the eigenvalues of A by this process. Which eigenvalues, then, does the Arnoldi iteration find? Typically, it finds *extreme* eigenvalues, that is, eigenvalues near the edge of the spectrum of A. Fortunately, these are precisely the eigenvalues of main interest in most applications.

For example, in problems of hydrodynamic stability, the aim is to determine whether small perturbations to a smooth fluid flow may grow unstably, in which case the flow is likely to break down into another form that is more complicated, perhaps turbulent. This determination of stability is conventionally achieved by linearizing the problem to obtain a Jacobian operator governing evolution of small perturbations, then calculating the rightmost eigenvalue of this operator in the complex plane, that is, the eigenvalue with maximal real part. If this eigenvalue is in the right half-plane, the flow is unstable, whereas if it is in the left half-plane, it is stable.

Many variations on the Arnoldi method have been devised, such as acceleration devices for problems in which some parts of the complex plane are of greater importance than others. Here we confine our attention to the unadorned Arnoldi method, attempting to give an idea of why it converges, why it tends to find extreme eigenvalues of A, and how quickly it finds them.

A Note of Caution: Nonnormality

First, however, we call the reader's attention to a cautionary fact. The physical significance of the eigenvalues of nonhermitian matrices is sometimes not as great as supposed. If a matrix is far from normal—that is, if its eigenvectors are far from orthogonal, which implies that its eigenvalues are ill-conditioned— then the eigenvalues may have little to do with how a physical system governed by the matrix actually behaves (see Exercises 24.3 and 26.2). In the problem of fluid flow through a circular pipe, for example, the Jacobian of the linearized equations has all its eigenvalues in the left half-plane, suggesting that the flow ought to be stable. In actuality, however, high-speed flows through pipes are invariably turbulent. The resolution of this paradox is rooted in the fact that although the linearized problem has all decaying eigenmodes, it may nevertheless amplify certain nonmodal flow perturbations by many orders of magnitude. Such phenomena cannot occur with normal matrices or operators—e.g., symmetric, hermitian, or self-adjoint.

If the answer is highly sensitive to perturbations, you have probably asked the wrong question. We urge anyone faced with nonhermitian eigenvalue computations involving highly sensitive eigenvalues to bear this principle in mind. If you are a numerical analyst, and the problem was given to you by a colleague in science or engineering, do not accept without scrutiny your colleague's assurances that it is truly the eigenvalues that matter physically, even though their condition numbers with respect to perturbations of the matrix entries are 10^4. Perhaps situations exist where highly sensitive eigenvalues of nonnormal operators are of genuine physical significance, but they are outnumbered by situations where eigenvalues are mistakenly investigated when a deeper analysis is properly called for.

Arnoldi and Polynomial Approximation

Let x be a vector in the Krylov subspace $\mathcal{K}_n$ (33.5). Such an x can be written as a linear combination of powers of A times b:

$$x = c_0 b + c_1 Ab + c_2 A^2 b + \cdots + c_{n-1} A^{n-1} b. \tag{34.1}$$

This expression has a compact description: it is a *polynomial in A* times b. That is, if q is the polynomial $q(z) = c_0 + c_1 z + \cdots + c_{n-1} z^{n-1}$, then we have

$$x = q(A)b, \tag{34.2}$$

as pointed out already in Exercise 33.1. Because every vector of the form (34.1) can be expressed in the form (34.2), Krylov subspace iterations can always be analyzed in terms of matrix polynomials.

One analysis of the Arnoldi iteration takes the following form. Define

$$P^n = \{\text{monic polynomials of degree } n\}.$$

(The word "monic" means that the coefficient of the term of degree n is 1.) Now consider the following approximation problem.

Arnoldi/Lanczos Approximation Problem. Find $p^n \in P^n$ such that

$$\| p^n(A)b \| = \text{minimum}. \tag{34.3}$$

As always in this part of the book, $\| \cdot \| = \| \cdot \|_2$. The reason we have used superscripts on P^n is that in the next lecture, we shall consider the space of degree-n polynomials normalized by $c_0 = 1$ rather than $c_n = 1$, and for this space we shall use a subscript.

The Arnoldi iteration has the remarkable property that it solves (34.3) exactly.

Theorem 34.1. *As long as the Arnoldi iteration does not break down (i.e., K_n is of full rank n), (34.3) has a unique solution p^n, namely, the characteristic polynomial of H_n.*

Proof. First we note that if $p \in P^n$, then the vector $p(A)b$ can be written $p(A)b = A^n b - Q_n y$ for some $y \in \mathbb{C}^n$, where Q_n is defined by (33.1). In other words, (34.3) is equivalent to a linear least squares problem: find the point in $\mathcal{K}_n$ closest to $A^n b$, or in matrix terms, find y such that

$$\|A^n b - Q_n y\| = \text{minimum}.$$

The solution is characterized by the orthogonality condition $p^n(A)b \perp \mathcal{K}_n$, illustrated in Figure 34.1, or equivalently, $Q_n^* p^n(A)b = 0$. Now consider the

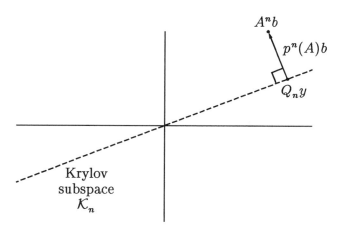

Figure 34.1. *The least squares polynomial approximation problem underlying the Arnoldi iteration.*

factorization $A = QHQ^*$ mentioned at the beginning of the last lecture. At step n of the Arnoldi process, we have computed the first n columns of Q and H; thus we know that a factorization of this kind exists with

$$Q = \left[\begin{array}{cc} Q_n & U \end{array} \right], \qquad H = \left[\begin{array}{cc} H_n & X_1 \\ X_2 & X_3 \end{array} \right]$$

for some $m \times (m-n)$ matrix U with orthonormal columns that are also orthogonal to the columns of Q_n and some matrices X_1, X_2, and X_3 of dimensions $n \times (m - n)$, $(m - n) \times n$, and $(m - n) \times (m - n)$, respectively, with all but the upper-right entry of X_2 equal to 0. The orthogonality condition becomes $Q_n^* Q p^n(H) Q^* b = 0$, which amounts to the condition that the first n entries of the first column of $p^n(H)$ are zero. Because of the structure of H, these are also the first n entries of the first column of $p^n(H_n)$. By the Cayley–Hamilton theorem, these are zero if p^n is the characteristic polynomial of H_n. Conversely, suppose there were another polynomial p^n with $p^n(A)b \perp K_n$. Taking the difference would give a nonzero polynomial q of degree $n - 1$ with $q(A)b = 0$, violating the assumption that K_n is of full rank.   □

   Theorem 34.1 gives us a new interpretation of the Ritz values generated by the Arnoldi iteration. They are the roots of the optimal polynomial (34.3).

Invariance Properties

Theorem 34.1 provides an easy way to remember some of the basic properties of the Arnoldi iteration. For example, since the family P^n of monic polynomials is invariant with respect to translations $z \mapsto z + \alpha$, the Arnoldi iteration is also translation-invariant. Unlike iterative algorithms for solving $Ax = b$, to be discussed later, such as GMRES, it does not "know where the origin is."

Here is a summary of this and other invariance properties. Each part of this theorem is a corollary of Theorem 34.1; we do not give proofs.

Theorem 34.2. *Let the Arnoldi iteration be applied to a matrix $A \in \mathbb{C}^{m \times m}$ as described above.*

Translation-invariance. *If A is changed to $A + \sigma I$ for some $\sigma \in \mathbb{C}$, and b is left unchanged, the Ritz values $\{\theta_j\}$ at each step change to $\{\theta_j + \sigma\}$.*

Scale-invariance. *If A is changed to σA for some $\sigma \in \mathbb{C}$, and b is left unchanged, the Ritz values $\{\theta_j\}$ change to $\{\sigma\theta_j\}$.*

Invariance under unitary similarity transformations. *If A is changed to UAU^* for some unitary matrix U, and b is changed to Ub, the Ritz values $\{\theta_j\}$ do not change.*

In all three cases the Ritz vectors, *namely the vectors $Q_n y_j$ corresponding to the eigenvectors y_j of H_n, do not change under the indicated transformation.*

By Theorem 24.9, every matrix A is unitarily similar to an upper-triangular matrix. Thus the property of invariance under unitary similarity transformations implies that for understanding convergence properties of the Arnoldi iteration, it is enough in principle to consider upper-triangular matrices. It is not, however, enough to consider diagonal matrices. In the nonhermitian (nonnormal) case, there is more to a matrix than its eigenvalues.

How Arnoldi Locates Eigenvalues

We can now begin to address the problem raised in the title of this lecture. Theorem 34.1 asserts that at bottom, the "goal" of the Arnoldi iteration is to solve a polynomial approximation problem, or equivalently, a least squares problem involving a Krylov subspace. If the Arnoldi iteration tends to find eigenvalues, it must be as a by-product of achieving this goal.

What does polynomial approximation have to do with the eigenvalues of A? A little thought shows that there is a connection between these two along the following lines. If one's aim is to find a polynomial p^n with the property that $p^n(A)$ is small, an effective means to that end may be to pick p^n to have zeros close to the eigenvalues of A.

Consider an extreme case. Suppose that A is diagonalizable and has only $n \ll m$ distinct eigenvalues, hence a minimal polynomial of degree n. Then from Theorem 34.1 it is clear that after n steps, all of these eigenvalues will be found exactly, at least if the vector b contains components in directions associated with every eigenvalue. Thus after n steps, the Arnoldi iteration has computed the minimal polynomial of A exactly.

In practical applications, much the same phenomenon takes place. Now, however, the agreement of Ritz values with eigenvalues is approximate instead of exact, and instead of a minimal polynomial, the result is a "pseudo-minimal polynomial," i.e., a polynomial p^n such that $\|p^n(A)\|$ is small.

Arnoldi Lemniscates

This convergence process can be illustrated graphically by plotting lemniscates in the complex plane. A *lemniscate* is a curve or collection of curves

$$\{z \in \mathbb{C} : |p(z)| = C\}, \tag{34.4}$$

where p is a polynomial and C is a real constant. If p is the Arnoldi polynomial p^n for an Arnoldi iteration with matrix A at step n, and C takes the value

$$C = \frac{\|p^n(A)b\|}{\|b\|}, \tag{34.5}$$

then the curves defined by (34.4) may be called *Arnoldi lemniscates*. As the iteration number n increases, components of these lemniscates typically appear which surround the extreme eigenvalues of A and then shrink rapidly to a point, namely the eigenvalue itself.

To illustrate these ideas, let A be a square matrix of dimension $m = 100$ whose entries are independent random numbers from the real normal distribution of mean 0 and standard deviation $m^{-1/2}$ (Exercise 12.3). Since A is real, its eigenvalues consist of real numbers and complex conjugate pairs. Our choice of the standard deviation is such that the eigenvalues are approximately uniformly distributed in the unit disk $|z| \leq 1$. But now we create an outlier eigenvalue by changing the corner entry a_{11} to 1.5. See Figure 34.2.

We apply the Arnoldi iteration to this matrix, beginning with a random vector q_1. At step n, the matrix H_n has been constructed and its characteristic polynomial reflects the iteration's current knowledge about the spectrum of A. Figure 34.3 plots Arnoldi lemniscates at steps $n = 5, 6, 7, 8$. At $n = 1$ (not shown), the lemniscate is an exact circle, little affected by the outlying eigenvalue. By $n = 5$, the circle has begun to bulge in the direction of λ. At $n = 6$, the bulge pinches off and a new component of the lemniscate appears. This component then proceeds to shrink with each subsequent iteration.

Geometric Convergence

Under certain circumstances, the convergence of some of the Arnoldi eigenvalue estimates to eigenvalues of A is geometric. These matters are incompletely understood at present, and we shall not cite theoretical results here. Instead, we shall just take a look at the convergence in the numerical example above, and give a partial explanation.

From a figure like Figure 34.3, one cannot assess the rate of convergence beyond one or two digits of accuracy. Figure 34.4 fills this gap by plotting $|\lambda^{(n)} - \lambda|$ as a function of n for the same example as before, where $\lambda^{(n)}$ is the Arnoldi eigenvalue estimate closest to λ at step n. The first thing we notice is that the convergence is clearly geometric. After fifty iterations, we have

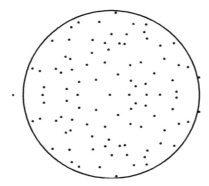

Figure 34.2. *Eigenvalues of a* 100×100 *matrix A, random except in the* $1,1$ *position. The circle is the unit circle in* $\mathbb{C}$. *The eigenvalues are approximately uniformly distributed in the unit disk except for the outlier* $\lambda \approx 1.4852$.

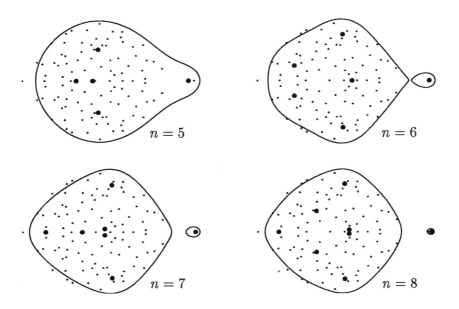

Figure 34.3. *Arnoldi lemniscates* (34.4)–(34.5) *at steps* $n = 5, 6, 7, 8$ *for the same matrix A. The small dots are the eigenvalues of A, and the large dots are the eigenvalues of* H_n, *i.e., the Ritz values. One component of the Arnoldi lemniscate first "swallows" the outlier eigenvalue, and in subsequent iterations it then shrinks to a point at a geometric rate.*

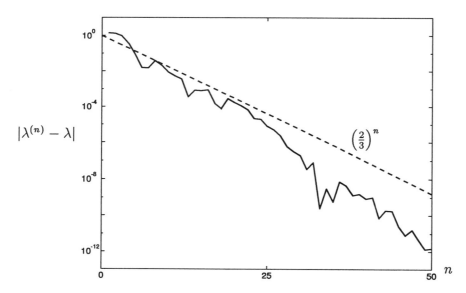

Figure 34.4. *Convergence of the rightmost Arnoldi eigenvalue estimate.*

achieved twelve digits of precision, and this figure would not have been much different if A had had dimension 1000 instead of 100.

We can be more quantitative. For the first few dozen iterations, at least, the convergence in Figure 34.3 approximates the rate

$$|\lambda^{(n)} - \lambda| \approx \left(\frac{2}{3}\right)^n. \tag{34.6}$$

An explanation of this behavior is as follows. To minimize (34.3), p^n must take a roughly minimal value at each of the eigenvalues of A. Consider for example the candidate polynomial $p(z) = z^{n-1}(z - \tilde{\lambda})$, where $\tilde{\lambda}$ is some number close to λ. At each of the eigenvalues of A in the unit disk, $|p(z)|$ is of order 1 or smaller. At $z = \lambda$, however, it has magnitude

$$|p(\lambda)| \approx \left(\frac{3}{2}\right)^n |\tilde{\lambda} - \lambda|$$

(this would be an equality if λ were exactly equal to $3/2$). When n is large, $(3/2)^n$ is huge. For this number also to be of order 1, $|\tilde{\lambda} - \lambda|$ must be small enough to balance it, that is, of order $(2/3)^n$, as in (34.6).

Another feature is apparent in Figure 34.4. After the initial few dozen steps, the convergence begins to accelerate, a phenomenon common in Krylov subspace iterations. What is happening here is that the iteration is beginning to resolve some of the other outer eigenvalues of A, near the unit circle. If the dimension of A had been $m = 300$, then the cloud of eigenvalues would have filled the unit disk sufficiently densely that no such acceleration would have been visible in the fifty iterative steps shown in Figure 34.4.

Exercises

34.1. Given $A \in \mathbb{C}^{m \times m}$, $b \in \mathbb{C}^m$, and $p \in P^n$, suppose we want to compute $p(A)b$. A natural place to start is with Horner's rule, which can be written

$$p(z) = c_0 + z(c_1 + z(c_2 + \cdots + z(c_{m-1} + z) \cdots)). \qquad (34.7)$$

(a) Write a **for** loop based on (34.7) (on paper, not on a computer) that computes $p(A)$ and then applies the result to b. Determine the number of flops required by this algorithm, to leading order.

(b) Write another **for** loop for computing $p(A)b$, a far better one, in which b is introduced into the process at the beginning. Again determine the number of flops required to leading order.

(c) In introductory numerical analysis texts, Horner's rule is recommended for evaluation of polynomials because it is faster than the obvious method of computing powers z^k, multiplying by coefficients c_k, and adding. Show that for computing $p(A)$ or $p(A)b$, by contrast, Horner's rule is not significantly faster than the obvious method.

34.2. We have seen that the Arnoldi polynomial p^n minimizes $\|p^n(A)b\|$. Another polynomial that might give cleaner information about eigenvalues would be the *ideal Arnoldi polynomial*, also known as the *Chebyshev polynomial of A*, defined as the unique $p^* \in P^n$ that minimizes $\|p^n(A)\|$. (This polynomial is not used in practice, because there is no fast way to compute it.)

(a) Prove that p^* exists.

(b) Prove that provided $p^*(A) \neq 0$, p^* is unique. (Hint: Suppose p_1 and p_2 are distinct ideal Arnoldi polynomials for given A and n, set $p = (p_1 + p_2)/2$, and consider the singular vectors of $p(A)$ corresponding to the maximal singular value. This is a hard problem.)

34.3. Let A be the $N \times N$ bidiagonal matrix with $a_{k,k+1} = a_{k,k} = k^{-1/2}$, $N = 64$. (In the limit $N \to \infty$, A becomes a non-self-adjoint compact operator.)

(a) Produce a plot showing the spectrum $\Lambda(A)$ and the boundaries of the ϵ-pseudospectra $\Lambda_\epsilon(A)$ (Exercises 26.1 and 26.2) for $\epsilon = 10^{-1}, 10^{-2}, 10^{-3}, 10^{-4}$.

(b) Starting from a random initial vector, run the Arnoldi iteration and compute Ritz values at steps $n = 1, 2, \ldots, 30$. Produce plots to indicate the rates of convergence to eigenvalues of A, and comment on your results.

(c) The Arnoldi iteration can also be used to approximate pseudospectra of A by those of H_n or $\tilde{H}_n$. (In the latter case, the boundary of $\Lambda_\epsilon(\tilde{H}_n)$ is defined by the condition $\sigma_{\min}(zI - \tilde{H}_n) = \epsilon$, or equivalently $\|(zI - \tilde{H}_n)^+\| = \epsilon^{-1}$, where I is a rectangular version of the identity.) Experiment with this idea by plotting the ϵ-pseudospectra of $\tilde{H}_n$ for $n = 5, 10, 15, 20$. How closely do they match the corresponding pseudospectra of A?

Lecture 35. GMRES

In the last lecture we showed how the Arnoldi process can be used to find eigenvalues. Here we show that it can also be used to solve systems of equations $Ax = b$. The standard algorithm of this kind is known as GMRES, which stands for "generalized minimal residuals."

Residual Minimization in $\mathcal{K}_n$

As in the last two lectures, let $A \in \mathbb{C}^{m \times m}$ be a square matrix and $b \in \mathbb{C}^m$ a vector, and let $\mathcal{K}_n$ denote the Krylov subspace $\langle b, Ab, \ldots, A^{n-1}b \rangle$ of (33.5). Now, however, we assume that A is nonsingular, for our goal is to solve a system of equations $Ax = b$. It will be convenient to have a notation for the exact solution of this problem: $x_* = A^{-1}b$.

The idea of GMRES is a one-liner. At step n, we shall approximate x_* by the vector $x_n \in \mathcal{K}_n$ that minimizes the norm of the residual $r_n = b - Ax_n$. In other words, we shall determine x_n by solving a least squares problem, illustrated in Figure 35.1.

The obvious way to solve this least squares problem would be as follows.

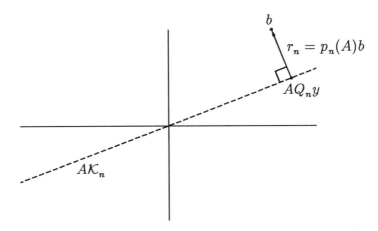

Figure 35.1. *The least squares polynomial approximation problem underlying GMRES: minimize the residual norm $\|r_n\|$. Compare Figure 34.1.*

Let K_n be the $m \times n$ Krylov matrix (33.6), so that we have

$$AK_n = \left[\begin{array}{c|c|c|c} Ab & A^2b & \cdots & A^nb \end{array} \right]. \tag{35.1}$$

The column space of this matrix is AK_n. Thus our problem is to find a vector $c \in \mathbb{C}^n$ such that

$$\|AK_nc - b\| = \text{minimum}, \tag{35.2}$$

where $\|\cdot\| = \|\cdot\|_2$, as always in this part of the book. This could be done by means of a QR factorization of AK_n, in analogy to (33.7). Once c is found, we would set $x_n = K_nc$.

The procedure just described is numerically unstable, however, and it constructs a factor R that is not needed. Here is what is actually done instead. We use the Arnoldi iteration (Algorithm 33.1) to construct a sequence of Krylov matrices Q_n whose columns $q_1, q_2, \ldots$ span the successive Krylov subspaces K_n. Thus we write $x_n = Q_ny$ instead of $x_n = K_nc$. Instead of (35.2), our least squares problem is to find a vector $y \in \mathbb{C}^n$ such that

$$\|AQ_ny - b\| = \text{minimum}. \tag{35.3}$$

Superficially, the problem (35.3) has dimensions $m \times n$. In actuality, however, because of the special structure of Krylov subspaces, it is essentially of dimensions $(n+1) \times n$. We can reveal this as follows. Applying (33.3) transforms the equation to

$$\|Q_{n+1}\tilde{H}_ny - b\| = \text{minimum}. \tag{35.4}$$

Now both vectors inside the norm are in the column space of Q_{n+1}. Therefore, multiplying on the left by Q_{n+1}^* does not change that norm. Thus a further equivalent problem is

$$\|\tilde{H}_n y - Q_{n+1}^* b\| = \text{minimum}. \qquad (35.5)$$

Finally, we note that by construction of the Krylov matrices $\{Q_n\}$, $Q_{n+1}^* b$ is equal to $\|b\|e_1$, where $e_1 = (1, 0, 0, \dots)^*$ as usual. Thus we reach at last the final form of the GMRES least squares problem:

$$\|\tilde{H}_n y - \|b\|e_1\| = \text{minimum}. \qquad (35.6)$$

At step n of GMRES we solve this problem for y, then set $x_n = Q_n y$.

Mechanics of GMRES

This completes our derivation of the GMRES algorithm. A high-level description is as follows.

Algorithm 35.1. GMRES

$q_1 = b/\|b\|$

for $n = 1, 2, 3, \dots$

    ⟨ *step n of Arnoldi iteration, Algorithm 33.1* ⟩

    Find y to minimize $\|\tilde{H}_n y - \|b\|e_1\|$ $(= \|r_n\|)$

    $x_n = Q_n y$.

At each step, GMRES minimizes the norm of the residual $r_n = b - Ax_n$ over all vectors $x_n \in \mathcal{K}_n$. The quantity $\|r_n\|$ is computed in the course of finding y; one need not calculate it explicitly from x_n.

The inner, "Find y" step of Algorithm 35.1 is an $(n+1) \times n$ matrix least squares problem with Hessenberg structure. It can be solved via QR factorization in the usual manner described in Lecture 11, at a cost of $O(n^2)$ flops, thanks to the Hessenberg structure. In addition, it is possible to save further work by a more specialized approach. Rather than construct QR factorizations of the successive matrices $\tilde{H}_1, \tilde{H}_2, \dots$ independently, one can use an updating process to get the QR factorization of $\tilde{H}_n$ from that of $\tilde{H}_{n-1}$. All that is required is a single Givens rotation (Exercise 10.4) and $O(n)$ work. See Exercise 35.4.

GMRES and Polynomial Approximation

In the last lecture, we showed that the calculation of eigenvalues by the Arnoldi iteration is related to the approximation problem (34.3): find $p^n \in P^n$ to minimize $\|p^n(A)b\|$, where P^n denotes the set of monic polynomials of degree n.

The GMRES iteration also solves an approximation problem, the only difference being that the space of polynomials is now

$$P_n = \{\text{polynomials } p \text{ of degree} \leq n \text{ with } p(0) = 1\}. \qquad (35.7)$$

Expressed in terms of polynomial coefficients, our normalization is now $c_0 = 1$ rather than $c_n = 1$. Notationally, we have changed from a superscript to a subscript as a reminder of the change from normalization at $z = \infty$ to normalization at $z = 0$.

Here is how GMRES can be reduced to polynomial approximation in P_n. The iterate x_n can be written

$$x_n = q_n(A)b, \qquad (35.8)$$

where q is a polynomial of degree $n - 1$; its coefficients are the entries of the vector c of (35.2). The corresponding residual $r_n = b - Ax_n$ is $r_n = (I - Aq_n(A))b$, where p_n is the polynomial defined by $p_n(z) = 1 - zq(z)$. In other words, we have

$$r_n = p_n(A)b \qquad (35.9)$$

for some polynomial $p_n \in P_n$. The GMRES process chooses the coefficients of p_n to minimize the norm of this residual. Thus we have shown that GMRES solves the following approximation problem successively for $n = 1, 2, 3 \ldots$.

GMRES Approximation Problem. Find $p_n \in P_n$ such that

$$\|p_n(A)b\| = \text{minimum}. \qquad (35.10)$$

Like the Arnoldi iteration, GMRES satisfies certain invariance properties. The following theorem is easily proved from (35.10).

Theorem 35.1. *Let the GMRES iteration be applied to a matrix $A \in \mathbb{C}^{m \times m}$ as described above.*

Scale-invariance. If A is changed to σA for some $\sigma \in \mathbb{C}$, and b is changed to σb, the residuals $\{r_n\}$ change to $\{\sigma r_n\}$.

Invariance under unitary similarity transformations. If A is changed to UAU^* for some unitary matrix U, and b is changed to Ub, the residuals $\{r_n\}$ change to $\{U^* r_n\}$.

GMRES is not invariant under translation, since the normalization $p(0) = 1$ involves the translation-dependent point 0. On the contrary, its behavior depends strongly on the position of the origin—loosely speaking, on the condition number of A.

Convergence of GMRES

How quickly does GMRES converge? How many steps n must be taken before $\|r_n\|/\|b\|$ is reduced to a satisfactory level such as 10^{-3} or 10^{-16}? As a practical matter, this often becomes the question of designing a good preconditioner (Lecture 40). Mathematically, the problem is to investigate what properties of A determine the size of $\|r_n\|$.

We begin with two observations. The first is that GMRES converges monotonically:

$$\|r_{n+1}\| \leq \|r_n\|. \tag{35.11}$$

The reason is that $\|r_n\|$ is as small as possible for the subspace $\mathcal{K}_n$; by enlarging $\mathcal{K}_n$ to the space $\mathcal{K}_{n+1}$, we can only decrease the residual norm, or at worst leave it unchanged. (Note that $P_n \subseteq P_{n+1}$, whereas $P^n \not\subseteq P^{n+1}$.) The second is that after at most m steps the process must converge, at least in the absence of rounding errors:

$$\|r_m\| = 0. \tag{35.12}$$

For generic data A and b, this will happen because $\mathcal{K}_m = \mathbb{C}^m$, and in special cases, if b happens to lie in $\mathcal{K}_n$ for some $n < m$, convergence will occur earlier. Equation (35.12) is useful as a reminder of the mathematics of GMRES, but it has little practical content, since a GMRES iteration must converge to satisfactory precision in $n \ll m$ steps if it is to be useful.

To obtain more useful information about convergence, we must turn to the polynomial approximation problem (35.10). We know that $\|r_n\| = \|p_n(A)b\| \leq \|p_n(A)\| \|b\|$ is minimal. Except for problems where the right-hand side b has special structure related to that of A, the critical factor that determines the size of this quantity is usually $\|p_n(A)\|$. That is, what determines the convergence rate of GMRES is generally the inequality

$$\frac{\|r_n\|}{\|b\|} \leq \inf_{p_n \in P_n} \|p_n(A)\|. \tag{35.13}$$

This brings us to the mathematically elegant question: given a matrix A and a number n, how small can $\|p_n(A)\|$ be? This question is the basis of almost all analyses of convergence of Krylov subspace iterations for solving systems of equations.

Polynomials Small on the Spectrum

Given A and n, how small can $\|p_n(A)\|$ be? The standard way of obtaining estimates is to look for polynomials $p(z)$ that are as small as possible on the spectrum $\Lambda(A)$, while still satisfying $p(0) = 1$. If p is a polynomial and S is a set in the complex plane, let us define the scalar $\|p\|_S$ by

$$\|p\|_S = \sup_{z \in S} |p(z)|. \tag{35.14}$$

Suppose A is diagonalizable, satisfying $A = V\Lambda V^{-1}$ for some nonsingular matrix V and diagonal matrix Λ. Then we have

$$\|p(A)\| \leq \|V\|\|p(\Lambda)\|\|V^{-1}\| = \kappa(V)\|p\|_{\Lambda(A)}. \tag{35.15}$$

Combining this result with (35.13) gives the following basic theorem on convergence of GMRES.

Theorem 35.2. *At step n of the GMRES iteration, the residual r_n satisfies*

$$\frac{\|r_n\|}{\|b\|} \leq \inf_{p_n \in P_n} \|p_n(A)\| \leq \kappa(V) \inf_{p_n \in P_n} \|p_n\|_{\Lambda(A)}, \tag{35.16}$$

where $\Lambda(A)$ is the set of eigenvalues of A, V is a nonsingular matrix of eigenvectors (assuming A is diagonalizable), and $\|p_n\|_{\Lambda(A)}$ is defined by (35.14).

This theorem can be summarized in words as follows. If A is not too far from normal in the sense that $\kappa(V)$ is not too large, and if properly normalized degree n polynomials can be found whose size on the spectrum $\Lambda(A)$ decreases quickly with n, then GMRES converges quickly.

Example 35.1. Here is a numerical example. Let A be a 200×200 matrix whose entries are independent samples from the real normal distribution of mean 2 and standard deviation $0.5/\sqrt{200}$. In MATLAB,

$$m = 200; \quad A = 2*\text{eye}(m) + 0.5*\text{randn}(m)/\text{sqrt}(m). \tag{35.17}$$

Figure 35.2 shows the eigenvalues of A, a set of points roughly uniformly distributed in the disk of radius $1/2$ centered at $z = 2$ (Exercise 12.3). Figure 35.3 shows the convergence curve for the GMRES iteration applied to the problem $Ax = b$, where $b = (1, 1, \ldots, 1)^*$. The convergence in this case is extraordinarily steady at a rate approximately 4^{-n}. The reason for this is not hard to spot. Since the spectrum of A approximately fills the disk indicated, $\|p(A)\|$ is approximately minimized by the choice $p(z) = (1 - z/2)^n$. Since $I - A/2$ is a random matrix scaled so that its spectrum approximately fills the disk of radius $1/4$ about 0, we have $\|p(A)\| = \|(I - A/2)^n\| \approx 4^{-n}$. This matrix A is well-conditioned, with condition number $\kappa(A) \approx 2.03$. The deviation from normality is modest, with $\kappa(V) \approx 141$.

Figure 35.3 illustrates the convergence of matrix iterations under favorable circumstances—when the matrix A is well behaved (which often means a good preconditioner has been applied). We see that six-digit accuracy is achieved after ten iterations, at a cost of approximately $10 \times 2m^2 = 8.0 \times 10^5$ flops, since the work is dominated by the matrix-vector multiplication at each step. Solving the same system by Gaussian elimination would require $\frac{2}{3}m^3 \approx 5.3 \times 10^6$ flops. This improvement by a factor close to 7 was achieved even though A has no sparsity to take advantage of and even though the dimension $m = 200$ is

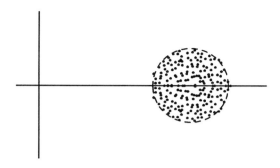

Figure 35.2. *Eigenvalues of the* 200×200 *matrix A of* (35.17). *The dashed curve is the circle of radius* $1/2$ *with center* $z = 2$ *in* $\mathbb{C}$. *The eigenvalues are approximately uniformly distributed within this disk.*

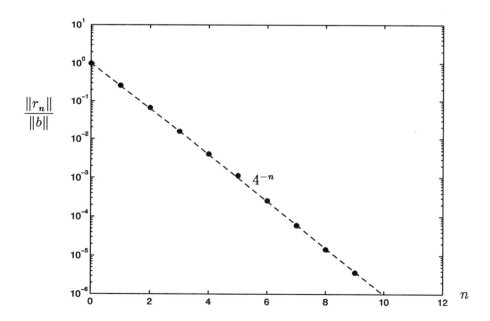

Figure 35.3. *GMRES convergence curve for the same matrix A. This rapid, steady convergence is illustrative of Krylov subspace iterations under ideal circumstances, when A is a well-behaved (or well-preconditioned) matrix.*

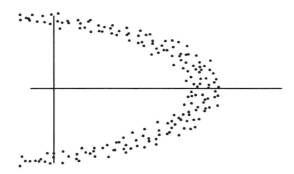

Figure 35.4. *Eigenvalues of a* 200×200 *matrix, like that of* (35.17) *except with a modified diagonal. Now the eigenvalues surround the origin on one side.*

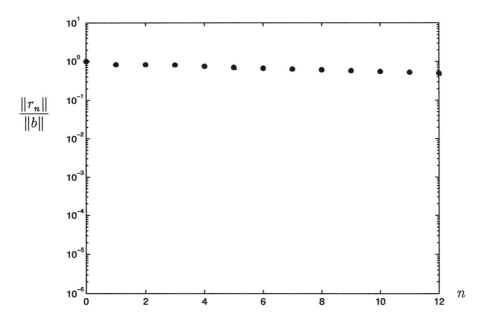

Figure 35.5. *GMRES convergence curve for the matrix of Figure 35.4. The convergence has slowed down greatly. When an iterative method stagnates like this, it is time to look for a better preconditioner.*

not high. For the same example with $m = 2000$, GMRES would beat Gaussian elimination by a factor more like 70. For a 2000×2000 matrix with similar spectral properties but 90% or 99% sparsity, the factor would improve to on the order of 700 or 7000, respectively, and the storage required by GMRES would also diminish dramatically. $\square$

Example 35.2. If the eigenvalues of a matrix "surround the origin," on the other hand, such rapid convergence cannot be expected. Figures 35.4–35.5 present an example. The matrix is now $A' = A + D$, where A is the matrix of (35.17) and D is the diagonal matrix with complex entries

$$d_k = (-2 + 2 \sin \theta_k) + i \cos \theta_k, \qquad \theta_k = \frac{k\pi}{m-1}, \ 0 \le k \le m-1.$$

As is evident in Figure 35.4, the eigenvalues now lie in a semicircular cloud that bends around the origin. The convergence rate is much worse than before, making the iterative computation no better than Gaussian elimination for this problem. The condition numbers are now $\kappa(A) \approx 4.32$ and $\kappa(V) \approx 54.0$, so the deterioration in convergence cannot be explained by conditioning alone; it is the locations of the eigenvalues, not their magnitudes (or those of the singular values) that are causing the trouble. If the arc extended much further around the spectrum, the convergence would worsen further. $\square$

Exercises

35.1. Show that if $S \subseteq \mathbb{C}$ contains infinitely many points, then (35.14) defines a norm on the vector space of all polynomials with complex coefficients. Explain what goes wrong if S has only finitely many points.

35.2. (a) Let $S \subseteq \mathbb{C}$ be a set whose convex hull contains 0 in its interior. That is, S is contained in no half-plane disjoint from the origin. Show that there is no $p \in P_1$ (i.e., no polynomial p of degree 1 with $p(0) = 1$) such that $\|p\|_S < 1$.

(b) Let A be a matrix, not necessarily normal, whose spectrum $\Lambda(A)$ has the property (a). Show that there is no $p \in P_1$ such that $\|p(A)\| < 1$.

(c) Though the convergence in Figure 35.5 is slow, it is clear that $\|r_1\| < \|r_0\|$. Explain why this does not contradict the result of (b). Describe what kind of polynomial $p_1 \in P_1$ GMRES has probably found to achieve $\|r_1\| < \|r_0\|$.

35.3. The recurrence $x_{n+1} = x_n + \alpha r_n = x_n + \alpha(b - Ax_n)$, where α is a scalar constant, is known as a *Richardson iteration*.

(a) What polynomial $p(A)$ at step n does this correspond to?

(b) What choice of α would you recommend for the matrix A of Figure 35.2, and what would you expect to be the corresponding convergence rate?

(c) Same questions for the matrix of Figure 35.4.

35.4. (a) Describe an $O(n^2)$ algorithm based on QR factorization by Givens rotations (Exercise 10.4) for solving the least squares problem of Algorithm 35.1.

(b) Show how the operation count can be improved to $O(n)$, as mentioned on p. 268, if the problem for step $n-1$ has already been solved.

35.5. Our statement of the GMRES algorithm (Algorithm 35.1) begins with the initial guess $x_0 = 0$, $r_0 = b$. (The same applies to CG and BCG, Algorithms 38.1 and 39.1.) Show that if one wishes to start with an arbitrary initial guess x_0, this can be accomplished by an easy modification of the right-hand side b.

35.6. For larger values of n, the cost of GMRES in operations and storage may be prohibitive. In such circumstances a method called *k-step restarted GMRES* or *GMRES(k)* is often employed, in which, after k steps, the GMRES iteration is started anew with the current vector x_k as an initial guess.

(a) Compare the asymptotic operation counts and storage requirements of GMRES and GMRES(k), for fixed k and increasing n.

(b) Describe an example in which GMRES(k) can be expected to converge in nearly as few iterations as GMRES (hence much faster in operation count).

(c) Describe another example in which GMRES(k) can be expected to fail to converge, whereas GMRES succeeds.

Lecture 36. The Lanczos Iteration

In the last three lectures we considered Krylov subspace iterations for non-hermitian matrix problems. We shall return to nonhermitian problems in Lecture 39, for there is more to this subject than Arnoldi and GMRES. But first, in this and the following two lectures, we specialize to the hermitian case, where a major simplification takes place.

Three-Term Recurrence

The Lanczos iteration is the Arnoldi iteration specialized to the case where A is hermitian. For simplicity of notation, we shall go a step further and assume, here and in the next two lectures, that A is real and symmetric.

Let us consider what happens to the Arnoldi process in this special case. Of course, all of the equations of Lectures 33 and 34 still apply, and in each formula we can replace $*$ by T. The first thing we notice is that it follows from (33.12) that the Ritz matrix H_n is symmetric. Therefore its eigenvalues, the Ritz values or Lanczos estimates (33.10), are also real. This seems natural enough, since the eigenvalues of A are real.

The second thing we notice is more dramatic. Since H_n is both symmetric and Hessenberg, it is tridiagonal. This means that in the inner loop of the Arnoldi iteration (Algorithm 33.1), the limits 1 to n can be replaced by $n-1$ to n. Thus instead of the $(n+1)$-term recurrence (33.4) at step n, the Lanczos iteration involves just a three-term recurrence. The result is that each step of the Lanczos iteration is much cheaper than the corresponding step of the

276

Arnoldi iteration. In Lecture 38 we shall see that analogously, for solving $Ax = b$, each step of the conjugate gradient iteration is much cheaper than the corresponding step of GMRES.

The fact that H_n is tridiagonal is so important that it is worth reviewing how it arises from the symmetry of A. The key equation is (33.12), which we can write entry-wise for real matrices A, H_n, and Q_n as

$$h_{ij} = q_i^T A q_j. \tag{36.1}$$

This implies that $h_{ij} = 0$ for $i > j + 1$, since $Aq_j \in \langle q_1, q_2, ..., q_{j+1} \rangle$ and the Krylov vectors are orthogonal. Taking the transpose gives

$$h_{ij} = q_j^T A^T q_i. \tag{36.2}$$

If $A = A^T$, then $h_{ij} = 0$ for $j > i + 1$ by the same reasoning as before. This simple argument leading to a three-term recurrence relation applies to arbitrary self-adjoint operators, not just to matrices.

The Lanczos Iteration

Since a symmetric tridiagonal matrix contains only two distinct vectors, it is customary to replace the generic notation a_{ij} by new variables. Let us write $\alpha_n = h_{nn}$ and $\beta_n = h_{n+1,n} = h_{n,n+1}$. Then H_n becomes

$$T_n = \begin{bmatrix} \alpha_1 & \beta_1 & & & \\ \beta_1 & \alpha_2 & \beta_2 & & \\ & \beta_2 & \alpha_3 & \ddots & \\ & & \ddots & \ddots & \beta_{n-1} \\ & & & \beta_{n-1} & \alpha_n \end{bmatrix}. \tag{36.3}$$

In this notation Algorithm 33.1 takes the following form.

Algorithm 36.1. Lanczos Iteration

$\beta_0 = 0$, $q_0 = 0$, $b = $ arbitrary, $q_1 = b/\|b\|$

for $n = 1, 2, 3, \ldots$

    $v = Aq_n$ [or $Aq_n - \beta_{n-1}q_{n-1}$ for greater stability]

    $\alpha_n = q_n^T v$

    $v = v - \beta_{n-1}q_{n-1} - \alpha_n q_n$

    $\beta_n = \|v\|$

    $q_{n+1} = v/\beta_n$

Each step consists of a matrix-vector multiplication, an inner product, and a couple of vector operations. If A has enough sparsity or other structure that matrix-vector products can be computed cheaply, then such an iteration can be applied without too much difficulty to problems of dimensions in the tens or hundreds of thousands.

The following theorem summarizes some of the properties of the Lanczos iteration (when carried out in exact arithmetic, of course, as with all such theorems in this book). Nothing here is new; these are restatements in the new notation of the results of Theorems 33.1 and 34.1 for the Arnoldi iteration.

Theorem 36.1. *The matrices Q_n of vectors q_n generated by the Lanczos iteration are reduced QR factors of the Krylov matrix (33.6),*

$$K_n = Q_n R_n. \tag{36.4}$$

The tridiagonal matrices T_n are the corresponding projections

$$T_n = Q_n^* A Q_n, \tag{36.5}$$

and the successive iterates are related by the formula

$$A Q_n = Q_{n+1} \tilde{T}_n, \tag{36.6}$$

which we can write in the form of a three-term recurrence at step n,

$$A q_n = \beta_{n-1} q_{n-1} + \alpha_n q_n + \beta_n q_{n+1}. \tag{36.7}$$

As long as the Lanczos iteration does not break down (i.e., K_n is of full rank n), the characteristic polynomial of T_n is the unique polynomial $p^n \in P^n$ that solves the Arnoldi/Lanczos approximation problem (34.3), i.e., that achieves

$$\| p^n(A) b \| = minimum. \tag{36.8}$$

Lanczos and Electric Charge Distributions

In practice, the Lanczos iteration is used to compute eigenvalues of large symmetric matrices just as the Arnoldi iteration is used for nonsymmetric matrices (Lecture 34). At each step n, or at occasional steps, the eigenvalues of the growing tridiagonal matrix T_n are determined by standard methods. These are the Ritz values or "Lanczos estimates" (33.10) for the given matrix A and starting vector q_1. Often some of these numbers are observed to converge geometrically to certain limits, which can then be expected to be eigenvalues of A.

As with the Arnoldi iteration, it is the outlying eigenvalues of A that are most often obtained first. This assertion can be made more precise by the following rule of thumb:

If the eigenvalues of A are more evenly spaced than Chebyshev
points, then the Lanczos iteration will tend to find outliers.

Here is what this statement means. Suppose the m eigenvalues $\{\lambda_j\}$ of A
are spread reasonably densely around an interval on the real axis. Since the
Lanczos iteration is scale- and translation-invariant (Theorem 34.2), we can
assume without loss of generality that this interval is $[-1, 1]$. The m *Chebyshev
points* in $[-1, 1]$ are defined by the formula

$$x_j = \cos \theta_j, \qquad \theta_j = \frac{(j - \frac{1}{2})\pi}{m}, \qquad 1 \le j \le m. \qquad (36.9)$$

The exact definition is not important; what matters is that these points cluster quadratically near the endpoints, with the spacing between points $O(m^{-1})$
in the interior and $O(m^{-2})$ near ± 1. The rule of thumb asserts that if the
eigenvalues $\{\lambda_j\}$ of A are more evenly distributed than this—less clustered at
the endpoints—then the Ritz values computed by a Lanczos iteration will tend
to converge to the outlying eigenvalues first. In particular, an approximately
uniform eigenvalue distribution will produce rapid convergence towards outliers. Conversely, if the eigenvalues of A are more than quadratically clustered
at the endpoints—a situation not so common in practice—then we can expect
convergence to some of the "inliers."

These observations can be given a physical interpretation. Consider m
point charges free to move about the interval $[-1, 1]$. Assume that the repul-
sive force between charges located at x_j and x_k is proportional to $|x_j - x_k|^{-1}$.
(For electric charges in 3D the force would be $|x_j - x_k|^{-2}$, but this becomes
$|x_j - x_k|^{-1}$ in 2D, where we can view each point as the intersection of an
infinite line in 3D with the plane.) Let these charges distribute themselves
in a minimal-energy equilibrium in $[-1, 1]$. Then this minimal-energy distribution and the Chebyshev distribution are approximately the same, and in
the limit $m \to \infty$, they both converge to a limiting continuous charge density
distribution proportional to $(1 - x^2)^{-1/2}$.

Think of the eigenvalues of A as point charges. If they are distributed approximately in a minimal-energy configuration in an interval, then the Lanczos
iteration will be useless; there will be little convergence before step $n = m$.
If the distribution is very different from this, however, then there is likely to
be rapid convergence to some eigenvalues, namely, the eigenvalues in regions
where there is "too little charge" in the sense that if the points were free
to move, more would tend to cluster here. The rule of thumb can now be
restated:

The Lanczos iteration tends to converge to eigenvalues in
regions of "too little charge" for an equilibrium distribution.

The explanation of this observation depends on the connection (36.8) of the
Lanczos iteration with polynomial approximation. Some of the details are
worked out in Exercise 36.2.

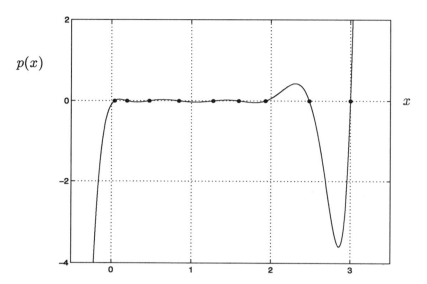

Figure 36.1. *Plot of the Lanczos polynomial at step 9 of the Lanczos iteration for the matrix (36.10). The roots are the Ritz values or "Lanczos eigenvalue estimates." The polynomial is small throughout* $[0, 2] \cup \{2.5\} \cup \{3.0\}$. *To achieve this, it must place one root near 2.5 and another very near 3.0.*

Example

The convergence of the Lanczos iteration is best illustrated by a numerical example. Let A be the 203×203 matrix

$$A = \operatorname{diag}(0, .01, .02, \ldots, 1.99, 2, 2.5, 3.0). \qquad (36.10)$$

The spectrum of A consists of a dense collection of eigenvalues throughout $[0, 2]$ together with two outliers, 2.5 and 3.0. We carry out a Lanczos iteration beginning with a random starting vector q_1.

Figure 36.1 shows the Ritz values and the associated Lanczos polynomial at step $n = 9$. Seven of the Ritz values lie in $[0, 2]$, and the polynomial is uniformly small on that interval; the beginnings of a tendency for the Ritz values to cluster near the endpoints can be detected. The other two Ritz values lie near the eigenvalues at 2.5 and 3.0. The leading three Ritz values are

$$1.93, 2.48, 2.999962.$$

Evidently we have little accuracy in the lower eigenvalues but five-digit accuracy in the leading one. A plot like this gives an idea of why outliers tend to be estimated accurately. The graph of $p(x)$ is so steep for $x \approx 3$ that if $p(3)$ is to be small, there must be a root of p very close to 3. This steepness of the graph is related to the presence of "too little charge" near this point. If the

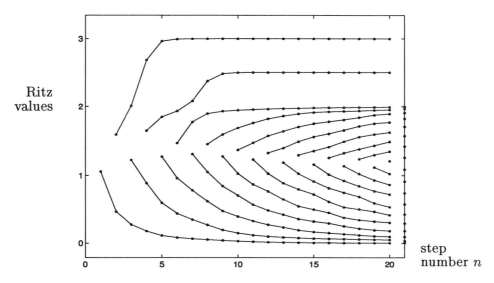

Figure 36.2. *Ritz values for the first* 20 *steps of the Lanczos iteration applied to the same matrix. The convergence to the eigenvalues* 2.5 *and* 3.0 *is geometric. Little useful convergence to individual eigenvalues occurs in the* [0, 2] *part of the spectrum. Instead, the Ritz values in* [0, 2] *approximate Chebyshev points in that interval, marked by dots on the right-hand boundary.*

charges were free to move about [0, 3] to minimize energy, more points would cluster near $x = 3$, and $p(x)$ would not be so steep there.

At step 20 the leading three Ritz values are

$$1.9906, \quad 2.499999999987, \quad 3.00000000000000.$$

Now we have about fifteen digits of accuracy in the leading eigenvalue and twelve digits in the second. A plot of $p(x)$ would be correspondingly steep near the points 2.5 and 3.0. Note that convergence to the third eigenvalue is also beginning to occur, a reflection of the fact that the eigenvalues in [0, 2] are distributed evenly rather than in a Chebyshev distribution.

An "aerial view" of the convergence process appears in Figure 36.2, which shows the Ritz values for all steps from $n = 1$ to $n = 20$. Each vertical slice of this plot corresponds to the Ritz values at one iteration; the lines connecting the dots help the eye follow what is going on but have no precise meaning. The plot shows pronounced convergence to the leading eigenvalue after about $n = 5$ and to the next one around $n = 10$. In the interval [0, 2] containing the other eigenvalues, they show a density of Ritz values approximately proportional to $(1 - x^2)^{-1/2}$, with very clear bunching at endpoints.

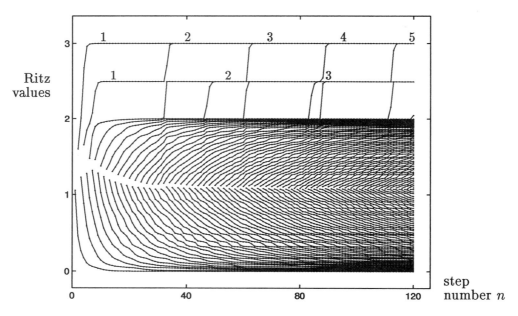

Figure 36.3. *Continuation to* 120 *steps of the Lanczos iteration. The numbers indicate multiplicities of the Ritz values. Note the appearance of four "ghost" copies of the eigenvalue* 3.0 *and two "ghost" copies of the eigenvalue* 2.5.

Rounding Errors and "Ghost" Eigenvalues

Rounding errors have a complex effect on the Lanczos iteration and, indeed, on all iterations of numerical linear algebra based on three-term recurrence relations. The source of the difficulty is easily identified. In an iteration based on an n-term recurrence relation, such as Arnoldi or GMRES, the vectors $q_1, q_2, q_3, \ldots$ are forced to be orthogonal by explicit Gram–Schmidt operations. Three-term recurrences like Lanczos and conjugate gradients, however, depend upon orthogonality of the vectors $\{q_j\}$ to arise "automatically" from a mathematical identity. In practice, such identities are not accurately preserved in the presence of rounding errors, and after a number of iterations, orthogonality is lost.

The loss of orthogonality in practical Lanczos iterations sounds wholly bad, but the situation is more subtle than that. As it happens, loss of orthogonality is connected closely with the convergence of Ritz values to eigenvalues of A. A great deal is known about this subject, though not as much as one might like; we shall not give details.

Because of complexities like these, no straightforward theorem is known to the effect that the Lanczos or conjugate gradient iterations is stable in the sense defined in this book. Nonetheless, these iterations are extraordinarily useful in practice. Figure 36.3 gives an idea of the way in which instability is often manifested in practice without preventing the iteration from being useful. The figure is a repetition of Figure 36.2, but for 120 instead of 20 steps of the iteration. Everything looks as expected until around step 30, when a second copy of the eigenvalue 3.0 appears among the Ritz values. A third copy appears around step 60, a fourth copy around step 90, and so on. Meanwhile, additional copies of the eigenvalue 2.5 also appear around step 40 and 80 and (just beginning to be visible) 120. These extra Ritz values are known as "ghost" eigenvalues, and they have nothing to do with the actual multiplicities of the corresponding eigenvalues of A.

A rigorous analysis of the phenomenon of ghost eigenvalues is complicated. Intuitive explanations, however, are not hard to devise. One idea is that in the presence of rounding errors, one should think of each eigenvalue of A not as a point but as a small interval of size roughly $O(\epsilon_{\text{machine}}\|A\|)$; ghost eigenvalues arise from the need for $p(z)$ to be small not just at the exact eigenvalues but throughout these small intervals. Another, rather different explanation is that convergence of a Ritz value to an eigenvalue of A annihilates the corresponding eigenvector component in the vector being operated upon; but in the presence of rounding errors, random noise must be expected to excite that component slightly again. After sufficiently many iterations, this previously annihilated component will have been amplified enough that another Ritz value is needed to annihilate it again—and then again, and again.

Both of these explanations capture some of the truth about the behavior of the Lanczos iteration in floating point arithmetic. The second one has perhaps more quantitative accuracy.

Exercises

36.1. In Lecture 27 it was pointed out that the eigenvalues of a symmetric matrix $A \in \mathbb{R}^{m \times m}$ are the stationary values of the Rayleigh quotient $r(x) = (x^T A x)/(x^T x)$ for $x \in \mathbb{R}^m$. Show that the Ritz values at step n of the Lanczos iteration are the stationary values of $r(x)$ if x is restricted to $\mathcal{K}_n$.

36.2. Consider a polynomial $p \in P^n$, i.e., $p(z) = \prod_{k=1}^{n}(z - z_k)$ for some $z_k \in \mathbb{C}$.

(a) Write $\log|p(z)|$ as a sum of n terms corresponding to the points z_k.

(b) Explain why the term involving z_k can be interpreted as the potential corresponding to a negative unit point charge located at z_k, if charges repel in inverse proportion to their separation. Thus $\log|p(z)|$ can be viewed as the potential at z induced by n point charges.

(c) Replacing each charge -1 by $-1/n$ and taking the limit $n \to \infty$, we get a continuous charge density distribution $\mu(\zeta)$ with integral -1, which we can expect to be related to the limiting density of zeros of polynomials $p \in P^n$ as

$n \to \infty$. Write an integral representing the potential $\varphi(z)$ corresponding to $\mu(\zeta)$, and explain its connection to $|p(z)|$.

(d) Let S be a closed, bounded subset of $\mathbb{C}$ with no isolated points. Suppose we seek a distribution $\mu(z)$ with support in S that minimizes $\max_{z \in S} \varphi(z)$. Give an argument (not rigorous) for why such a $\mu(z)$ should satisfy $\varphi(z) = $ constant throughout S. Explain why this means that the "charges" are in equilibrium, experiencing no net forces. In other words, S is like a 2D electrical conductor on which a quantity -1 of charge has distributed itself freely. Except for an additive constant, $\varphi(z)$ is the *Green's function* for S.

(e) As a step toward explaining the rule of thumb of p. 279, suppose that A is a real symmetric matrix with spectrum densely distributed in $[a, b] \cup \{c\} \cup [d, e]$ for $a < b < c < d < e$. Thus (b, d) is a region of "too little charge" for the set $S = [a, e]$. Explain why rapid convergence of a Ritz value to c can be expected, and estimate the rate of convergence in terms of the equilibrium potential $\varphi(z)$ associated with the set $S' = [a, b] \cup [c, d]$.

36.3. Let A be the 1000×1000 symmetric matrix whose entries are all zero except for $a_{ij} = \sqrt{i}$ on the diagonal, $a_{ij} = 1$ on the sub- and superdiagonals, and $a_{ij} = 1$ on the 100th sub- and superdiagonals, i.e., for $|i - j| = 100$. Determine the smallest eigenvalue of A to six digits of accuracy by the Lanczos iteration.

36.4. As a special case of the Arnoldi lemniscates of Lecture 34, "Lanczos lemniscates" can be employed to illustrate the convergence of the Lanczos iteration. Find a way to modify your program of Exercise 36.3 to plot the Lanczos lemniscates at each step. Your method need not be elegant, efficient, or numerically robust. Produce plots of Lanczos lemniscates at steps $n = 1, 2, \ldots, 12$ for the example of Figure 36.2 and for an example of your own choosing.

Lecture 37. From Lanczos to Gauss Quadrature

If discrete vectors become continuous functions on $[-1, 1]$, and the matrix A is taken to be the operator of pointwise multiplication by x, then the Lanczos iteration becomes the standard procedure for constructing orthogonal polynomials via a three-term recurrence relation. From here it is a short step to Gauss quadrature formulas, whose nodes and weights can be computed by solving a symmetric tridiagonal matrix eigenvalue problem.

Orthogonal Polynomials

In Lecture 7 we considered a continuous analogue of QR factorization. We now consider a continuous analogue of the Lanczos iteration, restricting attention, as in the last lecture, to real vectors (now functions) and real symmetric matrices (now linear operators).

The first thing we do is replace $\mathbb{R}^m$ by $L^2[-1, 1]$, a vector space of real-valued functions on $[-1, 1]$. The inner product of two functions $u, v \in L^2[-1, 1]$ is defined by

$$(u, v) = \int_{-1}^{1} u(x)\, v(x)\, dx, \tag{37.1}$$

and the norm of a function $u \in L^2[-1, 1]$ is $\|u\| = (u, u)^{1/2}$.

In Lecture 7 we took A to be a "$[-1, 1] \times n$ matrix" whose columns were powers of x. Here, for the Lanczos iteration, A should be square instead of rectangular. We shall take it to be the "$[-1, 1] \times [-1, 1]$ matrix" corresponding

285

to pointwise multiplication by x. That is, A is the linear operator defined by the equation

$$(Au)(x) = xu(x) \qquad (37.2)$$

for each $u \in L^2[-1, 1]$. This operator is analogous to a diagonal matrix, with the continuum of values $x \in [-1, 1]$ along the "diagonal." (One can make the diagonal matrix idea precise by describing A as an integral operator with kernel $k(x, y) = x\delta(x - y)$, where δ is the Dirac delta function.) In particular, A is symmetric, and because of the symmetry, the Arnoldi process specializes to the Lanczos process.

There is one more item to be pinned down. Let us agree that the initial function $b(x)$ will be a nonzero constant. The corresponding normalized initial function $q_1(x)$ will accordingly be $q_1(x) = 1/\sqrt{2}$.

The Lanczos iteration (Algorithm 36.1) now takes the following form.

Algorithm 37.1. Construction of Orthogonal Polynomials

$\beta_0 = 0$, $q_0(x) = 0$, $q_1(x) = 1/\sqrt{2}$
for $n = 1, 2, 3, \ldots$
$\qquad v(x) = xq_n(x)$
$\qquad \alpha_n = (q_n, v)$
$\qquad v(x) = v(x) - \beta_{n-1}q_{n-1}(x) - \alpha_n q_n(x)$
$\qquad \beta_n = \|v\|$
$\qquad q_{n+1}(x) = v(x)/\beta_n$

Note that the polynomials $q_1, q_2, q_3, \ldots$ are of degrees $0, 1, 2, \ldots$, and thus the polynomial of degree n in this sequence is q_{n+1}.

Algorithm 37.1 can be found in many books that have nothing ostensibly to do with linear algebra. It is precisely the usual three-term recurrence for constructing the sequence of *Legendre polynomials*, orthogonal on the interval $[-1, 1]$. Our statement of the algorithm is unusual, however, in notation and in normalization. The nth Legendre polynomial is usually written $P_n(x)$, of degree n, and normalized by $P_n(1) = 1$. Thus our $q_{n+1}(x)$ is a scalar multiple of the usual $P_n(x)$. Nevertheless we shall call it a Legendre polynomial.

The Legendre polynomials were considered already in (7.11) and Figure 7.1, where they were derived by Gram–Schmidt factorization of the "$[-1, 1]$ $\times n$ matrix" of monomials mentioned above. That matrix had the form of the Krylov matrix (33.6) generated by our present initial vector b and operator A, and that is why the Gram–Schmidt process in Lecture 7 had the same effect as the Lanczos process here.

We labeled Algorithm 37.1 "Construction of orthogonal polynomials" rather than "Construction of Legendre polynomials" because it is, in fact, more general. If (37.1) is modified by the inclusion of a nonconstant positive weight function $w(x)$ in the integrand, then one obtains other families of orthogonal polynomials such as Chebyshev polynomials and Jacobi polynomials. All of the developments of this lecture apply to these more general families, but we shall not give details.

Jacobi Matrices

All the formulas of the foregoing lectures on Arnoldi and Lanczos iterations remain valid for the polynomial orthogonalization process we have just described. Of course, they have to be interpreted properly, with $1, 2, \ldots, m$ replaced by $[-1, 1]$ and the usual vector inner product replaced by (37.1). For example, we now have " $[-1, 1] \times n$ matrices "

$$
K_n = \begin{bmatrix} 1 & x & \cdots & x^{n-1} \end{bmatrix}, \qquad Q_n = \begin{bmatrix} q_1(x) & q_2(x) & \cdots & q_n(x) \end{bmatrix}, \qquad (37.3)
$$

and they are related to one another exactly as in (36.4). (As just mentioned, K_n is the matrix that was called A in Lecture 7.)

The tridiagonal matrices $\{T_n\}$ described in the previous lecture are particularly important. There are still $n \times n$ discrete matrices, related to Q_n and A by (36.5) and (36.6). Their entries are given by the analogue of (36.1),

$$
t_{ij} = (q_i(x), xq_j(x)). \qquad (37.4)
$$

In the context of orthogonal polynomials, the matrices $\{T_n\}$ are known as *Jacobi matrices*. The three-term recurrence (36.7) takes the form

$$
xq_n(x) = \beta_{n-1}q_{n-1}(x) + \alpha_n q_n(x) + \beta_n q_{n+1}(x). \qquad (37.5)
$$

The statement of Algorithm 37.1 is perhaps misleading as written. It would appear that nontrivial computations are involved at each step of this algorithm: the evaluation of the inner product (q_n, v) and norm $\|v\|$ that define α_n and β_n. If (37.1) contained an arbitrary weight function $w(x)$, these computations would indeed be nontrivial. However, for the particular choice $w(x) = 1$ associated with Legendre polynomials, and also for various choices associated with other classical families of polynomials, the entries $\{\alpha_n\}$ and $\{\beta_n\}$ are known analytically. In the notation of (36.3), we have

$$
\alpha_n = 0, \qquad \beta_n = \frac{1}{2}(1 - (2n)^{-2})^{-1/2} \qquad (37.6)
$$

for Legendre polynomials. With the use of these formulas Algorithm 37.1 becomes a trivial mechanical procedure, a three-term recurrence relation and nothing more.

The Characteristic Polynomial

What becomes of the Arnoldi/Lanczos approximation problem (36.8) in this context of orthogonal polynomials? The answer comes upon noting that for our special choices of b and A, we have $p(A)\,b = p(x)/\sqrt{2}$. It follows that (36.8) can be written in the following way.

Orthogonal Polynomials Approximation Problem. Find $p^n \in P^n$ such that
$$\| \, p^n(x) \, \| = \text{minimum} . \qquad (37.7)$$

According to Theorem 36.1, the solution is the characteristic polynomial of the matrix T_n.

From here it is a short step to a remarkable conclusion. Note that any $p \in P^n$ can be written in the form $p(x) = Cq_{n+1}(x) + Q_n y$, where C is a constant—the inverse of the leading coefficient of $q_{n+1}(x)$. Note also that since the functions $\{q_n(x)\}$ are orthogonal, we have $\|p\| = (C^2 + \|y\|^2)^{1/2}$. It follows that the minimum in (37.7) is achieved by setting $y = 0$. In other words, $p^n(x)$ is the same as $q_{n+1}(x)$ up to that constant C. We express this conclusion as a theorem, the final assertion of which is proved in Exercise 37.2.

Theorem 37.1. *Let $\{q_n(x)\}$ be the sequence of orthogonal polynomials generated by Algorithm 37.1, let $\{T_n\}$ be the associated sequence of tridiagonal Jacobi matrices, and let p^n be the characteristic polynomial of T_n. Then for $n = 0, 1, 2, \ldots$,*
$$p^n(x) \; = \; C_n q_{n+1}(x), \qquad (37.8)$$
where C_n is a constant. In particular, the zeros of $q_{n+1}(x)$ are the eigenvalues of T_n. These n zeros are distinct and lie in the open interval $(-1, 1)$.

This theorem is of great computational importance. To determine the zeros of the Legendre polynomials, all one has to do is compute the eigenvalues of the associated Jacobi matrices, whose entries are given in closed form by (37.6). As we have seen in previous lectures, the eigenvalue problem for an $n \times n$ symmetric tridiagonal matrix is well-conditioned and can be solved very quickly, requiring only $O(n^2)$ flops. By contrast, computing the zeros of the Legendre polynomials directly, starting from the coefficients of the polynomials rather than the Jacobi matrices, is inefficient and numerically unstable.

Quadrature Formulas

There is a reason why the zeros of the Legendre polynomials are of computational interest: they are the nodes of the Gauss–Legendre quadrature formulas.

Let us briefly review the idea of numerical quadrature. Suppose $f(x)$ is a function defined on $[-1, 1]$ and we want to compute the integral

$$I(f) = \int_{-1}^{1} f(x)\, dx. \tag{37.9}$$

(If an interval of integration other than $[-1, 1]$ is of interest, this can be handled by a linear change of variables.) It is natural to consider approximating $I(f)$ by a finite sum

$$I_n(f) = \sum_{j=1}^{n} w_j f(x_j) \tag{37.10}$$

defined by a set of n *nodes* or *abscissas* $x_j \in [-1, 1]$ and corresponding *weights* w_j, chosen independently of f. This is an n-point *quadrature formula*, a notion studied by numerical analysts going back to Newton. Various forms of such formulas, often coupled with adaptive error estimation, interval subdivision, and order control, are the basis of most numerical integration carried out on computers today.

Any set of nodes is a candidate for a quadrature formula. The following result is a consequence of the nonsingularity of Vandermonde matrices (Exercise 37.3).

Theorem 37.2. *Let the nodes $\{x_j\}$ be an arbitrary set of n distinct points in $[-1, 1]$. Then there is a unique choice of weights $\{w_j\}$ with the property that the quadrature formula (37.10) has order of accuracy at least $n - 1$ in the sense that it is exact if $f(x)$ is any polynomial of degree $\leq n - 1$.*

If the nodes $\{x_j\}$ are taken equally spaced from -1 to 1, the quadrature formula provided by this theorem is known as a *Newton–Cotes* formula, the n-point generalization of the familiar trapezoid and Simpson rules. Newton–Cotes formulas have the order of accuracy guaranteed by this theorem but no higher. These formulas are simple and useful, especially for lower values of n. For larger n, their weights w_j have oscillating signs and huge amplitudes, of order 2^n, causing numerical instability.

Gauss Quadrature

The idea of Gauss quadrature is to pick not just the weights $\{w_j\}$ but also the nodes $\{x_j\}$ optimally, so as to raise the order of accuracy of (37.10) as high as possible. As it happens, there is a unique choice of nodes and weights that achieves this, and the resulting formula has order $2n - 1$. This is a dramatic improvement over order $n - 1$, a doubling of the number of digits of accuracy typically attainable for smooth functions and a fixed number of function evaluations. Moreover, the weights w_j are all positive, making these formulas stable even for high n.

What is the magic set of nodes $x_1, \ldots, x_n$ that doubles the order of accuracy of an n-point quadrature formula? It is nothing more than the set of zeros of the Legendre polynomial $q_{n+1}(x)$. The *Gauss* or *Gauss–Legendre quadrature formula* is defined as the quadrature formula (37.10) provided by Theorem 37.2 whose nodes $x_1, \ldots, x_n$ are the zeros of $q_{n+1}(x)$.

Theorem 37.3. *The n-point Gauss–Legendre quadrature formula has order of accuracy exactly $2n - 1$, and no quadrature formula (37.10) has order of accuracy higher than this.*

Proof. Given any set of distinct points $\{x_j\}$, let $f(x)$ be the polynomial $\prod_{j=1}^{n}(x - x_j)^2$ of degree $2n$. Then $I(f) > 0$, but $I_n(f) = 0$ since $f(x_j) = 0$ for each node x_j. Thus the quadrature formula is not exact for polynomials of degree $2n$.

On the other hand, suppose $f(x)$ is any polynomial of degree $\leq 2n - 1$, and take $\{x_j\}$ to be the Gauss quadrature nodes, the zeros of $\{q_{n+1}(x)\}$. The function $f(x)$ can be factored in the form

$$f(x) \;=\; g(x)q_{n+1}(x) + r(x),$$

where $g(x)$ is a polynomial of degree $\leq n-1$ and $r(x)$, the remainder term, is also a polynomial of degree $\leq n-1$. (In fact, $r(x)$ is the degree $n-1$ polynomial interpolant to f in the points $\{x_j\}$.) Now since $q_{n+1}(x)$ is orthogonal to all polynomials of lower degree, we have $I(g\,q_{n+1}) = 0$. At the same time, since $g(x_j)q_{n+1}(x_j) = 0$ for each node x_j, we have $I_n(g\,q_{n+1}) = 0$. Since I and I_n are linear operators, these identities imply $I(f) = I(r)$ and $I_n(f) = I_n(r)$. But since $r(x)$ is of degree $\leq n - 1$, we have $I(r) = I_n(r)$ by Theorem 37.2, and combining these results gives $I(f) = I_n(f)$, as claimed. $\square$

Gauss Quadrature via Jacobi Matrices

In six pages we have gone from the Lanczos iteration to Legendre polynomials and from Legendre polynomials to Gauss quadrature. We have even provided a fast and stable algorithm for determining the nodes of Gauss quadrature formulas: just set up the Jacobi matrices $\{T_n\}$ and compute their eigenvalues.

One final observation will finish the story. Not only the nodes but also the weights of Gauss quadrature formulas can be obtained from the eigenvalue problem for T_n. The jth Gauss weight turns out to be simply $w_j = 2(v_j)_1^2$, that is, twice the square of the first component of the jth eigenvector of T_n. We state this result without proof. An analogous result holds for Gauss quadrature formulas defined by inner products (37.1) with general weight functions $w(x)$.

Theorem 37.4. *Let T_n be the $n \times n$ Jacobi matrix (36.3) defined by Algorithm 37.1 or (36.5), with entries $\beta_1, \ldots, \beta_{n-1}$ given by (37.6). Let $T_n = VDV^T$ be an orthogonal diagonalization of T_n with $V = [v_1 | \cdots | v_n]$ and $D = \mathrm{diag}(\lambda_1, \ldots, \lambda_n)$. Then the nodes and weights of the Gauss–Legendre quadrature formula are given by*

$$x_j = \lambda_j, \qquad w_j = 2(v_j)_1^2, \qquad j = 1, \ldots, n. \tag{37.11}$$

Example

As an illustration of the power of Gauss quadrature for integrating smooth functions, suppose we wish to evaluate the integral

$$I(e^x) = \int_{-1}^{1} e^x \, dx = 2.35040239.$$

Taking $n = 4$, we find that the Jacobi matrix for four-point Gauss–Legendre quadrature is

$$T_4 = \begin{bmatrix} 0 & 0.577350269 & & \\ 0.577350269 & 0 & 0.516397779 & \\ & 0.516397779 & 0 & 0.507092553 \\ & & 0.507092553 & 0 \end{bmatrix}.$$

The eigenvalues of this matrix give the nodes

$$x_1 = -x_4 = 0.861136312, \qquad x_2 = -x_3 = 0.339981044,$$

and the first components of the eigenvectors give the corresponding weights

$$w_1 = w_4 = 0.347854845, \qquad w_2 = w_3 = 0.652145155.$$

Evaluating the sum (37.10) gives

$$I_n(e^x) = 2.35040209,$$

which agrees with the exact result to about seven digits. The four-point Newton–Cotes formula, by contrast, gives $I_n(e^x) \approx 2.3556$, accurate to only about three digits.

Exercises

37.1. The standard recurrence relation for Legendre polynomials is

$$P_n(x) = \frac{2n-1}{n} x P_{n-1}(x) - \frac{n-1}{n} P_{n-2}(x) \tag{37.12}$$

with initial values $P_0(x) = 1$, $P_1(x) = x$.

(a) Confirm that (37.12) gives the polynomials $P_2(x)$ and $P_3(x)$ of (7.11).

(b) Since $\{P_n(x)\}$ and $\{q_{n+1}(x)\}$ are normalized differently, (37.12) is not the same as the recurrence (37.5) with coefficients (37.6). Write down the two tridiagonal matrices corresponding to these formulas, and derive the relationship between them.

(c) Use the result of (b) to determine a formula for $q_{n+1}(1)$, or equivalently, for $\|P_n\|$.

37.2. Show based on the definition of orthogonality that $q_{n+1}(x)$ has n distinct zeros, all contained in the open interval $(-1, 1)$. (The fact that they are distinct also follows from Exercise 25.1, but here, use a direct argument.)

37.3. The problem of interpolating n data values $\{y_j\}$ in n distinct data points $\{x_j\}$ by a polynomial of degree $\leq n - 1$ was expressed in (11.4) as a square Vandermonde linear system of equations.

(a) Prove that this Vandermonde matrix is nonsingular by arguing that if the interpolation problem has a solution, it must be unique.

(b) Write down the analogous system of equations implicit in Theorem 37.2. Using the result of (a), prove this theorem.

37.4. (a) Write a six-line MATLAB program that computes the nodes and weights for the n-point Gauss–Legendre quadrature formula and applies these numbers to compute the approximate integral of the function f.

(b) Taking $f(x) = e^x$ and $n = 4$, confirm the example in the text. Then plot $|I(e^x) - I_n(e^x)|$ on a log scale for $n = 1, 2, \ldots, 40$ and comment on the results.

(c) Produce a similar plot for $f(x) = e^{|x|}$, and comment.

37.5. The program of Exercise 37.4 computes the zeros of Legendre polynomials, also known as *Legendre points* in $[-1, 1]$. The zeros of Chebyshev polynomials, *Chebyshev points*, are given by the explicit formula (36.9). Perform a sequence of calculations to generate numbers and plots illustrating as elegantly as you can that in the limit $n \to \infty$, both Legendre and Chebyshev points approach the limiting density distribution $\mu(x) = \pi^{-1}(1 - x^2)^{-1/2}$ (in the notation of Exercise 36.2). Produce further plots and numbers to explore the question: how close are Legendre points to Chebyshev points for various values of n?

Lecture 38. Conjugate Gradients

The conjugate gradient iteration is the "original" Krylov subspace iteration, the most famous of these methods and one of the mainstays of scientific computing. Discovered by Hestenes and Stiefel in 1952, it solves symmetric positive definite systems of equations amazingly quickly if the eigenvalues are well distributed.

Minimizing the 2-Norm of the Residual

As in the last two lectures, let $A \in \mathbb{R}^{m \times m}$ be real and symmetric, and suppose we wish to solve a nonsingular system of equations $Ax = b$, with exact solution $x_* = A^{-1}b$. Let $\mathcal{K}_n$ denote the nth Krylov subspace (33.5) generated by b,

$$\mathcal{K}_n = \langle b, Ab, \dots, A^{n-1}b \rangle. \tag{38.1}$$

One approach based on this Krylov subspace would be to solve the system by GMRES. As described in Lecture 35, this would mean that at step n, x_* is approximated by the vector $x_n \in \mathcal{K}_n$ that minimizes $\|r_n\|_2$, where $r_n = b - Ax_n$. Actually, the usual GMRES algorithm does more work than is necessary for minimizing $\|r_n\|_2$. Since A is symmetric, faster algorithms are available based on three-term instead of $(n+1)$-term recurrences at step n. One of these goes by the names of conjugate residuals or MINRES ("minimal residuals").

These methods, at least when constructed to apply to both definite and indefinite matrices, involve certain complications. Rather than describe them, we turn directly to the simpler and more important positive definite case.

Minimizing the A-Norm of the Error

Assume that A is not only real and symmetric but also *positive definite*. As discussed in Lecture 23, this means that the eigenvalues of A are all positive, or equivalently, that $x^T A x > 0$ for every nonzero $x \in \mathbb{R}^m$. Under this assumption, the function $\| \cdot \|_A$ defined by

$$\|x\|_A = \sqrt{x^T A x} \tag{38.2}$$

is a norm on $\mathbb{R}^m$, as can be verified from the definition (3.1). It is called the *A-norm*. (This is the same as the norm $\|x\|_W$ of (3.3), if W is a Cholesky factor of A or any other matrix satisfying $W^T W = A$.)

The vector whose A-norm will concern us is $e_n = x_* - x_n$, the error at step n. The conjugate gradient iteration can be described as follows. *It is a system of recurrence formulas that generates the unique sequence of iterates $\{x_n \in \mathcal{K}_n\}$ with the property that at step n, $\|e_n\|_A$ is minimized.*

We shall present the formulas for the CG iteration, without motivation at first, and derive some orthogonality properties (Theorem 38.1). From these, the claim about minimality of $\|e_n\|_A$ follows as a corollary (Theorem 38.2), and the motivation appears belatedly as we interpret CG as a nonlinear optimization algorithm.

The Conjugate Gradient Iteration

Here is the iteration that Hestenes and Stiefel made famous.

Algorithm 38.1. Conjugate Gradient (CG) Iteration

$x_0 = 0, \ r_0 = b, \ p_0 = r_0$
for $n = 1, 2, 3, \ldots$
$\qquad \alpha_n = (r_{n-1}^T r_{n-1})/(p_{n-1}^T A p_{n-1})$ step length
$\qquad x_n = x_{n-1} + \alpha_n p_{n-1}$ approximate solution
$\qquad r_n = r_{n-1} - \alpha_n A p_{n-1}$ residual
$\qquad \beta_n = (r_n^T r_n)/(r_{n-1}^T r_{n-1})$ improvement this step
$\qquad p_n = r_n + \beta_n p_{n-1}$ search direction

Before analyzing the mathematical properties of these formulas, let us examine them operationally. First we note that the CG iteration is extraordinarily simple—programmable in a few lines of MATLAB. Since it deals only with m-vectors, not with individual entries of vectors or matrices, it is simpler, for example, than Gaussian elimination with pivoting. The only complication—which we shall not address—is the choice of a convergence criterion.

At each step, the CG iteration involves several vector manipulations and one matrix-vector product, the computation of Ap_{n-1} (which appears twice in the listing but need be computed only once). If A is dense and unstructured, this matrix-vector product dominates the operation count, which becomes $\sim 2m^2$ flops for each step. If A is sparse or has other exploitable structure, Ap_{n-1} may be computable in as few as $O(m)$ operations, in which case the operation count may be as low as $O(m)$ flops per step.

From the five lines that define the algorithm, the following properties can be deduced. Like all the theorems in this book that do not explicitly mention rounding errors, this one assumes that the computation is performed in exact arithmetic. If there are rounding errors, these properties fail, and it becomes a subtle matter to explain the still very impressive performance of CG.

Theorem 38.1. *Let the CG iteration (Algorithm 38.1) be applied to a symmetric positive definite matrix problem $Ax = b$. As long as the iteration has not yet converged (i.e., $r_{n-1} \neq 0$), the algorithm proceeds without divisions by zero, and we have the following identities of subspaces:*

$$\mathcal{K}_n = \langle x_1, x_2, \ldots, x_n \rangle = \langle p_0, p_1, \ldots, p_{n-1} \rangle$$
$$= \langle r_0, r_1, \ldots, r_{n-1} \rangle = \langle b, Ab, \ldots, A^{n-1}b \rangle. \tag{38.3}$$

Moreover, the residuals are orthogonal,

$$r_n^T r_j = 0 \qquad (j < n), \tag{38.4}$$

and the search directions are "A-conjugate,"

$$p_n^T A p_j = 0 \qquad (j < n). \tag{38.5}$$

Proof. The proof is by induction on n; we sketch it informally. From the initial guess $x_0 = 0$ and the formula $x_n = x_{n-1} + \alpha_n p_{n-1}$, it follows by induction that x_n belongs to $\langle p_0, p_1, \ldots, p_{n-1} \rangle$. From $p_n = r_n + \beta_n p_{n-1}$ it follows that this is the same as $\langle r_0, r_1, \ldots, r_{n-1} \rangle$. From $r_n = r_{n-1} - \alpha_n A p_{n-1}$, finally, it follows that this is the same as $\langle b, Ab, \ldots, A^{n-1}b \rangle$. This establishes (38.3).

To prove (38.4), we apply the formula $r_n = r_{n-1} - \alpha_n A p_{n-1}$ and the identity $(A p_{n-1})^T = p_{n-1}^T A$ to compute

$$r_n^T r_j = r_{n-1}^T r_j - \alpha_n p_{n-1}^T A r_j.$$

If $j < n - 1$, both terms on the right are zero by induction. If $j = n - 1$, the difference on the right is zero provided $\alpha_n = (r_{n-1}^T r_{n-1})/(p_{n-1}^T A r_{n-1})$. Now this is the same as the line $\alpha_n = (r_{n-1}^T r_{n-1})/(p_{n-1}^T A p_{n-1})$ of Algorithm 38.1, except that $p_{n-1}^T A p_{n-1}$ has been replaced by $p_{n-1}^T A r_{n-1}$. Since p_{n-1} and r_{n-1} differ by $\beta_{n-1} p_{n-2}$, the effect of this replacement is to change the denominator by $\beta_{n-1} p_{n-1}^T A p_{n-2}$, which is zero by the induction hypothesis.

To prove (38.5), we apply the formula $p_n = r_n + \beta_n p_{n-1}$ to compute

$$p_n^T A p_j = r_n^T A p_j + \beta_n p_{n-1}^T A p_j.$$

If $j < n - 1$, both terms on the right are again zero by induction (since (38.4) has now been established for case n). If $j = n-1$, the sum on the right is zero provided $\beta_n = -(r_n^T A p_{n-1})/(p_{n-1}^T A p_{n-1})$, which we can write equivalently in the form $\beta_n = (-\alpha_n r_n^T A p_{n-1})/(\alpha_n p_{n-1}^T A p_{n-1})$. This is the same as the line $\beta_n = (r_n^T r_n)/(r_{n-1}^T r_{n-1})$ of Algorithm 38.1, except that $r_n^T r_n$ has been replaced by $r_n^T(-\alpha_n A p_{n-1})$ and $r_{n-1}^T r_{n-1}$ has been replaced by $p_{n-1}^T(\alpha_n A p_{n-1})$. By the induction hypothesis and lines 3 and 5 of Algorithm 38.1, these replacements can again readily be shown to have zero effect. $\square$

Optimality of CG

In deriving the orthogonality properties (38.4) and (38.5), we have finished the real work. It is now a straightforward matter to confirm that CG minimizes $\|e\|_A$ at each step.

Theorem 38.2. *Let the CG iteration be applied to a symmetric positive definite matrix problem $Ax = b$. If the iteration has not already converged (i.e., $r_{n-1} \neq 0$), then x_n is the unique point in $\mathcal{K}_n$ that minimizes $\|e_n\|_A$. The convergence is monotonic,*

$$\|e_n\|_A \leq \|e_{n-1}\|_A, \tag{38.6}$$

and $e_n = 0$ is achieved for some $n \leq m$.

Proof. From Theorem 38.1 we know that x_n belongs to $\mathcal{K}_n$. To show that it is the unique point in $\mathcal{K}_n$ that minimizes $\|e\|_A$, consider an arbitrary point $x = x_n - \Delta x \in \mathcal{K}_n$, with error $e = x_* - x = e_n + \Delta x$. We calculate

$$
\begin{aligned}
\|e\|_A^2 &= (e_n + \Delta x)^T A(e_n + \Delta x) \\
&= e_n^T A e_n + (\Delta x)^T A(\Delta x) + 2e_n^T A(\Delta x).
\end{aligned}
$$

The final term in this equation is $2r_n^T(\Delta x)$, an inner product of r_n with a vector in $\mathcal{K}_n$, and by Theorem 38.1, any such inner product is zero. This is the crucial orthogonality property that makes the CG iteration so powerful. It implies that we have

$$\|e\|_A^2 = e_n^T A e_n + (\Delta x)^T A(\Delta x).$$

Only the second of these terms depends on Δx, and since A is positive definite, that term is ≥ 0, attaining the value 0 if and only if $\Delta x = 0$, i.e., $x_n = x$. Thus $\|e\|_A$ is minimal if and only if $x_n = x$, as claimed.

The remaining statements of the theorem now follow readily. The monotonicity property (38.6) is a consequence of the inclusion $\mathcal{K}_n \subseteq \mathcal{K}_{n+1}$, and since $\mathcal{K}_n$ is a subset of $\mathbb{R}^m$ of dimension n as long as convergence has not yet been achieved, convergence must be achieved in at most m steps. $\qquad\square$

The guarantee that the CG iteration converges in at most m steps is void in floating point arithmetic. For arbitrary matrices A on a real computer, no decisive reduction in $\|e_n\|_A$ will necessarily be observed at all when $n = m$. In practice, however, CG is used not for arbitrary matrices but for matrices whose spectra, perhaps thanks to preconditioning, are well-enough behaved that convergence to a desired accuracy is achieved for $n \ll m$ (Lecture 32). The theoretical exact convergence at $n = m$ has no relevance to this use of the CG iteration in scientific computing.

CG as an Optimization Algorithm

We have just shown that the CG iteration has a certain optimality property: it minimizes $\|e_n\|_A$ at step n over all vectors $x \in \mathcal{K}_n$. In fact, as foreshadowed already by the use of such terms as "step length" and "search direction," this iteration can be interpreted as an algorithm of a standard form for minimizing a nonlinear function of $x \in \mathbb{R}^m$. At the heart of the iteration is the formula

$$x_n = x_{n-1} + \alpha_n p_{n-1}.$$

This is a familiar equation in optimization, in which a current approximation x_{n-1} is updated to a new approximation x_n by moving a distance α_n (the step length) in the direction p_{n-1} (the search direction). By a succession of such steps, the CG iteration attempts to find a minimum of a nonlinear function.

Which function? According to Theorem 38.2, the answer would appear to be $\|e\|_A$, or equivalently, $\|e\|_A^2$. However, although $\|e\|_A^2$ is indeed a function of x, it is not one we can evaluate without knowing x_*. It would not be very "standard" to interpret CG as an optimization process applied to a function that cannot be evaluated!

On the other hand, given A and b and $x \in \mathbb{R}^m$, the quantity

$$\varphi(x) = \tfrac{1}{2}x^T A x - x^T b \qquad (38.7)$$

can certainly be evaluated. A short computation now reveals

$$
\begin{aligned}
\|e_n\|_A^2 &= e_n^T A e_n = (x_* - x_n)^T A (x_* - x_n) \\
&= x_n^T A x_n - 2 x_n^T A x_* + x_*^T A x_* \\
&= x_n^T A x_n - 2 x_n^T b + x_*^T b = 2\varphi(x_n) + \text{constant}.
\end{aligned}
$$

Thus $\varphi(x)$ is the same as $\|e\|_A^2$ except for a factor of 2 and the (unknown) constant $x_*^T b$. Like $\|e\|_A^2$, it must achieve its minimum (namely, $-x_*^T b/2$) uniquely at $x = x_*$.

The CG iteration can be interpreted as an iterative process for minimizing the quadratic function $\varphi(x)$ of $x \in \mathbb{R}^m$. At each step, an iterate $x_n = x_{n-1} + \alpha_n p_{n-1}$ is computed that minimizes $\varphi(x)$ over all x in the one-dimensional space $x_{n-1} + \langle p_{n-1} \rangle$. (It is readily confirmed that the formula $\alpha_n = (r_{n-1}^T r_{n-1})/(p_{n-1}^T A p_{n-1})$ ensures that α_n is optimal in this sense among all step lengths α.) What makes the CG iteration remarkable is the choice of the search direction p_{n-1}, which has the special property that minimizing $\varphi(x)$ over $x_n + \langle p_{n-1} \rangle$ actually minimizes it over all of $\mathcal{K}_n$.

There is a close analogy between the CG iteration for solving $Ax = b$ and the Lanczos iteration for finding eigenvalues. The eigenvalues of A, as discussed in Lecture 27, are the stationary values for $x \in \mathbb{R}^m$ of the Rayleigh quotient, $r(x) = (x^T A x)/(x^T x)$. As pointed out in Exercise 36.1, the eigenvalue estimates (Ritz values) associated with step n of the Lanczos iteration are the stationary values of the same function $r(x)$ if x is restricted to the Krylov subspace $\mathcal{K}_n$. This is a perfect parallel of what we have shown in the last two pages, that the solution x_* of $Ax = b$ is the minimal point in $\mathbb{R}^m$ of the scalar function $\varphi(x)$, and the CG iterate x_n is the minimal point of the same function $\varphi(x)$ if x is restricted to $\mathcal{K}_n$.

CG and Polynomial Approximation

A theme of the last four lectures has been the connection between Krylov subspace iterations and polynomials of matrices. The Arnoldi and Lanczos iterations solve the Arnoldi/Lanczos approximation problem (34.3), and the GMRES iteration solves the GMRES approximation problem (35.10). For CG, the appropriate approximation problem involves the A-norm of the error.

CG Approximation Problem. Find $p_n \in P_n$ such that

$$\|p_n(A)e_0\|_A = \text{minimum}. \tag{38.8}$$

Here e_0 denotes the initial error, $e_0 = x_* - x_0 = x_*$, and P_n is again defined as in (35.7), the set of polynomials p of degree $\leq n$ with $p(0) = 1$. From Theorem 38.2 we may derive the following convergence theorem.

Theorem 38.3. *If the CG iteration has not already converged before step n (i.e., $r_{n-1} \neq 0$), then (38.8) has a unique solution $p_n \in P_n$, and the iterate x_n has error $e_n = p_n(A)e_0$ for this same polynomial p_n. Consequently we have*

$$\frac{\|e_n\|_A}{\|e_0\|_A} = \inf_{p \in P_n} \frac{\|p(A)e_0\|_A}{\|e_0\|_A} \leq \inf_{p \in P_n} \max_{\lambda \in \Lambda(A)} |p(\lambda)|, \tag{38.9}$$

where $\Lambda(A)$ denotes the spectrum of A.

Proof. From Theorem 38.1 it follows that $e_n = p(A)e_0$ for some $p \in P_n$. The equality in (38.9) is a consequence of this and Theorem 38.2. As for the inequality in (38.9), if $e_0 = \sum_{j=1}^m a_j v_j$ is an expansion of e_0 in orthonormal eigenvectors of A, then we have $p(A)e_0 = \sum_{j=1}^m a_j p(\lambda_j) v_j$ and thus

$$\|e_0\|_A^2 = \sum_{j=1}^m a_j^2 \lambda_j, \qquad \|p(A)e_0\|_A^2 = \sum_{j=1}^m a_j^2 \lambda_j (p(\lambda_j))^2.$$

These identities imply $\|p(A)e_0\|_A^2/\|e_0\|_A^2 \leq \max_{\lambda \in \Lambda(A)} |p(\lambda)|^2$, which implies the inequality in question. $\qquad\square$

Rate of Convergence

Theorem 38.3 establishes that the rate of convergence of the CG iteration is determined by the location of the spectrum of A. A good spectrum is one on which polynomials $p_n \in P_n$ can be very small, with size decreasing rapidly with n. Roughly speaking, this may happen for either or both of two reasons: the eigenvalues may be grouped in small clusters, or they may lie well separated in a relative sense from the origin. The two best-known corollaries of Theorem 38.3 address these two ideas in their extreme forms.

First, we suppose that the eigenvalues are perfectly clustered but assume nothing about the locations of these clusters.

Theorem 38.4. *If A has only n distinct eigenvalues, then the CG iteration converges in at most n steps.*

Proof. This is a corollary of (38.9), since a polynomial $p(x) = \prod_{j=1}^n (1 - x/\lambda_j) \in P_n$ exists that is zero at any specified set of n points $\{\lambda_j\}$. $\qquad\square$

At the other extreme, suppose we know nothing about any clustering of the eigenvalues but only that their distances from the origin vary by at most a factor $\kappa \geq 1$. In other words, suppose we know only the 2-norm condition number $\kappa = \lambda_{\max}/\lambda_{\min}$, where $\lambda_{\max}$ and $\lambda_{\min}$ are the extreme eigenvalues of A.

Theorem 38.5. *Let the CG iteration be applied to a symmetric positive definite matrix problem $Ax = b$, where A has 2-norm condition number κ. Then the A-norms of the errors satisfy*

$$\frac{\|e_n\|_A}{\|e_0\|_A} \leq 2 \left/ \left[\left(\frac{\sqrt{\kappa}+1}{\sqrt{\kappa}-1} \right)^n + \left(\frac{\sqrt{\kappa}+1}{\sqrt{\kappa}-1} \right)^{-n} \right] \right. \leq 2 \left(\frac{\sqrt{\kappa}-1}{\sqrt{\kappa}+1} \right)^n. \quad (38.10)$$

Proof. By Theorem 38.3, it is enough to find a polynomial $p \in P_n$ whose maximum value for $\lambda \in [\lambda_{\min}, \lambda_{\max}]$ is the middle expression of (38.10). The polynomial we choose is the scaled and shifted Chebyshev polynomial $p(x) = T_n(\gamma - 2x/(\lambda_{\max} - \lambda_{\min}))/T_n(\gamma)$, where T_n is the usual Chebyshev polynomial of degree n and γ takes the special value $\gamma = (\lambda_{\max} + \lambda_{\min})/(\lambda_{\max} - \lambda_{\min}) = (\kappa + 1)/(\kappa - 1)$. For $x \in [\lambda_{\min}, \lambda_{\max}]$, the argument of T_n in the numerator of $p(x)$ lies in $[-1, 1]$, which means the magnitude of that numerator is ≤ 1. Therefore, to prove the theorem, it will suffice to show

$$T_n(\gamma) = T_n\left(\frac{\kappa+1}{\kappa-1}\right) = \frac{1}{2}\left[\left(\frac{\sqrt{\kappa}+1}{\sqrt{\kappa}-1}\right)^n + \left(\frac{\sqrt{\kappa}+1}{\sqrt{\kappa}-1}\right)^{-n}\right]. \qquad (38.11)$$

We can do this by making the change of variables $x = \frac{1}{2}(z + z^{-1})$, $T_n(x) = \frac{1}{2}(z^n + z^{-n})$, standard in the study of Chebyshev polynomials. If $(\kappa+1)/(\kappa-1) = \frac{1}{2}(z + z^{-1})$, that is, $\frac{1}{2}z^2 - (\kappa+1)/(\kappa-1)z + \frac{1}{2} = 0$, then we have a quadratic equation with solution

$$z = \left(\frac{\kappa+1}{\kappa-1}\right) + \sqrt{\left(\frac{\kappa+1}{\kappa-1}\right)^2 - 1} = \frac{\kappa+1+\sqrt{(\kappa+1)^2-(\kappa-1)^2}}{\kappa-1}$$

$$= \frac{\kappa+1+\sqrt{4\kappa}}{\kappa-1} = \frac{(\sqrt{\kappa}+1)^2}{(\sqrt{\kappa}+1)(\sqrt{\kappa}-1)} = \frac{\sqrt{\kappa}+1}{\sqrt{\kappa}-1}.$$

Thus $T_n(\gamma) = \frac{1}{2}(z^n + z^{-n})$ for this value of z, which is (38.11), as claimed. $\square$

Theorem 38.5 is the most famous result about convergence of the CG iteration. Since

$$\frac{\sqrt{\kappa}-1}{\sqrt{\kappa}+1} \sim 1 - \frac{2}{\sqrt{\kappa}}$$

as $\kappa \to \infty$, it implies that if κ is large but not too large, convergence to a specified tolerance can be expected in $O(\sqrt{\kappa})$ iterations. One must remember that this is only an upper bound. Convergence may be faster for special right-hand sides (not so common) or if the spectrum is clustered (more common).

Example

For an example of the convergence of CG, consider a 500×500 sparse matrix A constructed as follows. First we put 1 at each diagonal position and a random number from the uniform distribution on $[-1, 1]$ at each off-diagonal position (maintaining the symmetry $A = A^T$). Then we replace each off-diagonal entry with $|a_{ij}| > \tau$ by zero, where τ is a parameter. For τ close to zero, the result is a well-conditioned positive definite matrix whose density of nonzero entries is approximately τ. As τ increases, both the condition number and the sparsity deteriorate.

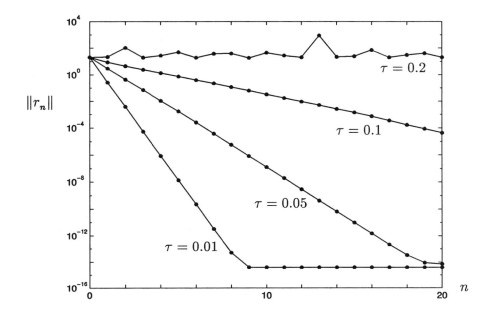

Figure 38.1. *CG convergence curves for the* 500×500 *sparse matrices A described in the text. For* $\tau = 0.01$, *the system is solved about* 700 *times faster by CG than by Cholesky factorization. For* $\tau = 0.2$, *the matrix is not positive definite and there is no convergence.*

Figure 38.1 shows convergence curves corresponding to 20 steps of the CG iteration for matrices of this kind with $\tau = 0.01, 0.05, 0.1, 0.2$. (The right-hand side b was taken to be a random vector.) For $\tau = 0.01$, A has 3092 nonzero entries and condition number $\kappa \approx 1.06$. Convergence to machine precision takes place in 9 steps, about 6×10^4 flops. For $\tau = 0.05$, there are 13,062 nonzeros with $\kappa \approx 1.83$, and convergence takes 19 steps, about 5×10^5 flops. For $\tau = 0.1$ we have 25,526 nonzeros and $\kappa \approx 10.3$, with only 5 digits of convergence after 20 steps and 10^6 flops. For $\tau = 0.2$, with 50,834 nonzeros, there is no convergence at all. The lowest eigenvalue is now negative, so A is no longer positive definite and the use of the CG iteration is inappropriate. (In fact, the CG iteration often succeeds with indefinite matrices, but in this case the matrix is not only indefinite but ill-conditioned.)

Note how closely the $\tau = 0.01$ curve of Figure 38.1 matches the schematic ideal depicted in Figure 32.1! For this example, the operation count of 6×10^4 flops beats Cholesky factorization (23.4) by a factor of about 700. Unfortunately, not every matrix arising in practice has such a well-behaved spectrum, even after the best efforts to find a good preconditioner.

Exercises

38.1. Based on the condition numbers κ reported in the text, determine the rate of convergence predicted by Theorem 38.5 for the matrices A of Figure 38.1 with $\tau = 0.01, 0.05, 0.1$. Draw lines on a copy of Figure 38.1 indicating how closely these predictions match the actual convergence rates.

38.2. Suppose A is a real symmetric 805×805 matrix with eigenvalues $1.00, 1.01, 1.02, \ldots, 8.98, 8.99, 9.00$ and also $10, 12, 16, 24$. How many steps of the conjugate gradient iteration must you take to be sure of reducing the initial error $\|e_0\|_A$ by a factor of 10^6?

38.3. The conjugate gradient is applied to a symmetric positive definite matrix A with the result $\|e_0\|_A = 1$, $\|e_{10}\|_A = 2 \times 2^{-10}$. Based solely on this data,

(a) What bound can you give on $\kappa(A)$?

(b) What bound can you give on $\|e_{20}\|_A$?

38.4. Suppose A is a dense symmetric positive definite 1000×1000 matrix with $\kappa(A) = 100$. Estimate roughly how many flops are required to solve $Ax = b$ to ten-digit accuracy by (a) Cholesky factorization, (b) Richardson iteration with the optimal parameter α (Exercise 35.3), and (c) CG.

38.5. We have described CG as an iterative minimization of the function $\varphi(x)$ of (38.7). Another way to minimize the same function—far slower, in general—is by the method of *steepest descent*.

(a) Derive the formula $\nabla \varphi(x) = -r$ for the gradient of $\varphi(x)$. Thus the steepest descent iteration corresponds to the choice $p_n = r_n$ instead of $p_n = r_n + \beta_n p_{n-1}$ in Algorithm 38.1.

(b) Determine the formula for the optimal step length α_n of the steepest descent iteration.

(c) Write down the full steepest descent iteration. There are three operations inside the main loop.

38.6. Let A be the 100×100 tridiagonal symmetric matrix with $1, 2, \ldots, 100$ on the diagonal and 1 on the sub- and superdiagonals, and set $b = (1, 1, \ldots, 1)^T$. Write a program that takes 100 steps of the CG and also the steepest descent iteration to approximately solve $Ax = b$. Produce a plot with four curves on it: the computed residual norms $\|r_n\|_2$ for CG, the actual residual norms $\|b - Ax_n\|_2$ for CG, the residual norms $\|r_n\|_2$ for steepest descent, and the estimate $2(\sqrt{\kappa} - 1)^n / (\sqrt{\kappa} + 1)^n$ of Theorem 38.5. Comment on your results.

Lecture 39. Biorthogonalization Methods

Not all Krylov subspace iterations for nonsymmetric systems involve recurrences of growing length and growing cost. Methods based on three-term recurrences have also been devised, and they are the most powerful nonsymmetric iterations available today. The price to be paid, at least for some of the iterations in this category, is that one must work with two Krylov subspaces rather than one, generated by multiplications by A^* as well as A.

Where We Stand

On p. 245 we presented a table of Krylov subspace matrix iterations:

	$Ax = b$	$Ax = \lambda x$
$A = A^*$	CG	Lanczos
$A \neq A^*$	GMRES CGN BCG et al.	Arnoldi

Our discussions of three of these boxes are now complete, and as for the fourth, lower-left position, we have already discussed GMRES. In this lecture we turn to the final two lines of the table. We spend just a moment on CGN,

303

a simple and easily analyzed algorithm, and then move to our main subject, the biorthogonalization methods represented by the entry "BCG et al."

CGN = CG Applied to the Normal Equations

Let $A \in \mathbb{C}^{m \times m}$ be nonsingular but not necessarily hermitian, so that $Ax = b$, for any $b \in \mathbb{C}^m$, is a nonsingular square system of equations. One of the simplest methods for solving such a system is to apply the CG iteration to the normal equations (11.9),

$$A^*Ax = A^*b. \tag{39.1}$$

(The matrix A^*A is not formed explicitly, which would require m^3 flops. Instead, each matrix-vector product A^*Av is evaluated in two steps as $A^*(Av)$.) Since A is nonsingular, A^*A is hermitian positive definite, or symmetric positive definite if A is real. Thus the theorems of the last lecture apply, and rapid convergence can be expected if the eigenvalues of A^*A are favorably distributed. This method goes by the name of CGN (also CGNR), which roughly stands for "CG applied to the normal equations."

Since we have already analyzed the behavior of CG, nothing new is needed to understand the behavior of CGN. If the initial guess is $x_0 = 0$, as in Algorithm 38.1, then from Theorem 38.1 we see that the later iterates belong to a Krylov subspace generated by A^*A:

$$x_n \in \langle A^*b, (A^*A)A^*b, \ldots, (A^*A)^{n-1}A^*b \rangle. \tag{39.2}$$

From Theorem 38.2 we know that the A^*A-norm of the error is minimized over this space at each step, and since $\|e_n\|_{A^*A}^2 = e_n^*A^*Ae_n = \|Ae_n\|_2^2 = \|r_n\|^2$, this is another way of saying that the 2-norm of the residual $r_n = b - Ax_n$ is minimized:

$$\|r_n\|_2 = \text{minimum}. \tag{39.3}$$

Thus CGN, like GMRES, is a minimal residual method, but since the Krylov subspaces (33.5) and (39.2) are different, these two methods are by no means equivalent.

According to Theorem 38.3, the convergence of CGN is controlled by the eigenvalues of A^*A. These numbers are equal to the squares of the singular values of A. Thus the convergence of CGN is determined by the singular values of A, and in principle has nothing to do with the eigenvalues of A. The fact that squares are involved is unfortunate, however. If A has condition number κ, then A^*A has condition number κ^2, and the analogue of (38.10) for CGN becomes

$$\frac{\|r_n\|_2}{\|r_0\|_2} \leq 2 \left(\frac{\kappa - 1}{\kappa + 1} \right)^n. \tag{39.4}$$

For large κ, this is far worse than (38.10); it implies that $O(\kappa)$ iterations are required for convergence to a fixed accuracy, not $O(\sqrt{\kappa})$.

This "squaring of the condition number" has given the CGN iteration a poor reputation, which, on balance, may be deserved. Nevertheless, for some problems CGN vastly outperforms alternative methods, since their convergence depends on eigenvalues rather than singular values. All one needs is a matrix whose singular values are well behaved but whose eigenvalues are not, such as a well-conditioned matrix whose spectrum surrounds the origin in the complex plane. An extreme example is provided by the $m \times m$ circulant matrix of the form

$$A = \begin{bmatrix} 0 & 1 & & & \\ & 0 & 1 & & \\ & & 0 & 1 & \\ & & & 0 & 1 \\ 1 & & & & 0 \end{bmatrix}. \tag{39.5}$$

The singular values of this matrix are all equal to 1, but the eigenvalues are the mth roots of unity. GMRES requires m steps for convergence for a general right-hand side b, while CGN converges in one step. (See Exercise 39.1.)

Another virtue of the CGN iteration is that since it is based on the normal equations, it applies without modification to least squares problems (cf. Algorithm 11.1), where A is no longer square. Alternatively, some iterative methods for least squares problems are based on the block system (19.4) of Exercise 19.1.

Tridiagonal Biorthogonalization

The Lanczos iteration, as we saw in Lecture 36, is a process of *tridiagonal orthogonalization*. If carried a full m steps (in exact arithmetic), it would produce a unitary reduction (36.5) of a hermitian matrix to tridiagonal form: $A = QTQ^*$.

If A is not hermitian, such a reduction is not possible in general: we must give up either the unitary transformations or the final tridiagonal form. The Arnoldi iteration, a process of *Hessenberg orthogonalization*, does the latter. If carried a full m steps, it would produce a unitary reduction (33.12) of an arbitrary square matrix to Hessenberg form: $A = QHQ^*$.

Biorthogonalization methods are based on the opposite choice. If we insist on a tridiagonal result but give up the use of unitary transformations, we have a process of *tridiagonal biorthogonalization*: $A = VTV^{-1}$, where V is nonsingular but generally not unitary (Figure 39.1). Taking the adjoint gives the equivalent equation $A^* = V^{-*}T^*(V^{-*})^{-1}$. (Recall from p. 12 that $V^{-*} = (V^*)^{-1} = (V^{-1})^*$.) The term "biorthogonal" refers to the fact that although the columns of V are not orthogonal to each other, they are orthogonal to the columns of V^{-*}, as follows trivially from the identity $(V^{-*})^*V = V^{-1}V = I$.

To begin to make this idea into an iterative algorithm, we must see what is involved for $n < m$. Let V be a nonsingular matrix such that $A = VTV^{-1}$

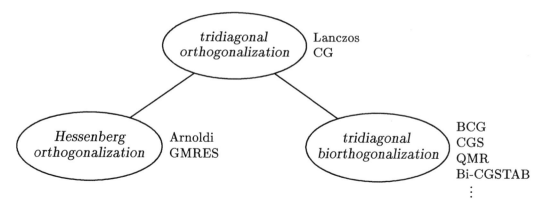

Figure 39.1. *Classification of Krylov subspace iterations. If the matrix is hermitian (top row), then it can be orthogonalized by a three-term recurrence relation—a tridiagonal matrix. If it is not hermitian, one must give up either the tridiagonal structure or the orthogonality.*

with T tridiagonal, and define $W = V^{-*}$. Let v_j and w_j denote the jth columns of V and W, respectively. These vectors are *biorthogonal* in the sense that

$$w_i^* v_j = \delta_{ij}, \tag{39.6}$$

where δ_{ij} is the Kronecker delta function (p. 14). For each n with $1 \le n \le m$, following (33.1), define the $m \times n$ matrices

$$V_n = \begin{bmatrix} \vert & & \vert \\ v_1 & \cdots & v_n \\ \vert & & \vert \end{bmatrix}, \qquad W_n = \begin{bmatrix} \vert & & \vert \\ w_1 & \cdots & w_n \\ \vert & & \vert \end{bmatrix}. \tag{39.7}$$

In matrix form, the biorthogonality condition can be written

$$W_n^* V_n = V_n^* W_n = I_n,$$

where I_n is the identity of dimension n.

We can now write down the key formulas that are the basis of biorthogonalization methods. For the Lanczos iteration, we had (36.5) and (36.6),

$$AQ_n = Q_{n+1}\tilde{T}_n, \qquad T_n = Q_n^* A Q_n.$$

For the Arnoldi iteration, we had (33.12) and (33.13),

$$AQ_n = Q_{n+1}\tilde{H}_n, \qquad H_n = Q_n^* A Q_n.$$

These are the corresponding formulas for biorthogonalization methods:

$$AV_n = V_{n+1}\tilde{T}_n, \tag{39.8}$$

$$A^*W_n = W_{n+1}\tilde{S}_n, \tag{39.9}$$

$$T_n = S_n^* = W_n^* A V_n. \tag{39.10}$$

Here V_n and W_n have dimensions $m \times n$, $\tilde{T}_{n+1}$ and $\tilde{S}_{n+1}$ are (nonhermitian) tridiagonal matrices with dimensions $(n+1) \times n$, and $T_n = S_n^*$ is the $n \times n$ matrix obtained by deleting the last row of $\tilde{T}_{n+1}$ or the last column of $\tilde{S}_{n+1}^*$.

In analogy to the developments of p. 252, (39.8) can be displayed as

$$\left[\begin{array}{c} A \end{array}\right]\left[\begin{array}{c|c|c} v_1 & \cdots & v_n \end{array}\right] = \left[\begin{array}{c|c|c} v_1 & \cdots & v_{n+1} \end{array}\right]\left[\begin{array}{ccccc} \alpha_1 & \gamma_1 & & & \\ \beta_1 & \alpha_2 & \gamma_2 & & \\ & \beta_2 & \alpha_3 & \ddots & \\ & & \ddots & \ddots & \gamma_{n-1} \\ & & & \beta_{n-1} & \alpha_n \\ & & & & \beta_n \end{array}\right],$$

which corresponds to the three-term recurrence relation

$$Av_n = \gamma_{n-1}v_{n-1} + \alpha_n v_n + \beta_n v_{n+1}. \tag{39.11}$$

Similarly, (39.9) takes the form

$$\left[\begin{array}{c} A^* \end{array}\right]\left[\begin{array}{c|c|c} w_1 & \cdots & w_n \end{array}\right] = \left[\begin{array}{c|c|c} w_1 & \cdots & w_{n+1} \end{array}\right]\left[\begin{array}{ccccc} \overline{\alpha}_1 & \overline{\beta}_1 & & & \\ \overline{\gamma}_1 & \overline{\alpha}_2 & \overline{\beta}_2 & & \\ & \overline{\gamma}_2 & \overline{\alpha}_3 & \ddots & \\ & & \ddots & \ddots & \overline{\beta}_{n-1} \\ & & & \overline{\gamma}_{n-1} & \overline{\alpha}_n \\ & & & & \overline{\gamma}_n \end{array}\right],$$

corresponding to

$$A^*w_n = \overline{\beta}_{n-1}w_{n-1} + \overline{\alpha}_n w_n + \overline{\gamma}_n w_{n+1}. \tag{39.12}$$

(We have not seen these bars for complex conjugation before, because in the last three lectures, we assumed that A was real.)

As usual with Krylov subspace iterations, these equations suggest an algorithm. Begin with vectors v_1 and w_1 that are arbitrary except for satisfying $v_1^* w_1 = 1$, and set $\beta_0 = \gamma_0 = 0$ and $v_0 = w_0 = 0$. Now, for each $n = 1, 2, \ldots$, set $\alpha_n = w_n^* A v_n$, as follows from (39.6) and (39.11) or (39.12). The vectors v_{n+1} and w_{n+1} are then determined by (39.11) and (39.12) up to scalar factors. These factors may be chosen arbitrarily, subject to the normalization $w_{n+1}^* v_{n+1} = 1$, whereupon β_{n+1} and γ_{n+1} are determined by (39.11) and (39.12).

The vectors generated by the procedure just described lie in the Krylov subspaces

$$v_n \in \langle v_1, Av_1, \dots, A^{n-1}v_1 \rangle, \quad w_n \in \langle w_1, A^*w_1, \dots, (A^*)^{n-1}w_1 \rangle. \quad (39.13)$$

For a generic matrix, in exact arithmetic, the procedure will run to completion after m steps, but for certain special matrices there may also be a breakdown of the process before this point. If $v_n = 0$ or $w_n = 0$ at some step, an invariant subspace of A or A^* has been found: the tridiagonal matrix T is reducible (cf. Exercise 33.2). Alternatively, it may also happen that $v_n \neq 0$ and $w_n \neq 0$ but $w_n^* v_n = 0$. The possibility of this more serious kind of breakdown is present in most biorthogonalization methods. As in other areas of numerical analysis, the fact that exact breakdown is possible for certain problems implies that near-breakdown may occur for many other problems, with potentially adverse consequences in floating point arithmetic. Some methods for coping with these phenomena are mentioned at the end of this lecture.

BCG = Biconjugate Gradients

One way to use the biorthogonalization process just described is to compute eigenvalues: as $n \to \infty$, some eigenvalues of T_n may converge rapidly to some eigenvalues of A. Another application, which we shall now briefly discuss, is the solution of nonsingular systems of equations $Ax = b$. The classic algorithm of this type is known as *biconjugate gradients* or *BCG*.

The principle of BCG is as follows. We take $v_1 = b$, so that the first Krylov subspace in (39.13) becomes $\mathcal{K}_n = \langle b, Ab, \dots, A^{n-1}b \rangle$. Recall that the principle of GMRES is to pick $x_n \in \mathcal{K}_n$ so that the orthogonality condition

$$\text{GMRES:} \quad r_n \perp \langle Ab, A^2b, \dots, A^nb \rangle = A\mathcal{K}_n \quad (39.14)$$

is satisfied, where $r_n = b - Ax_n$ is the residual corresponding to x_n (Figure 35.1). This choice has the effect of minimizing $\|r_n\|$, the 2-norm of the residual. The principle of the BCG algorithm is to pick x_n in the same subspace, $x_n \in \mathcal{K}_n$, but to enforce the orthogonality condition

$$\text{BCG:} \quad r_n \perp \langle w_1, A^*w_1, \dots, (A^*)^{n-1}w_1 \rangle. \quad (39.15)$$

Here $w_1 \in \mathbb{C}^m$ is an arbitrary vector satisfying $w_1^* v_1 = 1$; in applications one sometimes takes $w_1 = v_1/\|v_1\|_2$. Unlike (39.14), this choice does not minimize $\|r_n\|_2$, and it is not optimal from the point of view of minimizing the number of iterations. Its advantage is that it can be implemented with three-term recurrences rather than the $(n+1)$-term recurrences of GMRES.

Without giving details of the derivation, we now record the BCG algorithm in its standard form. What follows should be compared with Algorithm 38.1, the CG algorithm. The two are the same except that the sequence of search directions $\{p_n\}$ of CG has become two sequences $\{p_n\}$ and $\{q_n\}$, and the sequence of residuals $\{r_n\}$ of CG has become two sequences $\{r_n\}$ and $\{s_n\}$.

Algorithm 39.1. Biconjugate Gradient (BCG) Iteration

$x_0 = 0, \quad p_0 = r_0 = b, \quad q_0 = s_0 = \text{arbitrary}$

for $n = 1, 2, 3, \ldots$

$\quad \alpha_n = (s_{n-1}^* r_{n-1})/(q_{n-1}^* A p_{n-1})$

$\quad x_n = x_{n-1} + \alpha_n p_{n-1}$

$\quad r_n = r_{n-1} - \alpha_n A p_{n-1}$

$\quad s_n = s_{n-1} - \alpha_n A^* q_{n-1}$

$\quad \beta_n = (s_n^* r_n)/(s_{n-1}^* r_{n-1})$

$\quad p_n = r_n + \beta_n p_{n-1}$

$\quad q_n = s_n + \beta_n q_{n-1}$

As in Theorem 38.1, it is readily shown that $s_n^* r_j = 0$ and $q_n^* A p_j = 0$ for $j < n$.

Example

In Figure 38.1 we illustrated the convergence of the CG iteration for a 500×500 sparse symmetric positive definite matrix dependent on a parameter τ. To illustrate the convergence of BCG, consider the same matrix with one change: the signs of all the entries are randomized. This makes the matrix no longer hermitian, and it changes the dominant entries on the diagonal to 1 and -1 at random, rather than all 1, so that the eigenvalues are clustered around 1 and -1 instead of just 1.

Figure 39.2 shows the convergence of GMRES and BCG for such a matrix with $\tau = 0.01$. Considering first the GMRES curve, we note that the convergence is half as fast as in Figure 38.1, with essentially no progress at each odd-numbered step, but steady progress at each even step. This odd–even effect is a result of the approximate ± 1 symmetry of the matrix: a polynomial $p(z)$ of degree $2k + 1$ with $p(0) = 1$ can be no smaller at 1 and -1 than a corresponding polynomial of degree $2k$. Turning now to the BCG curve, we see that the convergence is comparable in an overall sense, but it is no longer monotonic, showing spikes of magnitude as great as about 10^2 at each odd-numbered step. The accuracy attained at the end has also suffered by more than a digit. All of these features are typical of BCG computations.

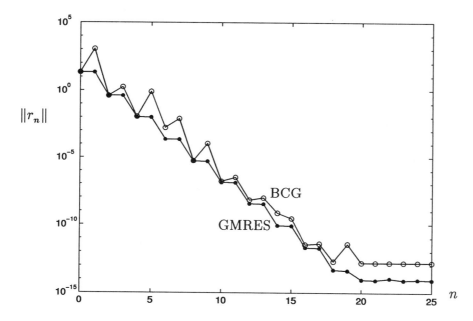

Figure 39.2. *Comparison of GMRES and BCG for the* 500×500 *matrix labeled* $\tau = 0.01$ *in Figure 38.1, but with the signs of the entries randomized.*

The horizontal axis in Figure 39.2 is the step number n, which is not the same as the computational cost. At each step, GMRES requires one matrix-vector multiplication involving A, whereas BCG requires multiplications involving both A and A^*. For problems where matrix-vector multiplications dominate the work and enough storage is available, GMRES may consequently be twice as fast as BCG or faster. Here, however, the matrix is sparse enough that the work associated with handling long recurrences is significant, and in fact, the BCG calculation of Figure 39.2 was faster than the GMRES calculation by better than a factor of 2.

QMR and Other Variants

BCG has one great advantage over GMRES: it involves three-term recurrences, enabling the work per step and the storage requirements to remain under control even when many steps are needed. On the other hand, it has two disadvantages. One is that in comparison to the monotonic and often rapid convergence of GMRES as a function of step number, its convergence is slower and often erratic, sometimes far more erratic than in Figure 39.2. Irregular convergence is unattractive, and it may have the consequence of reducing the ultimately attainable accuracy because of rounding errors (Exercise 39.4). In the extreme, it becomes the phenomenon of breakdown of the iteration, where an inner product becomes zero and no further progress is possible, even though the system of equations may be well-conditioned.

The other problem with BCG is that it requires multiplication by A^* as well as A. Depending on how these products are implemented both mathematically and in terms of computer architecture, this may be anything from a minor additional burden to effectively impossible.

In response to these two problems, beginning in the 1980s, more than a dozen variants of BCG have been proposed. Here are some of the best known of these; references are given in the Notes.

Look-ahead Lanczos (Parlett, Taylor, and Liu, 1985)

CGS = conjugate gradients squared (Sonneveld, 1989)

QMR = quasi-minimal residuals (Freund and Nachtigal, 1991)

Bi-CGSTAB = stabilized BCG (van der Vorst, 1992)

TFQMR = transpose-free QMR (Freund, 1993)

We shall say a few words about these methods but give no details.

The look-ahead Lanczos algorithm is based on the fact that when a breakdown is about to take place, it can be avoided by taking two or more steps of the iteration at once rather than a single step. The original idea of Parlett et al. has been developed extensively by later authors and is incorporated, for example, in the version of the QMR algorithm recommended by its authors. The phenomenon of breakdown can be shown to be equivalent to the phenomenon of square blocks of identical entries in the table of Padé approximants to a function, and the look-ahead idea amounts to a method of stepping across such blocks in one step. In practice, of course, one does not just test for exact breakdowns; a notion of near-breakdown defined by appropriate tolerances is involved.

The CGS algorithm is based on the discovery that if two steps of the BCG are combined into one in a different manner, so that the algorithm is "squared," then multiplication by A^* can be avoided. The result is a "transpose-free" method that sometimes converges up to twice as quickly as BCG, though the convergence is also twice as erratic.

The QMR algorithm is based on the observation that although three-term recurrences cannot be used to minimize $\|r_n\|$, they can be used to minimize a different, data-dependent norm that in practice is usually not so far from $\|r_n\|$. This may have a pronounced effect on the smoothness of convergence, significantly reducing the impact of rounding errors. The Bi-CGSTAB algorithm is another method that also significantly smooths the convergence rate of BCG, and TFQMR is a variant of QMR that combines its smooth convergence with the avoidance of the need for A^*.

Most recently, efforts have been directed at combining these three virtues of smoothed convergence curves, look-ahead to avoid breakdowns, and transpose-free operation. So far, all three have not yet been combined fully satisfactorily in a single algorithm, but this research area is young.

Exercises

39.1. Consider a problem $Ax = b$ for the matrix (39.5) of dimension m.

(a) Show that the singular values are all 1 and that this implies that CGN converges in one step.

(b) Show that the eigenvalues are the mth roots of unity and that this implies that GMRES requires m steps to converge for general b.

(c) This matrix A has so much structure that one does not need to consider eigenvalues or singular values to understand its convergence behavior. In particular, explain by an elementary argument why GMRES takes m steps to converge for the right-hand side $b = (1, 0, 0, \ldots, 0)^T$.

39.2. As a converse to Exercise 39.1, devise an example of a matrix of arbitrary dimension m with almost the opposite property: GMRES converges in two steps, but CGN requires m steps.

39.3. (a) If A is hermitian and s_0 is chosen appropriately, Algorithm 39.1 reduces to Algorithm 38.1. Confirm this statement and determine the appropriate s_0.

(b) Suppose A is a complex matrix that is symmetric but not hermitian. Show that with a different choice of s_0, Algorithm 39.1 again reduces to an iteration involving just one three-term recurrence.

39.4. Figure 39.2 illustrated that if the convergence curve for a biorthogonalization method has spikes in it, this may affect the attainable accuracy in floating point arithmetic. Without trying to be rigorous, explain why this is so, and comment on the analogy with growth factors in Gaussian elimination (Lecture 22).

39.5. Which of CG, GMRES, CGN, or BCG would you expect to be most effective for the following $m \times m$ problems $Ax = b$, and why?

(a) A dense nonhermitian matrix with $m = 10^4$, all but three of whose eigenvalues are approximately equal to -1.

(b) The same, but with all but three of the eigenvalues scattered about the region $-10 \le \text{Real}(\lambda) \le 10$, $-1 \le \text{Imag}(\lambda) \le 1$.

(c) A sparse nonhermitian matrix with $m = 10^6$ but only 10^7 nonzero entries, with eigenvalues as in (a).

(d) A sparse hermitian matrix with $m = 10^5$ whose eigenvalues are scattered through the interval $[1, 100]$.

(e) The same, except for outlying eigenvalues at 0.01 and 10,000.

(f) The same, but with additional outliers at -1, -10, and -100.

(g) A sparse, normal matrix with $m = 10^5$ whose eigenvalues are complex numbers scattered about the annulus $1 \le |\lambda| \le 2$.

Lecture 40. Preconditioning

The convergence of a matrix iteration depends on the properties of the matrix—the eigenvalues, the singular values, or sometimes other information. One of the developments that made it possible for these methods to take off in the 1970s and 1980s was the discovery that in many cases, the problem of interest can be transformed so that the properties of the matrix are improved drastically. This process of "preconditioning" is essential to most successful applications of iterative methods.

Preconditioners for $Ax = b$

In the abstract, the idea of preconditioning a system of equations is elementary. Suppose we wish to solve an $m \times m$ nonsingular system

$$Ax = b. \tag{40.1}$$

For any nonsingular $m \times m$ matrix M, the system

$$M^{-1}Ax = M^{-1}b \tag{40.2}$$

has the same solution. If we solve (40.2) iteratively, however, the convergence will depend on the properties of $M^{-1}A$ instead of those of A. If this *preconditioner* M is well chosen, (40.2) may be solved much more rapidly than (40.1).

For this idea to be useful, of course, it must be possible to compute the operation represented by the product $M^{-1}A$ efficiently. As usual in numerical

313

linear algebra, this will not mean an explicit construction of the inverse M^{-1}, but the solution of systems of equations of the form

$$My = c. \tag{40.3}$$

Two extreme cases come quickly to mind. If $M = A$, then (40.3) is the same as (40.1), so applying the preconditioner is as hard as solving the original problem, and nothing has been gained. If $M = I$, then (40.2) is the same as (40.1), so applying the preconditioner is a triviality, but it accomplishes nothing. Between these extremes lie the useful preconditioners, structured enough so that (40.3) can be solved quickly, but close enough to A in some sense that an iteration for (40.2) converges more quickly than an iteration for (40.1).

What does it mean for M to be "close enough to A?" Answering this question is the matter that has occupied our attention throughout this part of the book. If the eigenvalues of $M^{-1}A$ are close to 1 and $\|M^{-1}A - I\|_2$ is small, then any of the iterations we have discussed can be expected to converge quickly (Exercise 40.1). However, preconditioners that do not satisfy such a strong condition may also perform well. For example, the eigenvalues of $M^{-1}A$ could be clustered about a number other than 1, and there might be some outlier eigenvalues far from the others. For another example, if CGN is the iteration, it is enough for the singular values of $M^{-1}A$ to be clustered, not the eigenvalues. Detailed answers to questions of convergence rates depend, as always, on problems of polynomial approximation in the complex plane; all that changes for the analysis of preconditioners as opposed to basic iterations is that now it is the properties of $M^{-1}A$ rather than A that are of interest.

For most problems involving iterations other than CGN, fortunately, a simple rule of thumb is adequate. *A preconditioner M is good if $M^{-1}A$ is not too far from normal and its eigenvalues are clustered.*

Left, Right, and Hermitian Preconditioners

What we have described may be more precisely termed a *left preconditioner*. Another idea is to transform $Ax = b$ into $AM^{-1}y = b$, with $x = M^{-1}y$, in which case M is called a *right preconditioner*. Both left and right preconditioners are used in practice, and sometimes both are used at once. To keep the discussion simple, we shall confine our attention to the former.

If A is hermitian positive definite, then it is usual to preserve this property in preconditioning. Suppose M is also hermitian positive definite, with $M = CC^*$ for some C. Then (40.1) is equivalent to

$$[C^{-1}AC^{-*}]C^*x = C^{-1}b. \tag{40.4}$$

The matrix in brackets is hermitian positive definite, so this equation can be solved by CG or related iterations. At the same time we observe that since $C^{-1}AC^{-*}$ is similar to $C^{-*}C^{-1}A = M^{-1}A$, it is enough to examine the eigenvalues of the nonhermitian matrix $M^{-1}A$ to investigate convergence.

Example

Figure 40.1 presents an example of a preconditioned CG iteration for a symmetric positive definite matrix. The matrix A is adapted from Exercise 36.3: it is the 1000×1000 symmetric matrix whose entries are all zero except for $a_{ij} = 0.5 + \sqrt{i}$ on the diagonal, $a_{ij} = 1$ on the sub- and superdiagonals, and $a_{ij} = 1$ on the 100th sub- and superdiagonals, i.e., for $|i - j| = 100$. The right-hand side is $b = (1, 1, \ldots, 1)^T$. As the figure shows, a straight CG iteration for this matrix converges slowly, achieving about five-digit residual reduction after forty iterations. Since the matrix is very sparse, this is an improvement over a direct method, but one would like to do better.

As it happens, we can do much better with a simple diagonal preconditioner. Take $M = \text{diag}(A)$, the diagonal matrix with entries $m_{ii} = 0.5 + \sqrt{i}$. To preserve symmetry, set $C = \sqrt{M}$ and consider a new iteration preconditioned as in (40.4). The figure shows that thirty steps of the iteration now give convergence to fifteen digits.

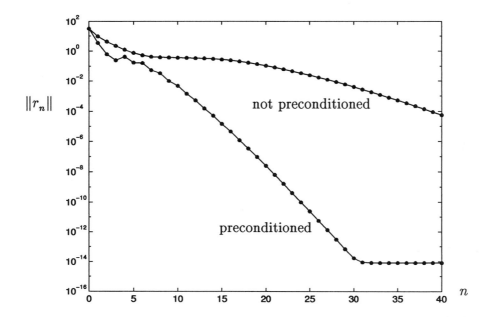

Figure 40.1. *CG and preconditioned CG convergence curves for the* 1000×1000 *sparse matrix A described in the text (the matrix of Exercise 36.3 plus* $0.5I$*).*

Survey of Preconditioners for $Ax = b$

The preconditioners used in practice are sometimes as simple as this one, but they are often far more complicated. Rather than consider one or two examples in detail, we shall take the opposite course and survey at a high level the wide range of preconditioning ideas that have been found useful over the years. Details can be found in the references listed in the Notes.

Diagonal scaling or Jacobi. Perhaps the most important preconditioner is the one just mentioned in the example: $M = \text{diag}(A)$, provided that this matrix is nonsingular. For certain problems, this transformation alone is enough to make a slow iteration into a fast one. More generally, one may take $M = \text{diag}(c)$ for a suitably chosen vector $c \in \mathbb{C}^m$. It is a hard mathematical problem to determine a vector c such that $\kappa(M^{-1}A)$ is exactly minimized, but fortunately, nothing like the exact minimum is needed in practice, and in any case, as the rule of thumb above shows, there is more to preconditioning than minimizing the condition number.

Incomplete Cholesky or LU factorization. Another star preconditioner is the one that made the idea of preconditioning famous in the 1970s. Suppose A is sparse, having just a few nonzeros per row. The difficulty with methods such as Gaussian elimination or Cholesky factorization is that these processes destroy zeros, so that if $A = R^*R$, for example, then the factor R will usually not be very sparse. However, suppose a matrix $\tilde{R}$ is computed by Cholesky-like formulas but allowed to have nonzeros only in positions where A has nonzeros, and we define $M = \tilde{R}^*\tilde{R}$. This *incomplete Cholesky* preconditioner may be highly effective for some problems; the acronym *ICCG* for incomplete Cholesky conjugate gradients is used. Similar *ILU* or *incomplete LU* preconditioners are useful in nonsymmetric cases. Numerous variants of the idea of incomplete factorization have been proposed and developed extensively.

These two examples of preconditioners are defined without reference to the origin of the underlying problem $Ax = b$. The best general advice one can give for designing preconditioners, however, is to examine that problem and take advantage of its structure. If it were simpler in a certain way, one asks, could it be solved quickly? If so, that simpler version of the problem may be an effective preconditioner. Most of our remaining examples are in this category.

Coarse-grid approximation. A discretization of a partial differential or integral equation on a fine grid may lead to a huge system of equations. The analogous discretization on a coarser grid, however, may lead to a small system that is easy to solve. If a method can be found to transfer solutions on the coarse grid to the fine grid and back again, e.g. by interpolation, then a powerful preconditioner may be obtained of the following schematic form:

$$M = \langle \text{transfer to fine grid} \rangle \circ A_{\text{coarse}} \circ \langle \text{transfer to coarse grid} \rangle. \quad (40.5)$$

Typically a preconditioner of this kind does a good job of handling the low-frequency components of the original problem, leaving the high frequencies to be treated by the Krylov subspace iteration. When this technique is iterated, resulting in a sequence of coarser and coarser grids, we obtain the idea of *multigrid iteration*.

Local approximation. A coarse-grid approximation takes into account some of the larger-scale structure of a problem while ignoring some of the finer structure. A kind of a converse to this idea is relevant to problems $Ax = b$ where A represents coupling between elements both near and far from one another. The elements may be physical objects such as particles, or they may be numerical objects such as the panels introduced in a boundary element discretization. In any case, it may be worth considering the operator M analogous to A but with the longer-range interactions omitted—a short-range approximation to A. In the simplest cases of this kind, M may consist simply of a few of the diagonals of A near the main diagonal, making this a generalization of the idea of a diagonal preconditioner.

Block preconditioners and domain decomposition. Throughout numerical linear algebra, most algorithms expressed in terms of the scalar entries of a matrix have analogues involving block matrices. An example is that a diagonal or Jacobi preconditioner may be generalized to block-diagonal or block-Jacobi form. This is another kind of local approximation, in that local effects within certain components are considered while connections to other components are ignored. In the past decade ideas of this kind have been widely generalized in the field of *domain decomposition*, in which solvers for certain subdomains of a problem are composed in flexible ways to form preconditioners for the global problem. These methods combine mathematical power with natural parallelizability.

Low-order discretization. Often a differential or integral equation is discretized by a higher-order method such as a fourth-order finite difference formula or a spectral method, bringing a gain in accuracy but making the discretization stencils bigger and the matrix less sparse. A lower-order approximation of the same problem, with its sparser matrix, may be an effective preconditioner. Thus, for example, one commonly encounters finite difference and finite element preconditioners for spectral discretizations.

Constant-coefficient or symmetric approximation. Special techniques, like fast Poisson solvers, are available for certain partial differential equations with constant coefficients. For a problem with variable coefficients, a constant-coefficient approximation implemented by a fast solver may make a good preconditioner. Analogously, if a differential equation is not self-adjoint but is close in some sense to a self-adjoint equation that can be solved more easily, then the latter may sometimes serve as a preconditioner.

Splitting of a multi-term operator. Many applications involve combinations of physical effects, such as the diffusion and convection that combine to make up the Navier–Stokes equations of fluid mechanics. The linear algebra result may be a matrix problem $Ax = b$ with $A = A_1 + A_2$ (or with more than two terms, of course), often embedded in a nonlinear iteration. If A_1 or A_2 is easily invertible, it may serve as a good preconditioner.

Dimensional splitting or ADI. Another kind of splitting takes advantage of the fact that an operator such as the Laplacian in two or three dimensions is composed of analogous operators in each of the dimensions separately. This idea may form the basis of a preconditioner, and in one form goes by the name of *ADI* or *alternating direction implicit* methods.

One step of a classical iterative method. In this book we have not discussed the "classical iterations" such as Jacobi, Gauss–Seidel, SOR, or SSOR, but one or more steps of these iterations—particularly Jacobi and SSOR—often serve excellently as preconditioners. This is also one of the key ideas behind multigrid methods.

Periodic or convolution approximation. Throughout the mathematical sciences, boundary conditions are a source of analytical and computational difficulty. If only there were no boundary conditions, so that the problem were posed on a periodic domain! This idea can sometimes be the basis of a good preconditioner. In the simplest linear algebra context, it becomes the idea of preconditioning a problem involving a Toeplitz matrix A (i.e., $a_{i,j} = a_{i-j}$) by a related *circulant matrix* M ($m_{i,j} = m_{(i-j)(\mathrm{mod}\,m)}$), which can be inverted in $O(m \log m)$ operations by a fast Fourier transform. This is a particularly well studied example in which $M^{-1}A$ may be far from the identity in norm but have highly clustered eigenvalues.

Unstable direct method. Certain numerical methods, such as Gaussian elimination without pivoting, deliver inaccurate answers because of instability. If the unstable method is fast, however, why not use it as a preconditioner? This is the "fly by wire" approach to numerical computation: solve the problem carelessly but quickly, and embed that solution in a robust control system. It is a powerful idea.

Polynomial preconditioners. Finally, we mention a technique that is different from the others in that it is essentially A^{-1} rather than A itself that is approximated by the preconditioner. A polynomial preconditioner is a matrix polynomial $M^{-1} = p(A)$ with the property that $p(A)A$ has better properties for iteration than A itself. For example, $p(A)$ might be obtained from the first few terms of the Neumann series $A^{-1} = I + (I - A) + (I - A)^2 + \cdots$, or from some other expression, often motivated by approximation theory in the complex plane. Implementation is easy, based on the same "black box" used for the Krylov subspace iteration itself, and the coefficients of the preconditioner may sometimes be determined adaptively.

Preconditioners for Eigenvalue Problems

Though the idea has been developed more recently and is not yet as famous, preconditioners can be effective for eigenvalue problems as well as systems of equations. Some of the best-known techniques in this area are *polynomial acceleration*, analogous to the polynomial preconditioning just described for systems of equations, *shift-and-invert Arnoldi* or the related *rational Krylov iteration*, which employ rational functions of A instead of polynomials, and the *Davidson* and *Jacobi–Davidson* methods, based on a kind of diagonal preconditioner. For example, shift-and-invert and rational Krylov methods are based on the fact that if $r(z)$ is a rational function and $\{\lambda_j\}$ are the eigenvalues of A, then the eigenvalues of $r(A)$ are $\{r(\lambda_j)\}$. If $r(A)$ can be computed with reasonable speed and its eigenvalues are better distributed for iteration than those of A, this may be a route to fast calculation of eigenvalues.

A Closing Note

In ending this book with the subject of preconditioners, we find ourselves at the philosophical center of the scientific computing of the future. The traditional view of computer scientists is that a computational problem is finite: after a short or long calculation, one obtains the solution exactly. Over the years, however, this view has come to be appropriate to fewer and fewer problems. The best methods for large-scale computational problems are usually approximate ones, methods that obtain a satisfactorily accurate solution in a short time rather than an exact one in a much longer or infinite time. Numerical analysis is indeed a branch of analysis, primarily, not algebra—even when the problems to be solved are from linear algebra. Further speculations on this phenomenon are presented in the Appendix.

Nothing will be more central to computational science in the next century than the art of transforming a problem that appears intractable into another whose solution can be approximated rapidly. For Krylov subspace matrix iterations, this is preconditioning. For the great range of computational problems, both continuous and discrete, we can only guess where this idea will take us.

Exercises

40.1. Suppose $A = M - N$, where M is nonsingular. Suppose $\|I - M^{-1}N\|_2 \leq 1/2$, and M is used as a preconditioner as in (40.2).

(a) Show that if GMRES is applied to this preconditioned problem, then the residual norm is guaranteed to be six orders of magnitude smaller, or better, after twenty steps.

(b) How many steps of CGN are needed for the same guarantee?

40.2. Show that if a matrix A and a preconditioner M are hermitian positive definite, then the same CG convergence rate is obtained whether M is used as a left preconditioner or a right preconditioner. Explain why this result does not hold for nonhermitian matrices and iterations such as GMRES, CGN, or BCG.

Appendix. The Definition of Numerical Analysis

by Lloyd N. Trefethen*

What is numerical analysis? I believe that this is more than a philosophical question. A certain wrong answer has taken hold among both outsiders to the field and insiders, distorting the image of a subject at the heart of the mathematical sciences.

Here is the wrong answer:

$$\text{Numerical analysis is the study of rounding errors.} \qquad \text{(D1)}$$

The reader will agree that it would be hard to devise a more uninviting description of a field. Rounding errors are inevitable, yes, but they are complicated and tedious and —*not fundamental*. If (D1) is a common perception, it is hardly surprising that numerical analysis is widely regarded as an unglamorous subject. In fact, mathematicians, physicists, and computer scientists have all tended to hold numerical analysis in low esteem for many years—a most unusual consensus.

---

*This essay is reprinted from the November 1992 issue of *SIAM News*. It was reprinted previously in the March/April 1993 issue of the *Bulletin of the Institute of Mathematics and Its Applications*.

Of course nobody believes or asserts (D1) quite as baldly as written. But consider the following opening chapter headings from some standard numerical analysis texts:

Isaacson & Keller (1966): 1. Norms, arithmetic, and well-posed computations.

Hamming (1971): 1. Roundoff and function evaluation.

Dahlquist & Björck (1974): 1. Some general principles of numerical calculation.
2. How to obtain and estimate accuracy....

Stoer & Bulirsch (1980): 1. Error analysis.

Conte & de Boor (1980): 1. Number systems and errors.

Atkinson (1987): 1. Error: its sources, propagation, and analysis.

Kahaner, Moler & Nash (1989): 1. Introduction.
2. Computer arithmetic and computational errors.

"Error" ... "roundoff" ... "computer arithmetic" — these are the words that keep reappearing. What impression does an inquisitive college student get upon opening such books? Or consider the definitions of numerical analysis in some dictionaries:

Webster's New Collegiate Dictionary (1973): "The study of quantitative approximations to the solutions of mathematical problems including consideration of the errors and bounds to the errors involved."

Chambers 20th Century Dictionary (1983): "The study of methods of approximation and their accuracy, etc."

The American Heritage Dictionary (1992): "The study of approximate solutions to mathematical problems, taking into account the extent of possible errors."

"Approximations" ... "accuracy" ... "errors" again. It seems to me that these definitions would serve most effectively to deter the curious from investigating further.

The singular value decomposition (SVD) affords another example of the perception of numerical analysis as the science of rounding errors. Although the roots of the SVD go back more than 100 years, it is mainly since the 1960s, through the work of Gene Golub and other numerical analysts, that it has achieved its present degree of prominence. The SVD is as fundamental an idea as the eigenvalue decomposition; it is the natural language for discussing all kinds of questions of norms and extrema involving nonsymmetric matrices or operators. Yet today, thirty years later, most mathematical scientists and even many applied mathematicians do not have a working knowledge of the SVD. Most of them have heard of it, but the impression seems to be widespread that the SVD is just a tool for combating rounding errors. A glance at a few numerical analysis textbooks suggests why. In one case after another, the SVD is buried deep in the book, typically in an advanced section on rank-deficient least squares problems, and recommended mainly for its stability properties.

I am convinced that consciously or unconsciously, many people think that (D1) is at least half true. In actuality, it is a very small part of the truth. And although there are historical explanations for the influence of (D1) in the past, it is a less appropriate definition today and is destined to become still less appropriate in the future.

I propose the following alternative definition with which to enter the new century:

$$\textit{Numerical analysis is the study of algorithms} \qquad \text{(D2)}$$
$$\textit{for the problems of continuous mathematics.}$$

Boundaries between fields are always fuzzy; no definition can be perfect. But it seems to me that (D2) is as sharp a characterization as you could come up with for most disciplines.

The pivotal word is *algorithms*. Where was this word in those chapter headings and dictionary definitions? Hidden between the lines, at best, and yet surely this is the center of numerical analysis: devising and analyzing algorithms to solve a certain class of problems.

These are the problems of *continuous mathematics*. "Continuous" means that real or complex variables are involved; its opposite is "discrete." A dozen qualifications aside, numerical analysts are broadly concerned with continuous problems, while algorithms for discrete problems are the concern of other computer scientists.

Let us consider the implications of (D2). First of all it is clear that since real and complex numbers cannot be represented exactly on computers, (D2) implies that part of the business of numerical analysis must be to approximate them. This is where the rounding errors come in. Now for a certain set of problems, namely the ones that are solved by algorithms that take a finite number of steps, that is all there is to it. The premier example is Gaussian elimination for solving a linear system of equations $Ax = b$. To understand Gaussian elimination, you have to understand computer science issues such as operation counts and machine architectures, and you have to understand the propagation of rounding errors—stability. That's all you have to understand, and if somebody claims that (D2) is just a more polite restatement of (D1), you can't prove him or her wrong with the example of Gaussian elimination.

But most problems of continuous mathematics cannot be solved by finite algorithms! Unlike $Ax = b$, and unlike the discrete problems of computer science, most of the problems of numerical analysis could not be solved exactly even if we could work in exact arithmetic. Numerical analysts know this, and mention it along with a few words about Abel and Galois when they teach algorithms for computing matrix eigenvalues. Too often they forget to mention that the same conclusion extends to virtually any problem with a nonlinear term or a derivative in it—zerofinding, quadrature, differential equations, integral equations, optimization, you name it.

Even if rounding errors vanished, numerical analysis would remain. Approximating mere numbers, the task of floating point arithmetic, is indeed a rather small topic and maybe even a tedious one. The deeper business of numerical analysis is approximating unknowns, not knowns. Rapid convergence of approximations is the aim, and the pride of our field is that, for many problems, we have invented algorithms that converge exceedingly fast.

These points are sometimes overlooked by enthusiasts of symbolic computing, especially recent converts, who are apt to think that the existence of Maple or Mathematica renders Matlab and Fortran obsolete. It is true that rounding errors can be made to vanish in the sense that in principle, any finite sequence of algebraic operations can be represented exactly on a computer by means of appropriate symbolic operations. Unless the problem being solved is a finite one, however, this only defers the inevitable approximations to the end of the calculation, by which point the quantities one is working with may have become extraordinarily cumbersome. Floating point arithmetic is a name for numerical analysts' habit of doing their pruning at every step along the way of a calculation rather than in a single act at the end. Whichever way one proceeds, in floating point or symbolically, the main problem of finding a rapidly convergent algorithm is the same.

In summary, it is a corollary of (D2) that numerical analysis is concerned with rounding errors and also with the deeper kinds of errors associated with convergence of approximations, which go by various names (truncation, discretization, iteration). Of course one could choose to make (D2) more explicit by adding words to describe these approximations and errors. But once words begin to be added it is hard to know where to stop, for (D2) also fails to mention some other important matters: that these algorithms are implemented on computers, whose architecture may be an important part of the problem; that reliability and efficiency are paramount goals; that some numerical analysts write programs and others prove theorems; and most important, that all of this work is *applied*, applied daily and successfully to thousands of applications on millions of computers around the world. "The problems of continuous mathematics" are the problems that science and engineering are built upon; without numerical methods, science and engineering as practiced today would come quickly to a halt. They are also the problems that preoccupied most mathematicians from the time of Newton to the twentieth century. As much as any pure mathematicians, numerical analysts are the heirs to the great tradition of Euler, Lagrange, Gauss and the rest. If Euler were alive today, he wouldn't be proving existence theorems.

* * *

Ten years ago, I would have stopped at this point. But the evolution of computing in the past decade has given the difference between (D1) and (D2) a new topicality.

Let us return to $Ax = b$. Much of numerical computation depends on linear algebra, and this highly developed subject has been the core of numerical analysis since the beginning. Numerical linear algebra served as the subject with respect to which the now standard concepts of stability, conditioning, and backward error analysis were defined and sharpened, and the central figure in these developments, from the 1950s to his death in 1986, was Jim Wilkinson.

I have mentioned that $Ax = b$ has the unusual feature that it can be solved in a finite sequence of operations. In fact, $Ax = b$ is more unusual than that, for the standard algorithm for solving it, Gaussian elimination, turns out to have extraordinarily complicated stability properties. Von Neumann wrote 180 pages of mathematics on this topic; Turing wrote one of his major papers; Wilkinson developed a theory that grew into two books and a career. Yet the fact remains that for certain $n \times n$ matrices, Gaussian elimination with partial pivoting amplifies rounding errors by a factor of order 2^n, making it a useless algorithm in the worst case. It seems that Gaussian elimination works in practice because the set of matrices with such behavior is vanishingly small, but to this day, nobody has a convincing explanation of why this should be so.[†]

In manifold ways, then, Gaussian elimination is atypical. Few numerical algorithms have such subtle stability properties, and certainly no other was scrutinized in such depth by von Neumann, Turing, and Wilkinson. The effect? Gaussian elimination, which should have been a sideshow, lingered in the spotlight while our field was young and grew into the canonical algorithm of numerical analysis. Gaussian elimination set the agenda, Wilkinson set the tone, and the distressing result has been (D1).

Of course there is more than this to the history of how (D1) acquired currency. In the early years of computers, it was inevitable that arithmetic issues would receive concerted attention. Fixed point computation required careful thought and novel hardware; floating point computation arrived as a second revolution a few years later. Until these matters were well understood it was natural that arithmetic issues should be a central topic of numerical analysis, and, besides this, another force was at work. There is a general principle of computing that seems to have no name: *the faster the computer, the more important the speed of algorithms.* In the early years, with the early computers, the dangers of instability were nearly as great as they are today, and far less familiar. The gaps between fast and slow algorithms, however, were narrower.

[†] This was written before the results of Lecture 22 were developed.

A development has occurred in recent years that reflects how far we have come from that time. Instances have been accumulating in which, even though a finite algorithm exists for a problem, an infinite algorithm may be better. The distinction that seems absolute from a logical point of view turns out to have little importance in practice—and in fact, Abel and Galois notwithstanding, large-scale matrix eigenvalue problems are about as easy to solve in practice as linear systems of equations. For $Ax = b$, iterative methods are becoming more and more often the methods of choice as computers grow faster, matrices grow larger and less sparse (because of the advance from 2D to 3D simulations), and the $O(N^3)$ operation counts of the usual direct (= finite) algorithms become ever more painful. The name of the new game is *iteration with preconditioning*. Increasingly often it is not optimal to try to solve a problem exactly in one pass; instead, solve it approximately, then iterate. Multigrid methods, perhaps the most important development in numerical computation in the past twenty years, are based on a recursive application of this idea.

Even direct algorithms have been affected by the new manner of computing. Thanks to the work of Skeel and others, it has been noticed that the expense of making a direct method stable—say, of pivoting in Gaussian elimination—may in certain contexts be cost-ineffective. Instead, skip that step—solve the problem directly but unstably, then do one or two steps of iterative refinement. "Exact" Gaussian elimination becomes just another preconditioner!

Other problems besides $Ax = b$ have undergone analogous changes, and the famous example is linear programming. Linear programming problems are mathematically finite, and for decades, people solved them by a finite algorithm: the simplex method. Then Karmarkar announced in 1984 that iterative, infinite algorithms are sometimes better. The result has been controversy, intellectual excitement, and a perceptible shift of the entire field of linear programming away from the rather anomalous position it has traditionally occupied towards the mainstream of numerical computation.

I believe that the existence of finite algorithms for certain problems, together with other historical forces, has distracted us for decades from a balanced view of numerical analysis. Rounding errors and instability are important, and numerical analysts will always be the experts in these subjects and at pains to ensure that the unwary are not tripped up by them. But our central mission is to compute quantities that are typically uncomputable, from an analytical point of view, and to do it with lightning speed. For guidance to the future we should study not Gaussian elimination and its beguiling stability properties, but the diabolically fast conjugate gradient iteration—or Greengard and Rokhlin's $O(N)$ multipole algorithm for particle simulations—or the exponential convergence of spectral methods for solving certain PDEs—or the convergence in $O(1)$ iteration achieved by multigrid methods for many kinds of problems—or even Borwein and Borwein's magical AGM iteration for determining 1,000,000 digits of π in the blink of an eye. *That* is the heart of numerical analysis.

Notes. Many people, too numerous to name, provided comments on drafts of this essay. Their suggestions led me to many publications that I would otherwise not have found.

I do not claim that any of the ideas expressed here are entirely new. In fact, 30 years ago, in his *Elements of Numerical Analysis*, Peter Henrici defined numerical analysis as "the theory of constructive methods in mathematical analysis." Others have expressed similar views; Joseph Traub (*Communications of the ACM*, 1972), for example, defined numerical analysis as "the analysis of continuous algorithms." For that matter, both the Random House and the Oxford English dictionaries offer better definitions than the three quoted here.

And should the field be called "numerical analysis," "scientific computing," or something else entirely? ("mathematical engineering?"). That is another essay.

Notes

There are a number of textbooks and monographs on numerical linear algebra, and a particularly notable group have been appearing in the second half of the 1990s. Rather than give a full survey, we highlight three current books that every reader who wishes to go further with this subject should be aware of:

- Golub and Van Loan, *Matrix Computations*, 3rd ed. [GoVa96],
- Higham, *Accuracy and Stability of Numerical Algorithms* [Hig96],
- Demmel, *Applied Numerical Linear Algebra* [Dem97].

The book by Golub and Van Loan, in its earlier editions, has long been the bible of this field—encyclopedic in its coverage and its references to the literature. The book by Higham is another encyclopedic treatment, exceedingly careful about details, with an emphasis on stability but full of algorithmic information and insights of all kinds. The book by Demmel has almost the same title as the present volume but is entirely different in style, being more focused on latest developments and considerations of computer architecture, less on mathematical foundations.

Other texts on numerical linear algebra include [Cia89], [Dat95], [GMW91], [Hag88], [Ste73], and [Wat91].

Excellent texts are also available on various more specialized subjects, including least squares, eigenvalue problems, and iterative methods. These are listed in the appropriate paragraphs below. For direct sparse matrix methods, not covered in this book, two standard texts are [GeLi81] and

[DER86]. For software, also not covered here, some of the landmark contributions are LAPACK [And95] and its predecessors EISPACK [Smi76] and LINPACK [DBMS79], the Basic Linear Algebra Subprograms (BLAS) developed for simplifying coding of linear algebra operations and maximizing efficiency on particular machines [DDDH90], the MATLAB repository managed by The MathWorks, Inc. (http://www.mathworks.com), and the Netlib automatic software distribution system (http://www.netlib.org), which has processed about 13 million requests as of this writing.

Finally, we mention that when it comes to matters of nonnumerical linear algebra, our own habit is always to turn first to the two remarkable volumes by Horn and Johnson, [HoJo85] and [HoJo91].

We turn now to notes on the individual Lectures of this text.

Lecture 1. Matrix-Vector Multiplication. It is impossible to understand the spirit of twentieth-century numerical linear algebra without learning to think in terms of operations on rows and columns of matrices. Virtually all the standard algorithms are normally conceived in this way, though modifications appear when it comes to exploiting sparsity.

In principle, the fastest algorithms for many problems may be recursive ones that involve manipulations of submatrices and thus require a different way of thinking. For example, Klyuyev and Kokovkin-Shcherbak showed in 1965 that solving an $m \times m$ system of equations solely by row and column operations requires $O(m^3)$ operations [KlKo65], but the subsequent work of Strassen and others (Lecture 32) improved this to $O(m^{2.81})$ and below by recursive fracturing of the matrix into smaller blocks [Str69]. The divide-and-conquer algorithm for computing eigenvalues (Lecture 30) is another example where row and column operations are not enough. It is possible that in the next century, the importance of such algorithms will grow to the point that new ways of thinking will come to prevail in numerical linear algebra, but we are not there yet.

Determinants were central to linear algebra in the nineteenth century, but their importance has diminished. For one perspective on the reasons, see [Axl95].

In one form or another, the material of this first lecture can be found in numerous textbooks, such as [Str88]. If there is another text that takes square matrices by default to have dimensions $m \times m$ rather than $n \times n$, however, we have not found it.

Lecture 2. Orthogonal Vectors and Matrices. The content of this lecture is standard material in linear algebra, which generalizes in the infinite-dimensional case to standard material in the theory of Hilbert spaces.

Algorithms based on orthogonal matrices became widespread in the early 1960s with the work of Householder, Francis, Givens, Wilkinson, Golub, and others, as it came to be recognized that such algorithms combine theoretical elegance with outstanding properties of numerical stability. The rapid spread of this point of view can be seen in Wilkinson's 1965 monograph [Wil65] and in the classic textbooks [Ste73] and [LaHa95] (first published in 1974).

Lecture 3. Norms. Though the use of norms has long been a feature of functional analysis, it has been slower to become standard in linear algebra, and even today, these ideas are often not emphasized in nonnumerical linear algebra texts and courses. The explanation for this is probably that linear algebra is historically rooted in algebra rather than in analysis, and hence makes sense in vector spaces more general than $\mathbb{R}^m$ and $\mathbb{C}^m$. Most scientific applications, however, lead to real or complex numbers, for which analysis is meaningful as well as algebra. In any application with a notion of "size," norms are probably useful. One certainly needs them if one wants to talk about convergence.

The importance of norms in numerical linear algebra was emphasized in the 1964 book by Householder [Hou64] and in the brief 1967 text on Gaussian elimination and related matters by Forsythe and Moler [FoMo67].

In infinite dimensions, the use of dual norms as in Exercise 3.6 becomes the Hahn–Banach theorem [Kat76].

Lectures 4 and 5. The Singular Value Decomposition. The SVD for matrices was discovered independently by Beltrami (1873) and Jordan (1874) and again by Sylvester (1889), and related work was done by Autonne (1915), Tagaki (1925), Williamson (1935), Eckart and Young (1939), and others. The infinite-dimensional generalization was developed in the context of integral equations by Schmidt (1907) and Weyl (1912); see [Smi70]. For historical discussions, see [HoJo91] and [Ste93].

Despite these deep roots, the SVD did not become widely known in applied mathematics until the late 1960s, when Golub and others showed that it could be computed effectively and used as the basis for many stable algorithms. Even after that time, perhaps because of numerical analysts' preoccupation with numerical stability, the mathematical world was slow to recognize the fundamental nature of the SVD. Again, the explanation may be the difference between algebra and analysis, for what makes the SVD so important is ultimately its analytic properties, as exemplified by Theorem 5.8. The importance of eigenvalues, by contrast, has been appreciated from the beginning, for eigenvalues are essentially algebraic in nature.

Closely related to the SVD is the *polar decomposition*, the representation of a matrix as a product of a symmetric positive definite matrix and a unitary matrix.

In the theory of Hilbert spaces, a compact operator is one that can be approximated by operators of finite rank, that is, one whose singular values decrease to zero.

Lecture 6. Projectors. Projectors are involved, explicitly or implicitly, whenever one expands a vector in a basis, and orthogonal projectors are one and the same as solutions of linear least squares problems. Perhaps it is unusual to make a discussion of projectors the starting point of a treatment of these matters, but only mildly so.

For a discussion of some relationships between norms of projectors and angles between complementary subspaces, see [IpMe95]. A full treatment of angles between subspaces is generally based on the *CS decomposition* [GoVa96].

Lecture 7. QR Factorization. The distinction between full and reduced QR factorizations appears wherever these ideas are applied, which means throughout numerical linear algebra, but this text is unusual in making the distinction explicit. More usually the QR factorization is defined in its full form, and columns of Q and rows of R are then stripped away as needed in applications. The same applies to the distinction between the full and reduced SVD.

The recognition of the importance of matrix factorizations for linear algebra computations is entirely a product of the computer age, beginning in the 1950s.

Concerning spectral methods for the numerical solution of partial differential equations, see [CHQZ88].

Lecture 8. Gram–Schmidt Orthogonalization. The idea of Gram (1883) and Schmidt (1907) is old and widely familiar, but its interpretation as a QR factorization is new to most students. In our view, this interpretation is an invaluable way to fix the Gram–Schmidt idea precisely in one's mind. The term QR factorization is due to Francis [Fra61].

The superiority of modified over classical Gram–Schmidt was first established by Rice (1966) and Björck (1967). Details and references are given in [Bjö96] and [Hig96].

Drawing pictures to calculate operation counts is nonstandard in respectable textbooks, since the same results are easily derived algebraically. But since we use the pictures in classroom teaching, we decided, why not include them in the book?

Lecture 9. MATLAB. As of 1996, about 150 textbooks in various fields of mathematics, science, and engineering have been published based on MATLAB, and the number is growing. Virtually all researchers in numerical linear algebra worldwide use MATLAB as their preferred programming language and environment, and in the Computer Science Department at Cornell, it is the principal language of all the numerical analysis courses. Information about MATLAB can be obtained from The MathWorks, 24 Prime Park Way, Natick, MA 01760, USA, tel. 508-647-7000, fax 508-647-7001, info@mathworks.com, http://www.mathworks.com.

Lecture 10. Householder Triangularization. Householder triangularization was introduced in a classic four-page paper in 1958 [Hou58]. (Householder reflectors themselves had been previously used as early as 1932, by

Turnbull and Aitken.) For thirty years, researchers in numerical linear algebra have gathered triennially for a conference on the state of their art, and these conferences are now known as Householder Symposia.

To make Householder reflections stable, it is not necessary to choose the sign as we have described. Alternative methods are described in [Par80] and [Hig96].

The symmetry between triangular orthogonalization and orthogonal triangularization is not novel mathematically, but as far as we are aware, it has not been stated in this epigrammatic form before.

For a beautiful and surprising connection between the modified Gram–Schmidt and Householder algorithms, see [BjPa92] or [Bjö96].

Lecture 11. Least Squares Problems. Who should get credit for the idea of least squares fitting? This question led to one of the great priority disputes in the history of mathematics, between Gauss, who invented the method in the 1790s, and Legendre, who first published it in 1805 (the same year in which Gauss invented the fast Fourier transform, which he also didn't publish). The honor was worth fighting over, as few ideas in mathematics have as far-reaching implications as least squares, but the fight brought honor to nobody; see [Sti86].

Troublesome square systems of equations, whose solutions may not seem to behave as they ought, arise frequently in discretization processes in scientific computing. The inverses of finite sections of an infinite matrix, for example, do not always converge as one might like to the sections of the infinite inverse matrix [Böt95]. Difficulties of this kind can often be avoided by looking at rectangular finite matrices instead and solving a least squares problem. This is just what was done in passing from Figure 11.1 to Figure 11.2.

The classic text by Lawson and Hanson gives a beautiful introduction to how numerical linear algebraists think about least squares problems; a lengthy appendix in the 1995 edition summarizes developments since the book's original publication in 1974 [LaHa95]. Other valuable introductions are presented in [Str88] and [GMW91]. A definitive work on numerical methods for least squares problems has recently been published by Björck [Bjö96], and it is here that one should turn for a full presentation of the state of the art.

Lecture 12. Conditioning and Condition Numbers. The idea of the condition number of a matrix was introduced in 1948 by Alan Turing [Tur48], the same Turing who founded theoretical computer science, who predicted the possibility of chemical waves long before they were discovered in the laboratory, and who contributed to the "Enigma" code-breaking effort that helped end the control of the Atlantic by German submarines in World War II.

A classic, more general paper on the subject of conditioning is [Ric66]; see also [Geu82].

The derivation of condition numbers is a special case of perturbation theory, and the definitive reference on perturbation theory for matrices and linear operators is [Kat76].

We have presented a simplified picture in that we only discuss normwise as opposed to componentwise condition numbers. For the latter increasingly important topic, see [Hig96] and [Dem97]. An example of a componentwise idea that we have omitted is that of the *Skeel condition number*, first proposed in [Ske79].

In many cases, the condition number of a well-posed problem is inversely related, at least approximately, to the distance to the nearest ill-posed problem. This point of view originated in a classic unpublished paper by Kahan [Kah72] and was developed in detail by Demmel [Dem87].

As mentioned in the text, Example 12.4 comes from Feynman [Fey85], who regrettably does not mention that the punch line of his story depends on ill-conditioning.

Concerning the ill-conditioning of roots of polynomials illustrated in Figure 12.1, two recent papers are [EdMu95] and [ToTr94], where pointers to earlier literature by Wilkinson and others can be found.

The result that Lebesgue constants for equispaced interpolation in n points grow asymptotically like $2^n/(e(n-1)\log n)$ was proved by Turetskii in 1940, but is not widely known. For historical comments, see [TrWe91].

Random matrices are of interest to statisticians and physicists as well as mathematicians, and the answers to the various parts of Exercise 12.3 can be found in [Ede88], [Gir90], [Meh91], and [TrVi97].

Lecture 13. Floating Point Arithmetic. Floating point arithmetic was first implemented as early as 1947, and from that point on, for many years, the details of the implementations by different manufacturers varied in ways hard to keep track of. The subject was simplified magnificently by the introduction and widespread adoption of the IEEE standard in the 1980s. For careful discussions of the issues involved, see [Gol91] and [Hig96].

Exercise 13.3 comes from Chapter 1 of [Dem97]. A similar plot for a sixth-order polynomial appears in Chapter 3 of [Code80]. Bob Lynch tells us that this example is due to Dave Dodson.

The results of Exercise 13.4, for which we thank Toby Driscoll, sometimes astonish people.

Lectures 14 and 15. Stability. The notion of backward stability is standard, and that of stability, reasonably so, but to define them formally via a precise interpretation of $O(\epsilon_{\text{machine}})$ is unusual. Most numerical analysts prefer to leave these ideas informal, so that they can be adapted to the particular features of different problems as needed. There are good reasons for this point of view, and we do not by any means claim that the course we have followed is the only proper one. Indeed, as mentioned in the text, for arbitrary problems of scientific computing, conditions involving $O(\epsilon_{\text{machine}})$ are probably too strict as a basis for definitions of stability.

Much the same formal definitions as ours can be found in [deJ77], a paper that has had less influence than it deserves.

Backward error analysis is one of the great ideas of numerical analysis, which made possible all the error estimates of numerical linear algebra that appear in this book. Credit for the development of this idea may be given to von Neumann and Goldstine, Turing, Givens, and Wilkinson. In recent years backward error analysis has been rediscovered by researchers in chaotic dynamical systems and developed under the name of *shadowing* [HYG88].

Lecture 16. Stability of Householder Triangularization. The astonishing difference between the low accuracy of the computed matrix factors Q and R individually and the high accuracy of their product exemplifies why backward error analysis is so powerful. In the 1950s and 1960s Wilkinson showed that similar effects occur in virtually every matrix algorithm. The first author was lucky enough to hear lectures by Wilkinson on these matters, which conveyed the wonder of such effects with unforgettable enthusiasm.

Theorems 16.1 and 16.2 are due to Wilkinson [Wil65], and proofs can also be found in §18.3 of [Hig96]. In the remainder of this book we state a number of stability theorems without proof. In most cases a proof, or a reference to another source containing a proof, can be found in [Hig96].

Lecture 17. Stability of Back Substitution. Carrying out a rounding error analysis in full detail can be deeply satisfying; some students have found this the most exciting lecture of the book. The results are originally due to Wilkinson; see [Wil61], [FoMo67], [Hig96].

Our remark at the end of this lecture indicates why we prefer to state results in terms of $O(\epsilon_{\text{machine}})$ rather than give explicit constants. Many numerical analysts feel differently, however, including N. J. Higham [Hig96], and we admit that it is reassuring to know that in most cases, explicit constants have been worked out and recorded in print.

Exercise 17.3, involving random matrices with entries ± 1, is based on [TrVi97] and subsequent developments from that paper.

Lecture 18. Conditioning of Least Squares Problems. The literature on this subject is not especially easy to read, partly because of the complication of rank-deficiency, which we have ignored. Several of the results in this area were first derived by Wedin [Wed73], and a paper by Stewart summarizes many of the key issues [Ste77]. The 1990 book by Stewart and Sun goes further, but is difficult reading [StSu90], and a good place to go for recent information is [Bjö96]. The papers [Geu82] and [Gra96] give exact condition numbers with respect to the Frobenius norm. For the 2-norm, the bottom row of Theorem 18.1 represents upper bounds; as far as we are aware, exact results are not known.

The geometric view of these conditioning questions is not always described explicitly, but one place where it is emphasized is [vdS75].

The differentiation of pseudoinverses is not useful just for stability analysis;

it also has algorithmic consequences. An influential paper in this area is [GoPe73].

Exercise 18.1 comes from [GMW91].

Lecture 19. Stability of Least Squares Algorithms. This is standard material, discussed in many books, including [Bjö96], [GoVa96], and [Hig96]. The subject of QR factorization with column pivoting is a large one belonging to the general area of *rank-revealing factorizations*; see [Bjö96] and [ChIp94].

Lecture 20. Gaussian Elimination. There is nothing unusual here except our deferral of this topic to the middle of the book. Gauss himself worked with positive definite systems around 1809; Jacobi extended the elimination idea to general matrices around 1857. The interpretation as a matrix factorization was first developed by Dwyer in 1944 [Dwy44].

Lecture 21. Pivoting. The terms "partial" and "complete" are due to Wilkinson in the 1950s, but pivoting was already being used as early as 1947 by von Neumann and Goldstine.

Numerous variants of the pivoting idea have found application in various computations of linear algebra. One example is the technique of *threshold pivoting*, in which one relaxes the pivot condition so that the pivot element need not be the largest in its column as long as it is within a prescribed factor of the largest. Though such a strategy may diminish the stability of the algorithm, it provides additional freedom that may be used to pick orderings that minimize fill-in in the treatment of sparse matrices. See [DER86].

Lecture 22. Stability of Gaussian Elimination. In the mid-1940s it was predicted by Hotelling and von Neumann and others that Gaussian elimination must be unstable because of exponentially compounding rounding errors, making the method unsuitable for problems of dimensions greater than a few dozen. By the early 1950s, computational experience had revealed that the algorithm was stable after all. Explaining this observation was a major theoretical challenge, and Wilkinson became famous for his contributions to the subject, which reduced the question of stability to the question of the size of the growth factor. Wilkinson's analysis was recorded in a landmark paper of 1961 [Wil61].

Wilkinson and his contemporaries did not address the problem of why, in practice, nothing like the worst-case growth factor is ever observed. In *The Algebraic Eigenvalue Problem* he comments, "experience suggests that though such a bound is attainable it is quite irrelevant for practical purposes" [Wil65], and similar remarks appear in texts from the 1960s to the 1990s. The first substantial paper on the behavior of growth factors was [TrSc90], which gave empirical evidence and other arguments that the phenomenon of practical stability is entirely statistical. The present lecture of this book, making the connection between large growth factors and exponentially skewed column spaces, represents the first explanation in print of this statistical phenomenon;

a fuller analysis is forthcoming.

Recently Wright [Wri93] and Foster [Fos94] have constructed examples of matrices for which Gaussian elimination is unstable which, though they apparently did not in fact arise in actual computations, plausibly might have done so.

Lecture 23. Cholesky Factorization. Cholesky factorization can be described, and programmed, in many different ways, and this lecture offers just one of the possibilities. As a method that takes advantage of a kind of structure of the matrix A (positive definiteness), Cholesky factorization is just the tip of an iceberg. Methods for all kinds of structured matrices have been devised, including symmetric indefinite, banded, arrowhead, Vandermonde, Toeplitz, Hankel, and other matrices; see [GoVa96].

As technology advances, the ingenious ideas that make progress possible tend to vanish into the inner workings of our machines, where only experts may be aware of their existence. So it often is with numerical algorithms, never of much interest to the public, yet hidden inside most of the appliances we use. Exercise 23.2 illustrates this phenomenon in a small way. Traditionally, an engineer wanting to solve a system of equations would choose the right method based on the properties of the system, but high-level tools like MATLAB's " \ " prefer to make these decisions by themselves. Still, by careful experimentation we can still deduce some of the advances in numerical analysis underlying those decisions.

Lecture 24. Eigenvalue Problems. This is all standard material, though the emphasis is different from what one would find in a nonnumerical text. For example, we mention the Schur factorization, which is important in computations, but not the Jordan canonical form, which usually is not, for reasons explained in [GoWi76].

Gerschgorin's theorem (Exercise 24.1) has many generalizations, some of which are reviewed in [BrRy91] and [BrMe94].

The abbreviations "ew" and "ev" (Exercise 24.1) are not standard, but perhaps they should be. We find them indispensable in the classroom.

Lecture 25. Overview of Eigenvalue Algorithms. Though more than thirty years old, Wilkinson's *The Algebraic Eigenvalue Problem* [Wil65] is still a valuable reference for details on all kinds of questions related to the computation of eigenvalues. For symmetric matrix problems, the 1980 book by Parlett is a standard reference and makes excellent reading [Par80]. For more recent developments, see [Dem97].

Though it is not mentioned in many textbooks, the $O(\log(|\log(\epsilon_{machine})|))$ iteration count of Exercise 25.2 applies to superlinearly converging algorithms all across scientific computing.

Lecture 26. Reduction to Hessenberg or Tridiagonal Form. The reduction of a matrix to Hessenberg form can also be carried out by nonunitary operations, and the asymptotic operation count is only half that of (26.1). In principle, nonunitary reductions are not always stable, but in practice they work very well. In the EISPACK software library of the 1970s [Smi76], nonunitary reduction was recommended as the default and unitary reduction was offered as an alternative. In the more recent LAPACK library [And95], only unitary reductions are provided for. Why is (nonunitary) Gaussian elimination the standard method for linear systems while unitary operations are standard for eigenvalue problems? Though unitary reductions are convenient for estimating eigenvalue condition numbers and related purposes, there seems to be no entirely compelling answer. The explanation may be that in view of the greater complexity of the eigenvalue problem, involving both a direct phase and an iterative one, numerical analysts have been less willing to take chances with stability.

For more on pseudospectra, including computed examples, see [Tre91] and [Tre97].

Lecture 27. Rayleigh Quotient, Inverse Iteration. Inverse iteration originated with Wielandt in the 1940s; for a history, see [Ips97]. For details on the phenomenon that an ill-conditioned matrix does not cause instability (Exercise 27.5), see [PeWi79], [Par80], or [GoVa96].

The convergence of the Rayleigh quotient iteration and its nonsymmetric generalization was analyzed in a sequence of papers by Ostrowski in the late 1950s [Ost59].

One of the best-known algorithms for computing zeros of polynomials is that of Jenkins and Traub. As pointed out in the original paper [JeTr70] and discussed also in the appendix of [ToTr94], the Jenkins–Traub iteration can be interpreted as a scheme for taking advantage of sparsity in a Rayleigh quotient iteration applied to a companion matrix, so that the work per step is reduced from $O(m^3)$ to $O(m)$.

Lectures 28 and 29. QR Algorithm. The QR algorithm was invented independently in 1961 by Francis [Fra61] and Kublanovskaya [Kub61], based on the earlier LR algorithm of Rutishauser, and came into worldwide use through the software package EISPACK [Smi76]. Our presentation is adapted from [Wat82]. Extensive discussions are given in [Par80] and [Wat91].

The computation of eigenvalues of matrices is one of the problems that has been most extensively studied by numerical analysts, and the amount of understanding incorporated in state-of-the-art software such as LAPACK [And95] is very great. Our "practical" Algorithm 28.2 certainly does not mention all the subtleties that must be addressed for robust computation. For example, when the QR algorithm is implemented in practice, the shifts are introduced in a more stable implicit manner by means of "chasing the bulge." See [Par80], [GoVa96], or [Dem97], where discussions of the properties of various shifts can also be found.

Lecture 30. Other Eigenvalue Algorithms. Jacobi's major paper on his eigenvalue algorithm appeared in 1846 [Jac46]; he used the method to find eigenvalues of a 7×7 matrix associated with the seven planets then known in the solar system. A classic modern reference is [FoHe60], and more recent developments, including the variant based on 4×4 blocks and quarternions, can be found in [Mac95] and the references therein. Because it avoids the tridiagonalization step, the Jacobi algorithm when carefully implemented is more accurate than the QR algorithm in a componentwise sense; see [DeVe92].

Divide-and-conquer algorithms were introduced by Cuppen in 1981 [Cup81] and made famous by Dongarra and Sorensen [DoSo87]. The literature since then is extensive. Some of the critical developments concerning stability, as well as the idea of acceleration via the fast multipole method, were introduced by Gu and Eisenstat; see [GuEi95] and [Dem97].

Lecture 31. Computing the SVD. The era of numerical computations of the SVD began in 1965 with the publication of a paper by Golub and Kahan [GoKa65], which recommended bidiagonalization by Householder reflections for Phase 1. The idea of applying the QR algorithm for Phase 2 is sometimes credited to the same paper, but in fact, the QR algorithm is not mentioned there, nor are the papers of Francis [Fra61] referenced. The key ideas developed very quickly in the late 1960s, however, through work by Golub, Kahan, Reinsch, and Businger.

Our discussion of alternative methods for Phase 1 is taken from [Bau94], where details concerning singular vectors as well as values can be found. For information about Phase II, see [GoVa96] and [Dem97].

Lecture 32. Overview of Iterative Methods. The history of the emergence of Krylov subspace iterative methods is fascinating. The foundations were laid in the early 1950s, but the machines of that era were too slow for these methods to be superior for most problems. Not only were they not extensively used, naturally enough, but their ultimate advantages concerning asymptotic complexity were not perceived very clearly. Nowadays, it is automatic to take note of the asymptotic complexity of algorithms; in the 1950s, it was not.

On the other hand, certain "classical iterations" such as Gauss–Seidel and SOR were used extensively in the 1950s for problems arising from discretizations of partial differential equations. We have given no attention to these methods here, as they are described in many books but are of diminishing practical importance today. A classic reference on this subject is [Var62].

For sparse direct matrix algorithms, see [GeLi81] and [DER86].

What is the dimension m of a "large" matrix, as a function of time? In recent years information on the subject has been collected by Edelman, who reported in 1994, for example, that he was unaware yet of any solutions of dense systems with $m > 100,000$, though matrices with $m = 76,800$ had been treated [Ede94].

A number of books have recently been written on iterative methods; we recommend in particular the monographs by Saad on eigenvalues [Saa92] and linear systems [Saa96] and the upcoming text on linear systems by Greenbaum [Gre97]. Other books on the subject include [Axe94], with extensive information on preconditioners, [Kel95], which emphasizes generalizations to nonlinear problems, and [Bru95], [Fis96], [Hac94], and [Wei96].

Since the 1950s it has been recognized that Krylov subspace methods are applicable to linear operators, not just matrices. An early reference in this vein is [Dan71], and a recent advanced one is [Nev93].

The Krylov idea of projection onto low-dimensional subspaces sounds analogous to one of the central ideas of numerical computation—discretization of a continuous problem so that it becomes finite-dimensional. One might ask whether this is more than an analogy, and if so, whether it might be possible to combine discretization and iteration into one process rather than separately replacing ∞ by m (discretization) and m by n (iteration). The answer is certainly yes, at least in some circumstances. However, many of the possibilities of this kind have not yet been explored, and at present, most scientific computations still keep discretization and iteration separate.

Strassen's famous paper appeared in 1969 [Str69], and pointers to the algorithms with still lower exponents represented in Figure 32.2 can be found in [Pan84] and [Hig96]. The current best exponent of 2.376 is due to Coppersmith and Winograd [CoWi90].

What we have called "the fundamental law of computer science" (p. 246) does not usually go by this name. This principle is discussed in [AHU74]; we do not know where it was first enunciated.

Lecture 33. The Arnoldi Iteration. Arnoldi's original paper was written in 1951, but his intentions were rather far from current ones [Arn51]. It took a long while for the various connections between the Arnoldi, Lanczos, CG, and other methods to be recognized.

Lecture 34. How Arnoldi Locates Eigenvalues. The convergence of the Lanczos iteration is reasonably well understood; some of the key papers are by Kaniel [Kan66], Paige [Pai71], and Saad [Saa80]. The convergence of the more general Arnoldi iteration, however, is not fully understood. For some of the results that are available, see [Saa92].

Our discussion in terms of lemniscates is nonstandard. The connection with polynomial approximation, including the notions of ideal Arnoldi and GMRES polynomials, is developed in [GrTr94]. An algorithm for computing these polynomials based on semidefinite programming is presented in [ToTr98], together with examples relating lemniscates to pseudospectra. The idea of estimating pseudospectra via the Arnoldi iteration comes from [ToTr96].

Concerning the "Note of Caution," see [TTRD93], [Tre91], and [Tre97].

Lecture 35. GMRES. The GMRES algorithm was proposed surprisingly recently, by Saad and Schultz in 1986 [SaSc86], though various related algorithms had appeared earlier.

Lecture 36. The Lanczos Iteration. The Lanczos iteration dates to 1950 [Lan50]. Though closely related to conjugate gradients, it was conceived independently. The Lanczos iteration was "rediscovered" in the 1970s, as tractable matrix problems grew to the size where it became competitive with other methods [Pai71]. A two-volume treatment was given in 1985 by Cullum and Willoughby [CuWi85].

The connection of Krylov subspace iterations with potential theory (electric charges) via polynomial approximation is well established. For a detailed analysis of what can and cannot be inferred about convergence from potential theory, see [DTT97].

Lecture 37. From Lanczos to Gauss Quadrature. Since 1969 it has been appreciated that the right way to compute Gauss quadrature nodes and weights is via tridiagonal matrix eigenvalue problems [GoWe69]. The brief presentation here describes the connection in full except for one omitted point: the relation of the weights to the first components of the eigenvectors, which can be derived from the Christoffel–Darboux formula. For information on this and other matters related to orthogonal polynomials, the classic reference is the book by Szegő [Sze75].

On p. 289 it is remarked that nth-order Newton–Cotes formulas have coefficients of order 2^n for large n. As Newton–Cotes formulas can be derived by interpolation, this is essentially the same factor 2^n mentioned in connection with Lebesgue constants in the notes on Lecture 12, above.

Lecture 38. Conjugate Gradients. The conjugate gradient iteration originated with Hestenes and Stiefel independently, but communication between the two men was established early enough (August 1951) for the original major paper on the subject, one of the great classics of numerical analysis, to be a joint effort [HeSt52]. Like the Lanczos iteration, CG was "rediscovered" in the 1970s, and soon became a mainstay of scientific computing. For the closely intertwined history of the CG and Lanczos iterations, see [GoOL89].

Much of what is known about the behavior of the CG iteration in floating point arithmetic is due to Greenbaum and her coauthors; see [Gre97].

Lecture 39. Biorthogonalization Methods. The biconjugate gradient iteration originated with Lanczos in 1952 [Lan52] and was revived (and christened) by Fletcher in 1976 [Fle76]. The other methods mentioned in the text are look-ahead Lanczos [PTL85], CGS [Son89], QMR [FrNa91], Bi-CGSTAB [vdV92], and TFQMR [Fre93]. For a survey as of 1991, see [FGN92], and for a description of the deep connections of these algorithms with orthogonal polynomials, continued fractions, Padé approximation, and other topics, see [Gut92].

For comparisons of the matrix properties that determine convergence of the various types of nonsymmetric matrix iterations, see [NRT92], where Exercises 39.1 and 39.2 are also addressed. For specific discussions of the relationships between BCG and QMR, see [FrNa91] and [CuGr96], where it is pointed out that spikes in the BCG convergence curve correspond in a precise way to flat (slow-progress) portions of the QMR convergence curve.

Lecture 40. Preconditioning. The word "preconditioning" originated with Turing in 1948, and some of the early contributions in the context of matrix iterations were due to Hestenes, Engeli, Wachspress, Evans, and Axelsson. The idea became famous in the 1970s with the introduction of incomplete factorization by Meijerink and van der Vorst [Meva77], and another influential paper of that decade was [CGO76]. For summaries of the current state of the art we recommend [Axe94] and [Saa96]. Domain decomposition is discussed in [SBG96], and the use of an unstable direct method as a preconditioner is considered in [Ske80]. The idea of circulant preconditioners for Toeplitz matrices originated with Strang [Str86] and has been widely generalized since then.

What about speeding up an iteration by changing the preconditioner adaptively at each step, just as the Rayleigh quotient shift speeds up inverse iteration from linear to cubic convergence? This idea is a promising one, and has recently been getting some attention; see [Saa96].

Preconditioners for eigenvalue problems have come into their own in the 1990s, though Davidson's original paper dates to 1975 [Dav75]; a good place to begin with these methods is [Saa92]. Polynomial acceleration devices have been developed by Chatelin [Cha93], Saad, Scott, Lehoucq and Sorensen [LeSo96], and others. Shift-and-invert Arnoldi methods have been developed by Saad and Spence, and rational Krylov iterations by Ruhe; for a recent survey see [MeRo96]. The Jacobi–Davidson algorithm was introduced by Sleijpen and van der Vorst [Slvd96].

Bibliography

[AHU74] A. V. Aho, J. E. Hopcroft, and J. D. Ullman, *The Design and Analysis of Computer Algorithms*, Addison-Wesley, Reading, MA, 1974.

[And95] E. Anderson et al., *LAPACK Users' Guide*, 2nd ed., SIAM, Philadelphia, 1995.

[Arn51] W. E. Arnoldi, *The principle of minimized iteration in the solution of the matrix eigenvalue problem*, Quart. Appl. Math. 9 (1951), 17–29.

[Axe94] O. Axelsson, *Iterative Solution Methods*, Cambridge U. Press, Cambridge, UK, 1994.

[Axl95] S. Axler, *Down with determinants*, Amer. Math. Monthly 102 (1995), 139–154.

[Bar94] R. Barrett et al., *Templates for the Solution of Linear Systems: Building Blocks for Iterative Methods*, SIAM, Philadelphia, 1994.

[Bau94] D. Bau, *Faster SVD for matrices with small m/n*, TR94-1414, Computer Science Dept., Cornell U., 1994.

[Bjö96] Å. Björck, *Numerical Methods for Least Squares Problems*, SIAM, Philadelphia, 1996.

[BjPa92] Å. Björck and C. C. Paige, *Loss and recapture of orthogonality in the modified Gram–Schmidt algorithm*, SIAM J. Matrix Anal. Appl. 13 (1992), 176–190.

[Böt95] A. Böttcher, *Infinite matrices and projection methods*, in P. Lancaster, ed., Lectures on Operator Theory and Its Applications, Amer. Math. Soc., Providence, RI, 1995.

[BrMe94] R. A. Brualdi and S. Mellendorf, *Regions in the complex plane containing the eigenvalues of a matrix*, Amer. Math. Monthly 101 (1994), 975–985.

[BrRy91] R. A. Brualdi and H. J. Ryser, *Combinatorial Matrix Theory*, Cambridge U. Press, Cambridge, UK, 1991.

[Bru95] A. M. Bruaset, *A Survey of Preconditioned Iterative Methods*, Addison-Wesley Longman, Harlow, Essex, UK, 1992.

[CHQZ88] C. Canuto, M. Y. Hussaini, A. Quarteroni, and T. A. Zang, *Spectral Methods in Fluid Dynamics*, Springer-Verlag, New York, 1988.

[ChIp94] S. Chandrasekaran and I. C. F. Ipsen, *On rank-revealing factorisations*, SIAM J. Matrix Anal. Appl. 15 (1994), 592–622.

[Cha93] F. Chatelin, *Eigenvalues of Matrices*, Wiley, New York, 1993.

[Cia89] P. G. Ciarlet, *Introduction to Numerical Linear Algebra and Optimisation*, Cambridge U. Press, Cambridge, UK, 1989.

[CGO76] P. Concus, G. H. Golub, and D. P. O'Leary, *A generalized conjugate gradient method for the numerical solution of elliptic partial differential equations*, in J. R. Bunch and D. J. Rose, eds., Sparse Matrix Computations, Academic Press, New York, 1976.

[Code80] S. D. Conte and C. de Boor, *Elementary Numerical Analysis: An Algorithmic Approach*, 3rd ed., McGraw-Hill, New York, 1980.

[CoWi90] D. Coppersmith and S. Winograd, *Matrix multiplication via arithmetic progressions*, J. Symbolic Comput. 9 (1990), 251–280.

[CuGr96] J. Cullum and A. Greenbaum, *Relations between Galerkin and norm-minimizing iterative methods for solving linear systems*, SIAM J. Matrix Anal. Appl. 17 (1996), 223–247.

[CuWi85] J. K. Cullum and R. A. Willoughby, *Lanczos Algorithms for Large Symmetric Eigenvalue Computations*, v. 1 and 2, Birkhäuser, Boston, 1985.

[Dan71] J. W. Daniel, *The Approximate Minimization of Functionals*, Prentice Hall, Englewood Cliffs, NJ, 1971.

[Dat95] B. N. Datta, *Numerical Linear Algebra and Applications*, Brooks/Cole, Pacific Grove, CA, 1995.

[Dav75] E. R. Davidson, *The iterative calculation of a few of the lowest eigenvalues and corresponding eigenvectors of large real symmetric matrices*, J. Comp. Phys. 17 (1975), 87–94.

[deJ77] L. S. de Jong, *Towards a formal definition of numerical stability*, Numer. Math. 28 (1977), 211–219.

[Dem87] J. W. Demmel, *On condition numbers and the distance to the nearest ill-posed problem*, Numer. Math. 51 (1987), 251–289.

[Dem97] J. W. Demmel, *Applied Numerical Linear Algebra*, SIAM, Philadelphia, 1997.

[DeVe92] J. Demmel and K. Veselić, *Jacobi's method is more accurate than QR*, SIAM J. Matrix Anal. Appl. 13 (1992), 1204–1245.

[DBMS79] J. J. Dongarra, J. R. Bunch, C. B. Moler, and G. W. Stewart, *LINPACK Users' Guide*, SIAM, Philadelphia, 1979.

[DDDH90] J. J. Dongarra, J. J. Du Croz, I. S. Duff, and S. J. Hammarling, *Algorithm 679. A set of level 3 basic linear algebra subprograms: Model implementation and test programs*, ACM Trans. Math. Software 16 (1990), 18–28.

[DoSo88] J. J. Dongarra and D. C. Sorensen, *A fully parallel algorithm for the symmetric eigenvalue problem*, SIAM J. Sci. Stat. Comput. 8 (1987), s139–s154.

[DTT97] T. A. Driscoll, K.-C. Toh, and L. N. Trefethen, *Matrix iterations: The six gaps between potential theory and convergence*, submitted to SIAM Review.

[DER86] I. S. Duff, A. M. Erisman, and J. K. Reid, *Direct Methods for Sparse Matrices*, Clarendon Press, Oxford, UK, 1986.

[Dwy44] P. S. Dwyer, *A matrix presentation of least squares and correlation theory with matrix justification of improved methods of solutions*, Ann. Math. Stat. 15 (1944), 82–89.

[Ede88] A. Edelman, *Eigenvalues and condition numbers of random matrices*, SIAM J. Matrix Anal. Appl. 9 (1988), 543–560.

[Ede94] A. Edelman, *Large numerical linear algebra in 1994: The continuing influence of parallel computing*, Proc. 1994 Scalable High Performance Computing Conf., IEEE Computer Soc. Press, Los Alamitos, CA, 1994, 781–787.

[EdMu95] A. Edelman and H. Murakami, *Polynomial roots from companion matrix eigenvalues*, Math. Comp. 64 (1995), 763–776.

[Fey85] R. P. Feynman, *Surely You're Joking, Mr. Feynman! Adventures of a Curious Character*, Norton, New York, 1985.

[Fis96] B. Fischer, *Polynomial Based Iteration Methods for Symmetric Linear Systems*, Wiley-Teubner, Chichester, UK, 1996.

[Fle76] R. Fletcher, *Conjugate gradient methods for indefinite systems*, in G. A. Watson, ed., Numerical Analysis Dundee 1975, Lec. Notes in Math. v. 506, Springer-Verlag, Berlin, 1976, 73–89.

[FoHe60] G. E Forsythe and P. Henrici, *The cyclic Jacobi method for computing the principal values of a complex matrix*, Trans. Amer. Math. Soc. 94 (1960), 1–23.

[FoMo67] G. E. Forsythe and C. B. Moler, *Computer Solution of Linear Algebraic Systems*, Prentice Hall, Englewood Cliffs, NJ, 1967.

[Fos94] L. V. Foster, *Gaussian elimination with partial pivoting can fail in practice*, SIAM J. Matrix Anal. Appl. 15 (1994), 1354–1362.

[Fra61] J. G. F. Francis, *The QR transformation: A unitary analogue to the LR transformation, parts* I *and* II, Computer J. 4 (1961), 256–72 and 332–45.

[Fre93] R. W. Freund, *A transpose-free quasi-minimal residual algorithm for non-hermitian linear systems*, SIAM J. Sci. Stat. Comput. 13 (1992), 425–448.

[FGN92] R. W. Freund, G. H. Golub, and N. M. Nachtigal, *Iterative solution of linear systems*, Acta Numerica 1 (1992), 57–100.

[FrNa91] R. W. Freund and N. M. Nachtigal, *QMR: A quasi-minimal residual method for non-Hermitian linear systems*, Numer. Math. 60 (1991), 315–339.

[GeLi81] A. George and J. W.-H. Liu, *Computer Solution of Large Sparse Positive Definite Systems*, Prentice Hall, Englewood Cliffs, NJ, 1981.

[Geu82] A. J. Geurts, *A contribution to the theory of condition*, Numer. Math. 39 (1982), 85–96.

[GMW91] P. E. Gill, W. Murray, and M. H. Wright, *Numerical Linear Algebra and Optimization*, Addison-Wesley, Redwood City, CA, 1991.

[Gir90] V. L. Girko, *Theory of Random Determinants*, Kluwer, Dordrecht, the Netherlands, 1990.

[Gol91] D. Goldberg, *What every computer scientist should know about floating-point arithmetic*, ACM Computing Surveys 23 (1991), 5–48.

[GoKa65] G. Golub and W. Kahan, *Calculating the singular values and pseudo-inverse of a matrix*, SIAM J. Numer. Anal. 2 (1965), 205–224.

[GoOL89] G. H. Golub and D. P. O'Leary, *Some history of the conjugate gradient and Lanczos methods*, SIAM Review 31 (1989), 50–100.

[GoPe73] G. H. Golub and V. Pereyra, *The differentiation of pseudoinverses and nonlinear least squares problems whose variables separate*, SIAM J. Numer. Anal. 10 (1973), 413–432.

[GoVa96] G. H. Golub and C. F. Van Loan, *Matrix Computations*, 3rd ed., Johns Hopkins U. Press, Baltimore, 1996.

[GoWe69] G. H. Golub and J. H. Welsch, *Calculation of Gauss quadrature rules*, Math. Comp. 23 (1969), 221–230.

[GoWi76] G. H. Golub and J. H. Wilkinson, *Ill-conditioned eigensystems and the computation of the Jordan canonical form*, SIAM Review 18 (1976), 578–619.

[Gra96] S. Gratton, *On the condition number of linear least squares problems in a weighted Frobenius norm*, BIT 36 (1996), 523–530.

[Gre97] A. Greenbaum, *Iterative Methods for Solving Linear Systems*, SIAM, Philadelphia, 1997.

[GrTr94] A. Greenbaum and L. N. Trefethen, *GMRES/CR and Arnoldi/Lanczos as matrix approximation problems*, SIAM J. Sci. Comput. 15 (1994), 359–368.

[GuEi95] M. Gu and S. C. Eisenstat, *A divide-and-conquer algorithm for the symmetric tridiagonal eigenproblem*, SIAM J. Matrix Anal. Appl. 16 (1995), 172–191.

[Gut92] M. H. Gutknecht, *A completed theory of the unsymmetric Lanczos process and related algorithms, part* I, SIAM J. Matrix Anal. Appl. 13 (1992), 594–639.

[Hac94] W. Hackbusch, *Iterative Solution of Large Sparse Linear Systems of Equations*, Springer-Verlag, Berlin, 1994.

[Hag88] W. Hager, *Applied Numerical Linear Algebra*, Prentice Hall, Englewood Cliffs, NJ, 1988.

[HYG88] S. M. Hammel, J. A. Yorke, and C. Grebogi, *Numerical orbits of chaotic processes represent true orbits*, Bull. Amer. Math. Soc. 19 (1988), 465–469.

[HeSt52] M. R. Hestenes and E. Stiefel, *Methods of conjugate gradients for solving linear systems*, J. Res. Nat. Bur. Stand. 49 (1952), 409–436.

[Hig96] N. J. Higham, *Accuracy and Stability of Numerical Algorithms*, SIAM, Philadelphia, 1996.

[HoJo85] R. A. Horn and C. R. Johnson, *Matrix Analysis*, Cambridge U. Press, Cambridge, UK, 1985.

[HoJo91] R. A. Horn and C. R. Johnson, *Topics in Matrix Analysis*, Cambridge U. Press, Cambridge, UK, 1991.

[Hou58] A. S. Householder, *Unitary triangularization of a nonsymmetric matrix*, J. Assoc. Comput. Mach. 5 (1958), 339–342.

[Hou64] A. S. Householder, *The Theory of Matrices in Numerical Analysis*, Blaisdell, New York, 1964.

[Ips97] I. C. F. Ipsen, *A history of inverse iteration*, in B. Huppert and H. Schneider, eds., Helmut Wielandt, Mathematische Werke, Mathematical Works, v. 2, Walter de Gruyter, Berlin, 1996, 453–463.

[IpMe95] I. C. F. Ipsen and C. D. Meyer, *The angle between complementary subspaces*, Amer. Math. Monthly 102 (1995), 904–911.

[Jac46] C. G. J. Jacobi, *Über ein leichtes Verfahren die in der Theorie der Säcularstörungen vorkommenden Gleichungen numerisch aufzulösen*, J. Reine Angew. Math. 30 (1846), 51–94.

[JeTr70] M. A. Jenkins and J. F. Traub, *A three-stage variable-shift iteration for polynomial zeros and its relation to generalized Rayleigh iteration*, Numer. Math. 14 (1970), 252–263.

[Kah72] W. M. Kahan, *Conserving confluence curbs ill-condition*, unpublished manuscript, 1972.

[Kan66] S. Kaniel, *Estimates for some computational techniques in linear algebra*, Math. Comp. 20 (1966), 369–378.

[Kat76] T. Kato, *Perturbation Theory for Linear Operators*, 2nd ed., Springer-Verlag, New York, 1976.

[Kel95] C. T. Kelley, *Iterative Methods for Linear and Nonlinear Equations*, SIAM, Philadelphia, 1995.

[KlKo65] V. V. Klyuyev and N. I. Kokovkin-Shcherbak, *On the minimization of the number of arithmetic operations for the solution of linear algebraic systems of equations*, Zh. Vychisl. Mat. i Mat. Fiz. 5 (1965), 21–33; translated from the Russian by G. J. Tee, Tech. Rep. CS24, Computer Science Dept., Stanford University, 1965.

[Koz92] D. C. Kozen, *The Design and Analysis of Algorithms*, Springer-Verlag, New York, 1992.

[Kry31] A. N. Krylov, *On the numerical solution of equations which in technical questions are determined by the frequency of small vibrations of material systems*, Izv. Akad. Nauk. S. S. S. R. Otd Mat. Estest. 1 (1931), 491–539.

[Kub61] V. N. Kublanovskaya, *On some algorithms for the solution of the complete eigenvalue problem*, USSR Comp. Math. Phys. 3 (1961), 637–657.

[Lan50] C. Lanczos, *An iteration method for the solution of the eigenvalue problem of linear differential and integral operators*, J. Res. Nat. Bur. Stand. 45 (1950), 255–282.

[Lan52] C. Lanczos, *Solution of systems of linear equations by minimized iterations*, J. Res. Nat. Bur. Stand. 49 (1952), 33–53.

[LaHa95] C. L. Lawson and R. J. Hanson, *Solving Least Squares Problems*, SIAM, Philadelphia, 1995 (reprinting with corrections and a new appendix of a 1974 Prentice Hall text).

[LeSo96] R. B. Lehoucq and D. C. Sorensen, *Deflation techniques for an implicitly restarted Arnoldi iteration*, SIAM J. Matrix Anal. Appl. 17 (1996), 789–821.

[Mac95] N. Mackey, *Hamilton and Jacobi meet again: Quaternions and the eigenvalue problem*, SIAM J. Matrix Anal. Appl. 16 (1995), 421–435.

[MeRo96] K. Meerbergen and D. Roose, *Matrix transformations for computing rightmost eigenvalues of large sparse non-symmetric eigenvalue problems*, IMA J. Numer. Anal. 16 (1996), 297–346.

[Meh91] M. L. Mehta, *Random Matrices*, 2nd ed., Academic Press, San Diego, 1991.

[Meva77] J. Meijerink and H. van der Vorst, *An iterative solution method for linear systems of which the coefficient matrix is a symmetric M-matrix*, Math. Comp. 31 (1977), 148–162.

[NRT92] N. M. Nachtigal, S. C. Reddy, and L. N. Trefethen, *How fast are nonsymmetric matrix iterations?*, SIAM J. Matrix Anal. Appl. 13 (1992), 778–795.

[Nev93] O. Nevanlinna, *Convergence of Iterations for Linear Equations*, Birkhäuser, Basel, 1993.

[Ost59] A. M. Ostrowski, *On the convergence of the Rayleigh quotient iteration for the computation of characteristic roots and vectors, IV. Generalized Rayleigh quotient for nonlinear elementary divisors*, Arch. Rational Mech. Anal. 3 (1959), 341–347.

[Pai71] C. C. Paige, *The Computation of Eigenvalues and Eigenvectors of Very Large Sparse Matrices*, PhD diss., U. of London, 1971.

[Pan84] V. Pan, *How to Multiply Matrices Faster*, Lec. Notes in Comp. Sci., v. 179, Springer-Verlag, Berlin, 1984.

[Par80] B. N. Parlett, *The Symmetric Eigenvalue Problem*, Prentice Hall, Englewood Cliffs, NJ, 1980.

[PTL85] B. N. Parlett, D. R. Taylor, and Z. A. Liu, *A look-ahead Lanczos algorithm for unsymmetric matrices*, Math. Comp. 44 (1985), 105–124.

[PeWi79] G. Peters and J. H. Wilkinson, *Inverse iteration, ill-conditioned equations and Newton's method*, SIAM Review 21 (1979), 339–360.

[Ric66] J. R. Rice, *A theory of condition*, SIAM J. Numer. Anal. 3 (1966), 287–310.

[Saa80] Y. Saad, *On the rates of convergence of the Lanczos and the block Lanczos methods*, SIAM J. Numer. Anal. 17 (1980), 687–706.

[Saa92] Y. Saad, *Numerical Methods for Large Eigenvalue Problems*, Manchester U. Press, Manchester, UK, 1992.

[Saa96] Y. Saad, *Iterative Methods for Sparse Linear Systems*, PWS Publishing, Boston, 1996.

[SaSc86] Y. Saad and M. H. Schultz, *GMRES: A generalized minimal residual algorithm for solving nonsymmetric linear systems*, SIAM J. Sci. Stat. Comput. 7 (1986), 856–869.

[Ske79] R. D. Skeel, *Scaling for numerical stability in Gaussian elimination*, J. Assoc. Comput. Mach. 26 (1979), 494–526.

[Ske80] R. D. Skeel, *Iterative refinement implies numerical stability for Gaussian elimination*, Math. Comp. 35 (1980), 817–832.

[Slvd96] G. L. G. Sleijpen and H. A. van der Vorst, *A Jacobi–Davidson iteration method for linear eigenvalue problems*, SIAM J. Matrix Anal. Appl. 17 (1996), 401–425.

[Smi76] B. T. Smith et al., *Matrix Eigensystem Routines—EISPACK Guide*, Springer-Verlag, Berlin, 1976.

[SBG96] B. Smith, P. Bjørstad, and W. Gropp, *Domain Decomposition: Parallel Multilevel Methods for Elliptic Partial Differential Equations*, Cambridge U. Press, Cambridge, UK, 1996.

[Smi70] F. Smithies, *Integral Equations*, Cambridge U. Press, Cambridge, UK, 1970.

[Son89] P. Sonneveld, *CGS, a fast Lanczos-type solver for nonsymmetric linear systems*, SIAM J. Sci. Stat. Comput. 10 (1989), 36–52.

[Ste73] G. W. Stewart, *Introduction to Matrix Computations*, Academic Press, New York, 1973.

[Ste77] G. W. Stewart, *On the perturbation of pseudo-inverses, projections, and linear least squares problems*, SIAM Review 19 (1977), 634–662.

[Ste93] G. W. Stewart, *On the early history of the singular value decomposition*, SIAM Review 35 (1993), 551–566.

[StSu90] G. W. Stewart and J. Sun, *Matrix Perturbation Theory*, Academic Press, Boston, 1990.

[Sti86] S. M. Stigler, *The History of Statistics*, Harvard U. Press, Cambridge, MA, 1986.

[Str86] G. Strang, *A proposal for Toeplitz matrix calculations*, Stud. Appl. Math. 74 (1986), 171–176.

[Str88] G. Strang, *Linear Algebra and Its Applications*, 3rd ed., Harcourt, Brace, and Jovanovich, San Diego, 1988.

[Str69] V. Strassen, *Gaussian elimination is not optimal*, Numer. Math. 13 (1969), 354–356.

[Sze75] G. Szegő, *Orthogonal Polynomials*, 4th ed., Amer. Math. Soc., Providence, RI, 1975.

[ToTr94] K.-C. Toh and L. N. Trefethen, *Pseudozeros of polynomials and pseudospectra of companion matrices*, Numer. Math. 68 (1994), 403–425.

[ToTr96] K.-C. Toh and L. N. Trefethen, *Computation of pseudospectra by the Arnoldi iteration*, SIAM J. Sci. Comput. 17 (1996), 1–15.

[ToTr98] K.-C. Toh and L. N. Trefethen, *The Chebyshev polynomials of a matrix*, SIAM J. Matrix Anal. Appl., to appear.

[Tre91] L. N. Trefethen, *Pseudospectra of matrices*, in D. F. Griffiths and G. A. Watson, eds., Numerical Analysis 1991, Longman Scientific and Technical, Harlow, Essex, UK, 1992, 234–266.

[Tre97] L. N. Trefethen, *Pseudospectra of linear operators*, SIAM Review 39 (1997), to appear.

[TrSc90] L. N. Trefethen and R. S. Schreiber, *Average-case stability of Gaussian elimination*, SIAM J. Matrix Anal. Appl. 11 (1990), 335–360.

[TTRD93] L. N. Trefethen, A. E. Trefethen, S. C. Reddy, and T. A. Driscoll, *Hydrodynamic stability without eigenvalues*, Science 261 (1993), 578–584.

[TrVi97] L. N. Trefethen and D. Viswanath, *The condition number of a random triangular matrix*, submitted to SIAM J. Matrix Anal. Appl.

[TrWe91] L. N. Trefethen and J. A. C. Weideman, *Two results on polynomial interpolation in equally spaced points*, J. Approx. Theory 65 (1991), 247–260.

[Tur48] A. M. Turing, *Rounding-off errors in matrix processes*, Quart. J. Mech. Appl. Math. 1 (1948), 287–308.

[vdS75] A. van der Sluis, *Stability of the solutions of linear least squares problems*, Numer. Math. 23 (1975), 241–254.

[vdV92] H. A. van der Vorst, *Bi-CGSTAB: A fast and smoothly convergent variant of Bi-CG for the solution of nonsymmetric linear systems*, SIAM J. Sci. Stat. Comput. 13 (1992), 631–644.

[Var62] R. S. Varga, *Matrix Iterative Analysis*, Prentice Hall, Englewood Cliffs, NJ, 1962.

[Wat82] D. S. Watkins, *Understanding the QR algorithm*, SIAM Review 24 (1982), 427–440.

[Wat91] D. S. Watkins, *Fundamentals of Matrix Computations*, Wiley, New York, 1991.

[Wed73] P.-Å. Wedin, *Perturbation theory for pseudo-inverses*, BIT 13 (1973), 217–232.

[Wei96] R. Weiss, *Parameter-Free Iterative Linear Solvers*, Akademie Verlag, Berlin, 1996.

[Wil61] J. H. Wilkinson, *Error analysis of direct methods of matrix inversion*, J. Assoc. Comput. Mach. 8 (1961), 281–330.

[Wil65] J. H. Wilkinson, *The Algebraic Eigenvalue Problem*, Clarendon Press, Oxford, UK, 1965.

[Wri93] S. J. Wright, *A collection of problems for which Gaussian elimination with partial pivoting is unstable*, SIAM J. Sci. Comput. 14 (1993), 231–238.

Index

~, 59

\ operator in MATLAB, 85, 138, 177, 337

Abel, Niels, 192, 324, 326
accuracy, 103, 111
A-conjugate vectors, 295
ADI (alternating direction implicit) splitting, 318
algorithm, formal definition, 102
angle between vectors or subspaces, 12, 214, 332
A-norm, 294
Arnoldi
    approximation problem, 259
    iteration, 245, 250–265, 340
    eigenvalue estimates, *see* Ritz values
    lemniscate, 262–263, 340
    polynomial, 262
    shift-and-invert, 319, 342
augmented matrix, 139, 141

back substitution, 121–128

backward
    error, 116
    error analysis, 108, 111–112, 334–335
    stability, 104, 334
banded matrix, 154, 161, 337
base, 98
basis, change of, 8, 15, 32–33, 182
Bauer–Fike theorem, 201
BCG (biconjugate gradients), 245, 303–312, 341
Bi-CGSTAB, 311, 341
biconjugate gradients, *see* BCG
bidiagonal
    matrix, 265
    reduction, 236–240
bilinear function, 12
biorthogonalization methods, 303–312
biorthogonal vectors, 305–306
bisection, 227–229, 233
BLAS (basic linear algebra subroutines), 330

block
   matrix, 143, 154, 230, 235, 249,
      317, 330
   power iteration, *see* simultane-
      ous iteration
boundary elements, 245, 248, 317
breakdown of Arnoldi iteration, 256

C, 63
cancellation error, 73, 91, 138
Cauchy–Schwarz inequality, 21
Cayley–Hamilton theorem, 260
Cayley transform, 16
Cayuga, Lake, 136
CG, *see* conjugate gradients
CGN or CGNR, 245, 303–305
CGS (conjugate gradients squared),
      311
chaos, 335
characteristic polynomial, 110, 183,
      184, 190
Chebyshev
   points, 79, 279, 292
   polynomials, 287, 292, 300
   polynomial of a matrix, 265, 340
χ^2 (chi-squared) distribution, 240
Cholesky factorization, 82, 141, 172–
      178, 301, 337
circulant matrix, 187, 305, 318, 342
column
   pivoting, 139–140, 143
   rank, 7
   space, 7
   spaces, sequence of, 48, 169, 245
communication, 59, 66
compact operator, 265, 331
companion matrix, 192, 338
complementary subspaces, 43, 332
complete pivoting, 161, 336
complex
   arithmetic, 59, 100
   conjugate, 11
   sign, 29, 72
   symmetric matrix, 312

componentwise analysis, 127, 227,
      334, 339
computers, speed of, 243–244, 339
conditioning, 89–96, 333
condition number
   absolute, 90
   computation of, 94
   of a matrix, 94, 333
   of an eigenvalue, 258
   relative, 90
   squaring of, 142, 235, 305
conjugate
   complex, 11
   gradients, 245, 293–302, 303, 341
   hermitian, 11
   residuals iteration, 293
convergence
   cubic, 195, 208, 212, 221–222
   linear or geometric, 195, 262–264
   quadratic, 195, 226
   superlinear, 195, 337
Coppersmith and Winograd, algo-
      rithm of, 247, 340
covariance matrix, 234
CS decomposition, 332
Cuppen, J. J. M., 229

data-fitting, *see* least squares prob-
      lem
Davidson method, 319
defective
   eigenvalue, 185
   matrix, 185
deflation, 212, 223, 232
deletion matrix, 9, 24
Demmel, James W., book by, 329
dense
   matrix, 244
   subset, 37
determinant, 8, 10, 34, 97, 161, 330
   computation of, 161
diagonalizable matrix, see nondefec-
      tive matrix
diagonalization, 188

diagonally dominant matrix, 162

diagonal matrix, 15, 18, 20, 32

dimensions, physical, 10, 107

direct algorithm, 190, 243, 247

divide-and-conquer algorithm, 212, 229–233, 239

domain decomposition, 317, 342

dual norm, 24, 95, 331

e_j, 7

eigenspace, 181, 183

eigenvalue decomposition, 33, 182

eigenvalue-revealing factorization, 188, 191

eigenvalues, 8, 15, 24, 181–189
  algebraic multiplicity of, 183–184
  computation of, 110, 190–233, 257–265
  defective, 185
  geometric multiplicity of, 183–184
  perturbation of, 188, 201, 258, 333
  simple, 184

eigenvectors, 15, 43, 181
  computation of, 202, 218, 227
  localization of, 232, 233

EISPACK, 257, 330, 337, 338

electric charge, 279, 283–284

error
  absolute, 103
  relative, 99, 103

Euclidean length, 12, 17, 78

ev and ew (abbreviations for eigenvector and eigenvalue), 188, 337

exponent, 98

exponential of a matrix, 33, 182, 189, 201

fast Fourier transform, 63

"fast matrix inverse", 248

fast Poisson solver, 317

Feynman, Richard, 91, 334

field of values, see numerical range

finite differences, 244, 317

finite elements, 254, 317

finite sections, 333

fixed point arithmetic, 98

fl, 99

floating point
  arithmetic, 66, 97–101, 334
  axioms, 99
  numbers, 98

flop (floating point operation), 58

Fortran, 63, 324

Forsythe and Moler, book by, 243, 331

forward error analysis, 108, 112, 177

4-norm, 18

fraction, 98

Frobenius norm, 22, 34

full rank, matrix of, 7

fundamental law of computer science, 246, 325, 340

Galois, Evariste, 192, 324, 326

gamma function, 85

Gaussian elimination, x, 35, 54, 61, 106, 147–171, 325
  stability, 152–154, 163–171, 325, 336

Gauss quadrature, 285–292, 341

Gauss–Seidel iteration, 318, 339

generalized minimal residuals, see GMRES

geometric interpretations, 12, 25, 36, 55, 59, 133, 201, 233, 332, 335

Gerschgorin's theorem, 189, 337

ghost eigenvalues, 282–283

Givens rotation, 76, 195, 218, 226, 268, 275

GMRES, 245, 266–275, 293, 303, 340
  approximation problem, 269
  restarted, 275

Golub, Gene H., 236, 330, 331, 339

Golub and Van Loan, book by, ix, 329

Golub–Kahan bidiagonalization, 236–237
gradient, 203, 302
Gram–Schmidt orthogonalization, 50–51, 56–62, 70, 148, 250–253, 332
    classical vs. modified, 51, 57, 65–66, 140, 332
graphics, 63
Green's function, 284
growth factor, 163–171, 312, 336
guard digit, 100

Hadamard
    inequality, 55
    matrix, 16
Hahn–Banach theorem, 331
Hein, Piet, 18
Henrici, Peter, 327
hermitian
    conjugate, 11
    matrix, 11, 15, 34, 44, 162, 172, 187
    positive definite matrix, 172, 294
Hessenberg
    matrix, 193, 198, 252
    orthogonalization, 305–306
    reduction, 193, 196–201, 250–251, 337–338
Hestenes, Magnus, 293, 341
Higham, Nicholas J., xii, 335
    book by, ix, 329
Hilbert space, 330, 331
Hilbert–Schmidt norm, see Frobenius norm
Hölder inequality, 21
Horn and Johnson, books by, 330
Horner's rule, 265
Householder
    Alston, 70, 330, 332
    reflector, 70–73
    Symposia, 333
    triangularization, 64, 69–76, 114–120, 147, 251, 332

tridiagonalization, 196–201, 251
hydrodynamic stability, 258
hyperellipse, 20, 25, 36, 95
hyperplane, 71

ICCG (incomplete Cholesky factorization), 316
ideal Arnoldi polynomial, see Chebyshev polynomial of a matrix
idempotent matrix, 41
identity, 8
IEEE arithmetic, 97, 334
ill-conditioned
    matrix, 94
    problem, 89, 91
ill-posed problem, 334
ILU (incomplete LU factorization), 316
image processing, 36, 68
incomplete factorization, 316, 342
infinitesimal perturbation, 90, 133, 135
∞-norm, 18, 20, 21
inner product, 12, 52, 109, 285
integral
    equation, 245, 331
    operator, 6, 53, 286
interlacing eigenvalues, 227–228
interpolation, 10, see also polynomial interpolation
intersection of subspaces, 36, 55
invariant subspace, 183
inverse, 8
    computation of, 161
    iteration, 206–207, 210, 219, 338
invertible matrix, see nonsingular matrix
irreducible matrix, 227
iterative methods, x, 69, 192, 243–249, 326, 339–340

Jacobi
    algorithm, 225–227, 233, 338–339
    Carl Gustav Jacob, 225

iteration, 318
matrix, 287–292
polynomial, 287
preconditioner, 316
rotation, 226
Jacobian, 90, 132–133, 258
Jacobi–Davidson methods, 319, 342
Jordan form, 337

Kahan, William M., 236, 334, 339
Karmarkar algorithm, 326
Kronecker delta function, 14
Krylov
    matrix, 253
    sequence, 245
    subspace iteration, 241–327
    subspaces, 245, 253

$L^2[-1,1]$, 52, 285
Lanczos
    iteration, 245, 250, 276–284, 298,
        303, 340
    lemniscate, 284
    polynomial, 280
LAPACK, 166, 205, 232, 243, 257,
    338
least squares problem, 36, 77–85,
    129–144, 305, 333
    rank-deficient, 143, 335
Lebesgue constants, 96, 334, 341
Legendre
    points, 292
    polynomial, 53, 54, 64, 68, 285–
        292
lemniscate, 262–263
LHC (Lawson–Hanson–Chan) bidi-
    agonalization, 237–239
LINPACK, 166, 243
look-ahead Lanczos, 311, 341
low-rank approximation, 35–36, 331
    computation of, 36
LU factorization, 147, 154, 160

machine epsilon, 66, 98, 100
mantissa, 98

mass–spring system, 9
MathWorks, Inc., The, 63, 330, 332
MATLAB, 31, 62, 63–68, 166, 205,
    257, 324, 332
matrix
    augmented, 139, 141
    banded, 154, 161, 337
    bidiagonal, 265
    block, 143, 154, 230, 235, 249,
        317, 330
    circulant, 187, 305, 318, 342
    companion, 192, 338
    complex symmetric, 312
    covariance, 234
    defective, 185
    deletion, 9, 24
    dense, 244
    diagonal, 15, 18, 20, 32
    diagonalizable, see nondefective
        matrix
    diagonally dominant, 162
    Hadamard, 16
    hermitian, 11, 15, 34, 44, 162,
        172, 187
    hermitian positive definite, 172,
        294
    Hessenberg, 193, 198, 252
    idempotent, 41
    identity, 8
    ill-conditioned, 94
    irreducible, 227
    nondefective, 185–186
    nonnormal, 186, 258
    nonsingular, 7
    normal, 92, 173, 187, 201
    orthogonal, 14, 218
    permutation, 34, 157, 220
    positive definite, see hermitian
        positive definite matrix
    random, 96, 114, 167–171, 189,
        233, 240, 244, 262, 271, 334
    random orthogonal, 65, 114, 120
    random sparse, 300, 309
    random triangular, 96, 128, 167

skew-hermitian, 16, 187
sparse, 232, 244, 300–301
symmetric, 11, 172
Toeplitz, 68, 318, 337, 342
triangular, 10, 15, 49, 240
tridiagonal, 194, 218
unitarily diagonalizable, *see* nor-
        mal matrix
unitary, 14–16, 119, 163, 187
unit triangular, 62, 148
Vandermonde, 4, 53, 64, 78, 137,
        289, 292, 337
well-conditioned, 94
matrix-matrix multiplication, 5
matrix-vector multiplication, 3, 93,
        330
memory hierarchy, 59
MINRES, 293
multigrid methods, 317, 326
multiplicity of an eigenvalue
        algebraic, 183
        geometric, 183
multipole methods, 232, 245, 326,
        339

nested dissection, 245
Netlib, 330
Newton–Cotes quadrature formula,
        289, 341
Newton's method, 101, 231
nondefective matrix, 185–186
nonnormal matrix, 186, 258
nonsingular matrix, 7
normal
        distribution, 96, 171, 240
        equations, 81, 82, 130, 137, 141,
                204
        matrix, 92, 173, 187, 201
norms, 17–24, 331
        1-, 2-, 4-, ∞-, p-, 18
        equivalence of, 37, 106, 117
        induced, 18
        matrix, 18, 22
        vector, 17

weighted, 18, 24, 294
normwise analysis, 127, 334
nullspace, 7, 33
        computation of, 36
numerical
        analysis, definition of, 321–327
        range, 209

O ("big O"), 103–106
$O(\epsilon_{\mathrm{machine}})$, 104
1-norm, 18, 20
one-to-one function, 7
operation count, 58–60
orthogonal
        matrix, 14, 218
        polynomials, 285–292, 341
        polynomials approximation prob-
                lem, 288
        projector, 43–47, 56, 81, 83, 129
        triangularization, 69–70, 148
        vectors, 13
orthogonality, loss of, 66–67, 282–
        283, 295
orthonormal
        basis, 36
        vectors, 13
outer product, 6, 22, 24, 109, *see
        also* rank-one matrix
overdetermined system, 77
overflow, 97

Padé approximation, 311, 341
panel methods, 245
parallel computer, 66, 233
partial differential equations, 53, 244,
        248, 316–318, 332
partial pivoting, 156, 160, 336
Pentium^{TM} microprocessor, 100
permutation matrix, 34, 157, 220
π, calculation of, 327
pivot element, 155
pivoting in Gaussian elimination, 155–
        162, 336
p-norm, 18

polar decomposition, 331
polynomial, 4, 101, 181, 283
    approximation, 246, 258, 268–269,
        298–299, 340–341
    Chebyshev, 292, 300
    interpolation, 78, 96, 292
    Legendre, 53, 54, 64, 68, 285–292
    monic, 183, 259
    of a matrix, 259, 265, 318
    orthogonal, 285–292
    preconditioner, 318
    quintic, 192
    roots, 92, 101, 110, 190, 191, 227,
        338
positive definite matrix, see hermi-
    tian positive definite matrix
potential theory, 279, 283–284, 341
power iteration, 191, 204–206
powers of a matrix, 33, 120, 182,
    189
precision, 98
preconditioning, 274, 297, 313–319,
    326, 342
principal minors, 154, 214
problem
    formal definition, 89, 102
    instance, 89
problem-solving environment, 63
projector, 41, 331–332
    complementary, 42
    oblique, 41
    orthogonal, 43–47, 56, 81, 83, 129
    rank-one, 14, 46
pseudoinverse, 81–85, 94, 129, 335
pseudo-minimal polynomial, 261
pseudospectra, 201, 265, 338, 340
    computation of, 201, 265, 340
Pythagorean theorem, 15, 81

QMR (quasi-minimal residuals), 310–
    311, 341
Q portrait, 169–170
QR algorithm, 211–224, 239, 253–
    254, 338

QR factorization, x, 36, 48–55, 48–
    55, 83, 253, 332
    full, 49
    reduced, 49
    with column pivoting, 49, 143
quadrature, 285–292
quasi-minimal residuals, see QMR

radix, 98
random matrix, 96, 114, 167–171,
    189, 233, 240, 244, 262, 271,
    334
    orthogonal, 65, 114, 120
    sparse, 300, 309
    triangular, 96, 128, 167
range, 6, 33
    computation of, 36
    sensitivity to perturbations, 133–
        134
rank, 7, 33, 55
    computation of, 36
rank-deficient matrix, 84, 143
rank-one
    matrix, 35, see also outer prod-
        uct
    perturbation, 16, 230
    projector, 14, 46
rank-revealing factorization, 336
rank-two perturbation, 232
Rayleigh–Ritz procedure, 254
Rayleigh quotient, 203, 209, 217, 254,
    283
    iteration, 207–209, 221, 338
    shift, 221, 342
recursion, 16, 230, 249
reflection, 15, 29, see also House-
    holder reflector
    of light, 136
regression, 136
regularization, 36
residual, 77, 116
resolvent, 201
resonance, 182
Richardson iteration, 274, 302

Ritz
    matrix, 276
    values, 255, 257, 278
rootfinding, *see* polynomial roots
rotation, 15, 29, 31, *see also* Givens
        rotation
rounding, 99
    errors, 321–327
row
    rank, 7
    vector, 21

Schur
    complement, 154
    factorization, 187, 193, 337
secular equation, 231
self-adjoint operator, 258
shadowing, 335
shifts in QR algorithm, 212, 219–
        224
similarity transformation, 34, 184
similar matrices, 184
simultaneous
    inverse iteration, 219
    iteration, 213–218, 253–254
singular
    value, 8, 26
    value decomposition, *see* SVD
    vector, 26
Skeel
    condition number, 334
    Robert D., 326
skew-hermitian matrix, 16, 187
software, 330
SOR (successive over-relaxation), 318,
        339
sparse
    direct methods, 339
    matrix, 232, 244, 300–301
spectral
    abscissa, 189, 258
    methods, 53, 255, 317, 326, 332
    radius, 24, 189
spectrum, 181, 201

splitting, 317–318
square root, 58, 91, 127
SSOR (symmetric SOR), 318
stability, 57, 66, 72, 84, 89, 102–
        113, 326
    formal definition, 104
    physical, 182, 258
stable algorithm, *see* stability
stationary point, 203, 283
steepest descent iteration, 302
Stiefel, Eduard, 293, 341
Strassen's algorithm, 247, 249, 330,
        340
Sturm sequence, 228
submatrix, 9, 333
subtraction, 91, 108
superellipse, 18
SVD (singular value decomposition),
        25–37, 83, 113, 120, 142, 201,
        322, 331
    computation of, 36, 113, 234–240,
        339
    full, 28
    reduced, 27
symbolic computation, 101, 324
symmetric matrix, 11, 172

TFQMR (transpose-free QMR), 311,
        341
three-step bidiagonalization, 238–240
three-term recurrence relation, 229,
        276, 282, 287, 291
threshold pivoting, 336
tilde ($\tilde{\ }$), 103
Toeplitz matrix, 68, 318, 337, 342
trace, 23
translation-invariance, 261, 269
transpose, 11
transpose-free iterations, 311
Traub, Joseph, 327
triangle inequality, 17
triangular
    matrix, 10, 15, 49, 240 *see also*
        random matrix, triangular

orthogonalization, 51, 70, 148

triangularization, 148

system of equations, 54, 82–83, 117, 121–128

tridiagonal

biorthogonalization, 305–306

matrix, 194, 218

orthogonalization, 305–306

reduction, 194, 196–201, 212

Turing, Alan, 325, 333, 335, 342

2-norm, 18, 20, 34

computation of, 36

underdetermined system, 143

underflow, 97

unit

ball, 20

sphere, 25

triangular matrix, 62, 148

unitarily diagonalizable matrix, *see* normal matrix

unitary

diagonalization, 187–188

equivalence, 31

matrix, 14–16, 119, 163, 187

triangularization, 188

unstable algorithm, *see* stability

Vandermonde matrix, 4, 53, 64, 78, 137, 289, 292, 337

Von Neumann, John, 325, 335, 336

wavelets, 245

weighted norm, 18, 24, 294

well-conditioned

matrix, 94

problem, 89, 91

Wilkinson, James H., 115, 325, 330, 335, 336

book by, 331, 337

polynomial, 92

shift, 222, 224

zerofinding, *see* polynomial roots

ziggurat, 75

從 2016年 1月16日 至 22年 9月 19日. 或 22年 12月. 我在整体上都保持着一種上升態勢吧.

　但從 22年暑假開始. 便陷入了一種

412
433
454
555
475

Winter

MATH 525
     526
     551
     572
STATS 489
ECON 490

Summer 555
       433
       454
       451 (?)